Family-Making

This volume explores the ethics of making or expanding families through adoption or technologically assisted reproduction. For many people, these methods are separate and distinct: they can choose either adoption or assisted reproduction. But for others, these options blend together. For example, in some jurisdictions, the path of assisted reproduction for same-sex couples is complicated by the need for the partner who is not genetically related to the resulting child to adopt this child if she wants to become the child's legal parent.

The essays in this volume critically examine moral choices to pursue adoption, assisted reproduction, or both, and highlight the social norms that can distort decision-making. Among these norms are those that favour people having biologically related children ('bionormativity') or that privilege a traditional understanding of family as a heterosexual unit with one or more children where both parents are the genetic, biological, legal, and social parents of these children.

As a whole, the book looks at how adoption and assisted reproduction are morally distinct from one another, but also emphasizes how the two are morally similar. Choosing one, the other, or both of these approaches to family-making can be complex in some respects, but ought to be simple in others, provided that one's main goal is to become a parent.

Françoise Baylis is Professor and Canada Research Chair in Bioethics and Philosophy at Dalhousie University, and founder of the NovelTechEthics research team.

Carolyn McLeod is Professor of Philosophy and an Affiliate Member of Women's Studies and Feminist Research at Western University.

ISSUES IN BIOMEDICAL ETHICS

General Editors

John Harris and Søren Holm

Consulting Editors

Raanan Gillon and Bonnie Steinbock

The late twentieth century witnessed dramatic technological developments in biomedical science and in the delivery of healthcare, and these developments have brought with them important social changes. All too often ethical analysis has lagged behind these changes. The purpose of this series is to provide lively, up-to-date, and authoritative studies for the increasingly large and diverse readership concerned with issues in biomedical ethics—not just healthcare trainees and professionals, but also philosophers, social scientists, lawyers, social workers, and legislators. The series will feature both single-author and multi-author books, short and accessible enough to be widely read, each of them focused on an issue of outstanding current importance and interest. Philoshophers, doctors, and lawyers from a number of countries feature among the authors lined up for the series.

Family-Making

Contemporary Ethical Challenges

EDITED BY

Françoise Baylis
and Carolyn McLeod

OXFORD
UNIVERSITY PRESS

OXFORD
UNIVERSITY PRESS

Great Clarendon Street, Oxford, OX2 6DP,
United Kingdom

Oxford University Press is a department of the University of Oxford.
It furthers the University's objective of excellence in research, scholarship,
and education by publishing worldwide. Oxford is a registered trade mark of
Oxford University Press in the UK and in certain other countries

First published 2014
First published in paperback 2017

Published in the United States of America by Oxford University Press
198 Madison Avenue, New York, NY 10016, United States of America

British Library Cataloguing in Publication Data

Data available

Library of Congress Cataloging in Publication Data

Data available

ISBN 978–0–19–965606–6 (Hbk.)
ISBN 978–0–19–877658–1 (Pbk.)

Acknowledgements

This work was supported, primarily, through a Canadian Institutes of Health Research grant FRN 119620, "Family-Making: Contemporary Ethics Challenges." As well, additional partial funding was provided through Baylis's Canada Research Chair in Bioethics and Philosophy, McLeod's Graham and Gale Wright Distinguished Scholar Award, and a Canadian Institutes of Health Research grant FRN 102516, "Let Conscience Be Their Guide? Conscientious Refusals in Reproductive Health Care." Special thanks are owed to Timothy Krahn for assistance with references and the index, and to Sandra Moore for administrative support.

Contents

Part V. Special Responsibilities of Parents

Part VI. Contested Practices

Notes on Contributors

FRANÇOISE BAYLIS is Professor and Canada Research Chair in Bioethics and Philosophy at Dalhousie University, Canada, and founder of the NovelTechEthics research team. She is an elected Fellow of the Royal Society of Canada and the Canadian Academy of Health Sciences. Her research interests are many and varied, and she has a love of thought-provoking questions. Her publications on the ethics of assisted human reproduction and human embryo research span more than thirty years. In addition to her academic research on reproductive ethics, she contributes to national policy via government research contracts, membership on national committees and public education. This work – all of which is informed by a strong commitment to the common good – focuses largely on issues of social justice. Her website is: www.noveltechethics.ca. Her blog is impactethics.ca.

LUCY BLAKE is a Post-Doctoral Research Associate at the Centre for Family Research, University of Cambridge, UK. Her main areas of interest are family relationships and psychological well-being in non-traditional families. Her Ph.D. examined family functioning in donor insemination and egg donation families in the UK, with particular reference to the issue of disclosure. She is currently investigating family functioning in families in which gay fathers have had a child using a surrogate and an egg donor. She is particularly interested in research that explores family relationships from the perspectives of the different members of the family, especially those voices that are not typically heard. Her most recent publications have explored family relationships and donor conception from the perspective of the child.

ANDREW BOTTERELL is Associate Professor at Western University, Canada, where he is jointly appointed to the Department of Philosophy and the Faculty of Law. His main teaching and research areas are philosophy of law and metaphysics, and his work has appeared in, among other places, *Legal Theory, Law and Philosophy, Criminal Justice Ethics*, and *Philosophy Compass*. A former Supreme Court of Canada clerk, he currently serves as Associate Editor of the *Canadian Journal of Law and Jurisprudence*.

SAMANTHA BRENNAN is Professor of Philosophy at Western University, Canada. She is also a member of the Rotman Institute of Philosophy, an affiliate member of the Department of Women's Studies and Feminist Research, and a member of the graduate faculty of the Department of Political Science. Brennan received her Ph.D. from the University of Illinois at Chicago. Brennan's B.A. in Philosophy is from Dalhousie University, Halifax, Nova Scotia. Brennan has broad-ranging research interests in contemporary normative theory, including applied ethics, bioethics, political philosophy, children's rights and family justice, gender and sexuality, death, and fashion. In recent years Brennan was a visiting faculty fellow at the Australian National University and a

Taylor Fellow in Philosophy at the University of Otago. Her website is http://publish.uwo.ca/~sbrennan.

HARRY BRIGHOUSE is Professor of Philosophy and of Educational Policy Studies at the University of Wisconsin, Madison, US. He has written several books, including *School Choice and Social Justice* (Oxford, 2000) and *On Education* (Routledge, 2005) and, with Adam Swift, *Family Values* (Princeton University Press, 2014). He is co-director, with Mike McPherson, of the Spencer Foundation's Initiative on Philosophy in Educational Policy and Practice.

JULIE CRAWFORD is Associate Professor of English and Comparative Literature at Columbia University, US. She is the author of a book on cheap print and post-Reformation English culture called *Marvelous Protestantism* (Johns Hopkins, 2005) and her essays on sixteenth- and seventeenth-century literary culture have appeared in a wide range of journals and edited collections including *PMLA, ELH, SEL, Renaissance Drama* and *The Blackwell Companion to Shakespeare.* She has just finished writing a series of essays on the history of women's reading and another on Shakespeare and the history of sexuality. Her new book, *Mediatrix: Women and the Politics of Literary Production in Early Modern England,* will be published by Oxford University Press.

JURGEN DE WISPELAERE is an occupational therapist turned political philosopher, currently a Post-Doctoral Fellow with the Montreal Health Equity Research Consortium (MHERC) at McGill University, Canada. Previously he worked at University College Dublin, Trinity College Dublin, Université de Montréal, and the Universitat Autònoma de Barcelona. His publications span a wide variety of topics in ethics and politics, including unconditional basic income, disability, organ procurement, and family policy. He is a founding editor of the journal *Basic Income Studies,* co-editor of three collections, and currently completing a book on republicanism (Continuum Press, forthcoming).

HEATH FOGG DAVIS is Associate Professor of Political Science at Temple University, US. His research and teaching are in the areas of anti-discrimination law and public policy, democratic political theory, African-American political thought, and feminist and gender theory. Publications include "Theorizing Black Lesbians within Black Feminism: A Critique of Same-Race Street Harassment" in *Politics & Gender, The Ethics of Transracial Adoption* (Cornell University Press), "Navigating Race in the Market for Human Gametes" in the *Hastings Center Report,* and "The Racial Retreat of Contemporary Political Theory" in *Perspectives on Politics.* He is a member of the Arcus Foundation-funded working group on the points of convergence and divergence between the black American civil rights movement and black lesbian, gay, bisexual, and transgender activism at Emory University.

SUSAN GOLOMBOK is Professor of Family Research and Director of the Centre for Family Research at the University of Cambridge, UK. Her research examines the impact of new family forms on parenting and child development. She has pioneered research on lesbian mother families, gay father families, single mothers by choice, and families created by assisted reproductive technologies, including in vitro fertilization (IVF), donor insemination, egg donation, and surrogacy. Her research has not only challenged commonly held assumptions about these families but also has contested widely held theories of child development by demonstrating that structural aspects of the family, such as the number, gender, sexual orientation, and genetic relatedness of parents, is less important for children's psychological well-being than the quality of family relationships. In addition to academic papers she is the author of *Parenting: What Really Counts?* and co-author of *Bottling it Up, Gender Development,* and *Growing up in a Lesbian Family.*

KIMBERLY LEIGHTON is an Assistant Professor of Philosophy at American University, US, where she teaches applied ethics, ethical theory, and bioethics. Topics of her research include adoption, assisted reproduction, and theories of bioethics, with a particular focus on how public debates, regulatory policy, and ethical theory together contribute to understandings of norms. Her current manuscript project, "Genetic Bewilderment: What we think genes can tell us about how we should—and shouldn't— be related", offers an intervention into debates about anonymous gamete donation. Drawing insights from current and historical concerns about the genetic background of adoptees and from the discourse of eugenics, the book shows the risks of claiming we have a right to know our genetic origins. Her articles include "Being Adopted and Being a Philosopher: An Exploration of the 'Desire to Know' Differently" and "Addressing the Harms of not Knowing One's Heredity: Lessons from Genealogical Bewilderment."

MIANNA LOTZ is a Senior Lecturer in the Department of Philosophy at Macquarie University, Australia. She teaches in applied ethics and ethical theory, and her main areas of research are in applied ethics and bioethics, with a particular focus on questions of parental liberty, the family, adoption, and children's rights and interests. Her work has appeared in a variety of edited volumes and journals, including *Bioethics, Journal of Social Philosophy,* and *Journal of Applied Philosophy.* She is Chair of the Macquarie University Arts Faculty Research Ethics Committee.

CAROLYN MCLEOD is Professor of Philosophy and an Affiliate Member of Women's Studies and Feminist Research at Western University, Canada. Her research interests lie at the intersection of health care ethics, feminist theory, and moral theory. She has had a long-standing interest in reproductive ethics, beginning with her book, *Self-Trust and Reproductive Autonomy* (MIT Press, 2002). She has published on various topics in this area, including reproductive autonomy, the commodification of women's reproductive labour, and conscientious refusals by health care professionals to provide reproductive health services such as abortions. Recently, her knowledge of the ethics of having children has broadened to include issues that concern adoption.

JAMIE LINDEMANN NELSON is Professor of Philosophy and Faculty Associate, Center for Ethics and Humanities in the Life Sciences, at Michigan State University,

US; she was formerly the Associate Dean for Graduate Studies in MSU's College of Arts and Letters as well. Nelson is a Fellow of the Hastings Center, a Member of the Ethics Committee of the United Network for Organ Sharing, and a co-editor of the *International Journal for Feminist Approaches to Bioethics*. The author or co-author of three books and well over a hundred articles and chapters, Nelson's work has been supported by the National Science Foundation, the National Institutes of Health, and the Greenwall Foundation, *inter alia*; in 2000, she was the co-director of a Summer Seminar for College and University Teachers, funded by the National Endowment for the Humanities. In her off time, she sails small boats on very small lakes, and knits (awkwardly) for her grandchildren.

CHRISTINE OVERALL is Professor Emerita of Philosophy and holds a University Research Chair at Queen's University, Canada. She is an elected Fellow of the Royal Society of Canada. Her research interests include feminist philosophy, applied ethics, social philosophy, and philosophy of religion. She is the author of more than 110 journal articles and book chapters, and has published eleven books, including *Why Have Children? The Ethical Debate* (MIT Press, 2012). Her book, *Aging, Death, and Human Longevity: A Philosophical Inquiry* (University of California Press, 2003), won awards from the Canadian Philosophical Association and the Royal Society of Canada. Her most recent book is an edited anthology, *Pets and People: The Ethics of Our Relationships with Companion Animals* (Oxford University Press, 2017). She is the mother of two children and the grandmother of four grandsons.

JENNIFER A. PARKS is Professor of Philosophy, Associate Faculty Member in the Women's Studies and Gender Studies Program, and Director of the Bioethics Minor Program at Loyola University of Chicago, US. She works in the area of feminist bioethics with specific interests in reproduction and reproductive technologies and ageing and long-term care. She is the author of three books: *No Place Like Home? Feminist Ethics and Home Health Care; Ethics, Aging and Society: The Critical Turn* (co-authored with Martha Holstein and Mark Waymack); and *The Complete Idiot's Guide to Understanding Ethics* (co-authored with David Ingram). She has also published a textbook with Prentice Hall entitled *Bioethics in a Changing World*. Articles by Parks appear in journals such as the *Hastings Center Report, Bioethics, Hypatia*, and the *Journal of Medicine and Philosophy*.

MARTIN RICHARDS is Emeritus Professor of Family Research at the Centre for Family Research, University of Cambridge, UK. After spending most of his research career working on parent–child relationships and the development of children, he turned his attention to genetic and reproductive technologies and families, including the social history of these technologies and practices. He also has interests in bioethics, especially the process of consent and the return of results to research participants. He has served on the Ethics and Law Committee of the Human Fertilisation and Embryology Authority and the UK Human Genetic Commission. He is vice chair of the UK Biobank Ethics and Governance

Council. Recent books include *Reproductive Donation: Practice, Policy and Bioethics*, co-edited with G. Pennings and J. Appleby (Cambridge University Press, 2012).

TINA RULLI is Assistant Professor of Philosophy at the University of California, Davis, US. She works on issues related to the choice between adoption and procreation and the moral significance of the genetic relationship. Her work on adoption comes out of her dissertation, where she argued for a duty of prospective parents to adopt children rather than procreate. A main theme in Tina's work is that we ought to prioritize making the worst-off among us happy. This includes making the worst-off happy rather than making happy people. From 2011 to 2013, Tina was a Postdoctoral Fellow at the National Institutes of Health Bioethics Department in Bethesda, Maryland. Her research at NIH focused on the duty to rescue in medical and research settings. Tina received her Ph.D. in Philosophy from Yale University in 2011.

ADAM SWIFT is Professor of Political Theory in the Department of Politics and International Studies at the University of Warwick, UK. For twenty-five years, he taught political theory and sociology at the University of Oxford, where he was Founding Director of the Centre for the Study of Social Justice. He co-authored *Liberals and Communitarians* (2nd edn, Blackwell, 1996), *Against the Odds? Social Class and Social Justice in Industrial Societies* (Oxford University Press, 1997), and *Family Values* (Princeton University Press, 2014) but wrote *How Not to Be a Hypocrite: School Choice for the Morally Perplexed Parent* (Routledge, 2003) and *Political Philosophy: A Beginners' Guide for Students and Politicians* (3rd edn, Polity, 2013) on his own.

DANIEL WEINSTOCK is a Professor in the Faculty of Law and in the Department of Philosophy of McGill University, Canada, where he is also the Director of the Institute for Health and Social Policy. For twenty years, he taught in the Department of Philosophy of the Université de Montréal, where he was also the Founding Director of the Centre de recherche en éthique de l'Université de Montréal. He has published widely on a large range of topics in political philosophy. Current projects involve investigating the moral and political foundations of family policy, and developing a political philosophy of the city. He is also the Principal Investigator in a major collaborative research initiative looking into philosophical and empirical dimensions of health equity.

CHARLOTTE WITT is Professor of Philosophy and Humanities at the University of New Hampshire, US. She is the author of *Substance and Essence in Aristotle* and *Ways of Being in Aristotle's Metaphysics,* both published by Cornell University Press. She is the co-editor of *A Mind of One's Own: Feminist Essays on Reason and Objectivity* and three other collections including *Adoption Matters: Philosophical and Feminist Essays.* Her most recent work includes a monograph *The Metaphysics of Gender* (Oxford University Press, 2011) and an edited volume *Feminist Metaphysics: Explorations in the Ontology of Sex, Gender and the Self* (Springer, 2011).

Introduction

Françoise Baylis and Carolyn McLeod

Many people want children but cannot have them through sexual intercourse, because they and/or their partner are infertile or they are not heterosexual. Other people want children and could have them through sexual intercourse, but choose not to because they do not have a sexual partner, they are at risk of having a child with a serious genetic condition, or they believe that there is a moral obligation not to create more children when there are existing children who are in need of families. In the not-so-distant past, these people had limited options in terms of how or whether they had children; for example, *in vitro* fertilization was not available to those who wanted to reproduce, and those who could not create a "wholesome" environment for a child (based on socioeconomic status, marital status, sexual orientation, age, and so on) were denied the ability to adopt children. Now, particularly in Western democratic states, many people who cannot, or do not want to, have children through sexual intercourse—including heterosexual or homosexual couples and single people—have more options available to them than would have been the case in the past.

This book is about the ethics of having children by adoption and technologically assisted reproduction (or "assisted reproduction" for short). For many people, these methods are separate and distinct from one another; they can choose *either* adoption *or* assisted reproduction. By contrast, for others these options blend together. Consider, for example, same-sex female couples for whom, in some jurisdictions, the path of assisted reproduction is complicated by the need for the partner who is not genetically related to the resulting child to adopt this child, if she wants to become the child's legal parent (see Crawford, Chapter 9). The situation is similar (again in some jurisdictions) for same-sex male couples and heterosexual couples using contract pregnancy where neither, or only one, partner is genetically related to the offspring. These couples do not choose between adoption and assisted reproduction (see Baylis, Chapter 14). Rather, they choose between adoption *without* assisted reproduction and assisted reproduction *with* adoption. (For simplicity's sake, we—and the authors in

this collection—sometimes refer to choosing between adoption and assisted reproduction, accepting that, in some cases, the option of assisted reproduction is coupled
with adoption.)

Some people confronted with the alternatives of adoption or assisted reproduction
struggle between them. Others do not agonize in this way. In the latter group are people who have a profound desire for a genetic link to the child(ren) they will parent
or a substantial interest in experiencing pregnancy and childbirth. For these people,
assisted reproduction is the preferred alternative. Also in this latter group are people
who have a profound desire to adopt a child (or children) and thus see adoption as
the best alternative. For them, adoption might be the only morally decent choice in a
world that is already overcrowded or in which some children simply do not have parents (see Rulli, Chapter 6).

This book critically examines moral choice situations that involve adoption and
assisted reproduction as ways of making families with children, and highlights the
social norms that can distort decision-making. Examples of such norms are those that
favour people having biologically related children ("bionormativity") or those that
privilege a traditional understanding of family as a heterosexual unit with one or more
children where both parents are the genetic, biological, legal, and social parents of
these children (see Witt, Chapter 3). Factors that could legitimately tip the balance in
favour of adoption or assisted reproduction are also discussed, such as cost, genetics,
scrutiny of prospective parents by the state, and discrimination against certain types
of families.

As a whole, the book looks at how adoption and assisted reproduction are morally distinct from one another, but also emphasizes how the two are morally similar.
Choosing one, the other, or both of these approaches to family-making can be complex in some respects, but ought to be simple in others, provided one's main goal is
to become a responsible, caring, and loving parent who willingly takes on the moral
obligations of parenting—obligations that involve both protecting and promoting the
rights and interests of one's children (see Brennan, Chapter 2).

Context

The title of the book, *Family-Making*, speaks to the fact that people use assisted reproduction and adoption to make, and sometimes expand, families that include children.
Families are composed of relationships in which people have moral responsibilities for
each other. In creating a family with children, one is acquiring responsibilities, serious
ones that go along with being a parent, rather than simply getting a baby or a child.

The fact that assisted reproduction is a form of family-making, not just baby-making, is barely evident, however, from the literature in reproductive ethics and indeed
from the experience of being a fertility-treatment patient. Few people who write about
ethics and assisted reproduction focus on the moral dimensions of becoming a parent

using assisted reproductive technologies. Instead, many focus on whether people have a right to children or a right to procreate. In addition, the state does no screening of prospective parents who choose to attempt assisted reproduction. Moreover, insofar as there is mandatory counselling before fertility treatment, it has little if anything to do with the moral responsibilities that one incurs in becoming a parent. Rather, the counselling focuses on what one can expect during treatment.

In sharp contrast, one can neither read the ethics literature on adoption nor be someone who is in the process of becoming an adoptive parent without realizing that adoption is a family-making enterprise that comes with considerable moral responsibilities. For example, the adoption process routinely includes mandatory screening and counselling that focuses on whether one is morally competent to be a parent, especially, though not exclusively, to an adopted child.

The different approaches to mandatory screening and counselling for assisted reproduction and adoption suggest different background assumptions about the role of the state in constraining family-making in the best interest of children. One of these assumptions has been that children who are adopted experience significantly more problems (psychological or otherwise) than children who are created using assisted reproduction: hence the need for increased scrutiny of prospective adoptive parents. The thought here—that families created using assisted reproduction as opposed to adoption are more likely to function well and therefore would not benefit from mandatory screening or counselling—is empirically questionable, however (see Blake, Richards, and Golombok, Chapter 4). Another assumption that could explain differences in the processes of becoming a parent through assisted reproduction and adoption is that the state has different moral and legal obligations towards its dependent members (e.g. live-born children in need of parents through adoption or foster care) than it does towards children who are but a twinkle in their parents' eye. However, it is not obvious that the state has any less of a moral obligation towards children born of assisted (or unassisted) reproduction, compared to children who are available for adoption, to ensure that these children have good or good enough parents (see McLeod and Botterell, Chapter 8).

This book challenges a number of morally questionable ideas, held by individuals and states, about how families formed through assisted reproduction or adoption differ from one another. At the same time, the book confirms that moral differences do exist between these families; in particular, differences that concern the kinds of responsibilities that parents have towards their children. Indeed, some chapters emphasize that, in choosing adoption or assisted reproduction, one incurs special responsibilities as a parent (Chapters 10, 11, and 12).

Overview

The book begins with two chapters that address basic moral questions about all families, with no particular attention to the means of family-making. In the opening chapter, Harry Brighouse and Adam Swift explain why it is worthwhile for most, though not all, adults to have children. In their view, parent–child relationships are not interchangeable with other intimate relationships that adults might have; being a parent makes a unique and important contribution to many people's flourishing; and whether the parenting is biological or adoptive is irrelevant to the kinds of benefits that it can provide. Brighouse and Swift argue that the goods of parenting exist, in part, because of the moral role that parents play in the lives of their child(ren). In Chapter 2, Samantha Brennan discusses the moral obligations attached to this role of promoting and protecting a child's well-being. She focuses on what parents are obligated to do to ensure the well-being not just of the future adult that their child will become, but also of the child herself. She presents a theory of children's well-being that is part of a theory of what rights they hold.

The next section of the book takes a critical look at the concept of "bionormativity" from both a philosophical and an empirical perspective. According to the bionormative conception of the family, families formed through biological reproduction are superior to other families. In her chapter, Charlotte Witt argues that because philosophical arguments supporting this vision of the family fail and the vision itself is stigmatizing, we ought to get rid of it. In the next chapter, Lucy Blake, Martin Richards, and Susan Golombok discuss empirical evidence about family functioning and child well-being in adoptive families, as well as families formed through assisted reproduction. They conclude that the results are good for the majority of families in both of these categories. Hence, their chapter represents a significant challenge to bionormativity.

Next, the focus is on the value of procreation and adoption, with particular attention to moral and pragmatic reasons for choosing one or the other family-making strategy. Christine Overall argues that, generally speaking, prospective parents do not have good moral or pragmatic reasons to prefer procreation, while Tina Rulli argues that they have strong moral reasons to choose adoption. According to Rulli, adoption has unique moral value for all prospective parents, not just for those who cannot have children without assistance from others.

The choices people make in becoming parents through assisted reproduction, adoption, or both are, in important respects, shaped by state policies and practices. The next few chapters question the legitimacy of specific state-sanctioned or imposed restrictions on family-making. Jurgen De Wispelaere and Daniel Weinstock discuss the legitimacy of financial barriers to assisted reproduction in jurisdictions where relevant technologies are not publicly funded. Arguably, such barriers are unjustified if people have a positive right to reproduce. De Wispelaere and Weinstock do not argue in favour of a positive right to reproduce. They do, however, defend the claim that people have a right to become parents. Further, since this right can be satisfied through adoption, they argue that financial barriers to assisted reproduction are justified in circumstances where there are many children available for adoption and where the obstacles to becoming an adoptive parent are not severe. An example they give of such an obstacle is onerous state-imposed licensing of adoptive parents. The next chapter, by Carolyn McLeod and Andrew Botterell, focuses squarely on parental licensing and what they call the "status quo" of requiring that only adoptive parents undergo licensing (i.e. by having to have a home study and possibly also having to take parental classes). By analyzing the arguments one might give in favour of the status quo and showing how they fail, McLeod and Botterell argue that the status quo is morally objectionable. In making their argument, they reflect briefly on their personal experience in being licensed to adopt children. Likewise, Julie Crawford, who has the last chapter in this section, discusses her personal experience of using assisted reproduction and adoption to become a non-genetic, in some ways biological, adoptive legal mother to a child born by her same-sex partner. A social background of heteronormativity profoundly shaped this experience, as the subtitle of her chapter suggests ("My Daughter is Going to be a Father!"). Crawford is appropriately critical of regulatory regimes for assisted reproductive technologies that automatically extend legal parenthood to non-genetic fathers, but withhold it from non-biological mothers until they go through (and pay for) an adoption.

The next three chapters look at the special responsibilities of parents incurred both in becoming and being a parent through assisted reproduction or adoption. In her chapter, Jamie Lindemann Nelson explains that all parental responsibilities are special or unique; however, these responsibilities can be "extra special" for some people who engage or participate in assisted reproduction. Included within this group, for Nelson, are people who reproduce using donor gametes and people who use assisted reproductive techniques that cause the birth of multiple children. The other two chapters on the special responsibilities of parents focus on adoptive parents. Mianna Lotz discusses the duties incurred by adoptive parents post-adoption. According to her, people who become parents through adoption have special obligations towards their children, who are likely to experience unique challenges given the prevailing bionormative conception of the family. The vulnerabilities identified by Lotz concern identity, development of a healthy sense of self and sense of belonging, and emotional independence. The main post-adoptive, parental obligation that she defends is that of initiating "communicative openness" with one's children about their adoption. In his chapter, Heath Fogg Davis looks specifically at the challenges likely to be experienced by black children who are adopted into white families in the United States (a country plagued by a very particular history of slavery and racism). Existing residential segregation in the United States (and elsewhere) between black and white communities raises both challenges and responsibilities for racially integrated families. Fogg Davis contends that adoptive parents in these circumstances have an obligation to select a racially diverse community for their biracial family. But rather than being entirely special to them, this responsibility is—as Fogg Davis puts it—"a more magnified version of the general moral responsibility that we all have to make residential decisions that do not perpetuate longstanding patterns of racially segregated housing" (pp. 222).

The book ends with three chapters on contested practices: anonymous gamete donation, transnational commercial contract pregnancy in India, and advanced age parenting. Kimberley Leighton critically examines the view that children born of anonymous gamete donation are fundamentally harmed by not having access to information about their genetic heritage and that they therefore have a right to this information. Arguments in favour of this view typically draw an analogy with adoption; they insist that children in closed adoptions are harmed by not knowing their genetic origins, and that children born through anonymous gamete donation must therefore be similarly harmed. According to Leighton, such arguments are flawed insofar as they ignore important differences between what it means to be adopted as opposed to donor-conceived. Along the way, she is critical of claims in favour of donor-conceived people having a "right to know" their genetic origins. In the next chapter, on transnational commercial contract pregnancy in India, Françoise Baylis looks at the harms of this practice for gestating Indian women and Indian women as a group, with a particular focus on the harm of exploitation. She then turns her attention to the potential harms for children born of commercial contract pregnancy where neither, or only one, partner is genetically related to the offspring. Here the focus is on identity formation. Baylis is more sympathetic than Leighton to the experiences of donor-conceived persons who report harms to identity formation from the withholding of information about biological parentage and about kinship relations. For children born of commercial contract pregnancy in India, Baylis is concerned about this harm and about the stigma associated with not knowing biological relatives (including one's birth mother), with being birthed by a woman who was exploited, and with being commodified. In the closing chapter of the book, Jennifer Parks discusses new fertility preservation technologies and the opportunity they provide for some women to become biological parents at an advanced age. She argues, from a feminist perspective, that the use of cryopreserved oocytes by older women (e.g. 60-year-old women) is morally permissible. However, she also counsels in favour of the removal of restrictions on older women or men becoming parents through other means, including adoption.

In summary, each chapter of this book contributes to our understanding of the moral and practical challenges of contemporary family-making practices. These challenges are considerable. They include whether people should rely on others' reproductive labour in having children, whether they should ensure that they will have a genetic tie to their children or that their children will have some connection to genetic relatives, whether they should bring a new child into the world at all, whether they should agree to what the government would require of them for an adoption, where they should live if the family they make is multi-racial, at what age they should forgo having children, and the list goes on. Together, the chapters shed considerable light on how individuals or governments should respond to the many ethical challenges involved in making families through adoption or assisted reproduction.

Addendum

This book explores morally relevant and irrelevant differences between families formed through adoption and assisted reproduction. In such a project, careful attention to the terminology used to describe these families or their members is essential. Terminology that supports or reifies differences that do not actually exist should be avoided, although that's easier said than done. Consider the terms commonly used to distinguish adoptive parents from other kinds of parents, all of which can be problematic:

- "Biological parents," which can be ambiguous when applied to women, who can contribute gestationally, genetically, or both to the creation of a child, and which is troubling when applied to men and women at the same time, as though they necessarily contribute in the same way biologically to their offspring.
- "Genetic parents," which, some argue, implies that genetic ties matter more than they actually do in being a parent; on this view, one is never *merely* a genetic parent.
- "Birth parents," which, when applied to men, is inappropriate if they were not present during the gestation and birth of the child, and which arguably is morally loaded because of its connection to the birth mother movement.
- "Natural parents," which is problematic in its suggestion that parents who have not adopted their children are naturally parents, while the same is not true of adoptive parents.

Contributors to this book have navigated as carefully as possible through the difficult terrain of naming in morally appropriate ways the parents, families, and children that result from adoption and assisted reproduction. They do not all use the same terms, because, as we have found, no terms are perfect. However, usually their use of specific terms is intentional and well thought out. There are some terms they have avoided or use only in a critical fashion, such as a "child of one's own," which does not refer just to biological or genetic children, although normally that is its intended meaning. Throughout, the goal has been to choose the best terms to distinguish among families that, morally speaking, have a lot in common with one another.

PART I

Families: Of Parents and Children

1

The Goods of Parenting*

Harry Brighouse and Adam Swift

Introduction

This chapter aims to identify the distinctive contribution that parent–child relationships make to the well-being or flourishing of adults.[1] The claim that those relationships are very important for children—perhaps especially for their emotional development—is widely accepted; we subscribe to that consensus. But the idea that adults benefit from parenting children, while no less familiar, warrants more careful attention than it has generally received.[2] By giving it that attention, we hope to challenge some conventional ways—often so taken-for-granted as to be unstated—in which parents think about their children. In particular, we query the significance of the biological connection between parent and child.

Though rarely conceived in such terms, it is widely believed that adults who get to parent children enjoy goods in their lives that are not realizable through alternative relationships, however intimate or loving, such as those with lovers, friends, or pets. Certainly many adults who desire strongly to become parents would reject the view that these other relationships can be adequate substitutes. They could be wrong: people can want things that do not in fact make their lives go better. This is not just a matter of their discovering, with hindsight, that something they wanted turns out to be something they would rather not have had. People can spend their whole lives believing things in it were good for them when, in fact, those things made their lives worse. So some of those who want to be parents may be mistaken about what will be good for them—perhaps, for them, other relationships would be as good or better—and some who are parents and think that being a parent is good for them may be mistaken about that too. Also, and perhaps more interestingly, people can misunderstand what is good

---

* This chapter draws on material in Brighouse and Swift (forthcoming).
[1] In this chapter we treat "well-being" and "flourishing" as synonymous, varying our usage only to avoid repetition. For us, anything that "benefits" a person makes her life better for her and should be understood as contributing to her well-being or flourishing.
[2] But, in addition to the works cited later, see Austin (2007) and Richards (2010).

about the things they are right to value. Parenting is indeed special, and especially valuable. But what makes it special is not necessarily what those who want to be parents think is special about it; some, we suggest, value parenting for the wrong reasons.

Why Parents?

It's easy to see why children should be looked after by *adults*, but we could imagine a system in which different adults were in charge of them at different ages—specialists in dealing with young babies being replaced by experts on toddlers, who in turn would cede authority to those with advanced qualifications on the development of 4–5 year olds, and so on. Or if we thought continuity of care was important, new-born babies could be handed over to state-run childrearing institutions staffed by well-qualified professionals. Or perhaps groups of twenty or thirty adults living together in communes could share the tasks of childrearing between them, with no particular child being the particular responsibility of any particular adult. In none of these alternatives would children have *parents*, as we will understand that term, and societies that reared their children those ways would not have *families*.

How does one go about evaluating childrearing arrangements? Some philosophers think that there are things that societies must (or must not) do to or for people irrespective of whether doing (or not doing) those things will make people's lives better. But we focus on the well-being interests of the different parties who have a stake in the matter. First, and most obviously, there are children; their vulnerability, and the fact that, however they are raised, they cannot be thought to have had any say in the matter, are so glaring that it is hard to hold that their interests play no role. Second, there are adults; adults too may flourish less or more depending on their society's rules about how they may and may not be involved in the process of childrearing. Third, there are third parties; whether or not an individual is herself directly involved in raising children, she will surely be affected by the way her society goes about it, since childrearing arrangements are bound to have what economists call externalities.

Though useful for analytical purposes, this tripartite division doesn't identify distinct people. Not all children become adults, alas, but all adults were once children; and all people, both children and adults, suffer or enjoy the negative or positive externalities of other people's childrearing arrangements. This framework is an intellectual tool for thinking about the distinct ways in which we are all affected by decisions about how children should be raised. Any individual, thinking just about what is best for herself, will seek to combine these different perspectives and come up with an all things considered judgement about which childrearing practices would be, or would have been, best for her overall. We can approach the social decision in essentially the same way.

This chapter focuses on the value of parenting to parents because that is relatively unexplored territory, not because we think adults' interests are more important than children's, or because we think the interests of third parties are irrelevant. If the kind of relationship we are going to describe were not also good for children, then it could not justify the practice of parenting. If childrearing arrangements that were valuable for parents and children were damaging to third parties, then that too would count importantly against them. But the idea that, generally speaking, children are better raised if they experience this kind of relationship is well established: basic attachment theory and other staples of child development all point in that direction (Waldfogel, 2006). It is conventional also to regard parent–child relationships as crucial for turning children into law-abiding, cooperative fellow citizens. (Witness the popular concern that young people's lack of discipline is due to parental failure: Morse, 1999.)

The fact that people want something doesn't mean they should be allowed, or helped, to get it. Perhaps, instead, the activity of parenting should be distributed only to those who would do it best. Would there be anything wrong with a system that distributed children to adults in a way that maximized the realization of children's interests, even if it left out some adults who would be willing, and adequately good, parents? We think there would. To be a parent is to have a certain kind of relationship with a child, and in our view many adults have a weighty interest in enjoying that kind of relationship. The relationship contributes extremely valuable and non-substitutable benefits to adults' lives—goods that we call "familial relationship goods." For many, parenting a child makes a distinctive and weighty contribution to their well-being as adults. It is distinctive in that it cannot be substituted by other forms of relationship, and, we claim, the goods in question are important enough to impose a duty on others to allow, and indeed to enable, adults to enjoy them.

What's Special about Parenting?

For most people, intimate relationships with others are essential if their lives are to have meaning for them. Rather than being alone in the world, seeking to fulfil their own pleasures, people thrive when they are connected to other human beings with whom they enjoy deep and close relationships. These relationships are challenging—in an intimate relationship one does not fully control the response of the other person; one has to discern her interests even when she does not necessarily articulate them well, and act to further those interests and come to share some of them as one's own. The love and voluntary compliance of others in a relationship, when recognized, results in a sense of well-being and self-worth, as does successful attendance to the well-being of those others. A life without such relationships, or in which they all fail, is usually an unsuccessful life.

But our intimate relationships are not all the same—they are not substitutable one for another. People need more than one kind. Most need, usually, a romantic lover, someone to whom we can bare our raw emotions and whom we are confident will love us anyway, with whom we share sexual love. We need close friendships that last, if not a whole lifetime then some long part of it, with people on whom we can rely for support when in need and who we know can rely on us, with whom we can share our joys and interests. We also need more casual relationships—relationships of trust with people whose lives we do not know intimately but with whom we form bonds around some particular shared interest, project, or adversity. A successful life is a life with a variety of successful relationships, including a variety of successful intimate relationships.

We believe that many, perhaps most, adults need to be involved in an intimate relationship of a very particular kind in order to have a fully flourishing life. The parent–child relationship is not, in our view, just another intimate relationship, valuable to both sides but substitutable for the adult by an additional relationship with a consenting adult. The relationship is, on the contrary, *sui generis*, a relationship that involves the adult in a quite unique combination of joys and challenge; experiencing and meeting these makes a distinctive set of demands, and produces a distinctive contribution to well-being. Other intimate relationships have their own value, but they are not substitutes for a parenting relationship with a child.[3]

The parent is charged with responsibility for both the immediate well-being of the child and the development of the child's capacities. The child has immediate interests in being kept safe, enjoying herself, being sheltered and well nourished, having loving relationships with others, etc. She has future interests in many of these same things, but also in becoming the kind of person who is not entirely dependent on others for having her interests met, and the kind of person who can make her own judgements about her interests, and act on them. The parent's fiduciary duties are to guarantee the child's immediate well-being, including assuring to her the intrinsic goods of childhood (see Brennan, in this volume), and to oversee her cognitive, emotional, physical, and moral development. Four broad features of this relationship combine to make the joys and challenges of parenting different from those that attend other kinds of relationship, including other kinds of fiduciary relationship.

First, obviously, parents and children cannot have equal power. Children are not in the relationship voluntarily and, unlike adults, they lack the power to exit the relationship at least until they reach sufficient age to escape (which age will be culturally sensitive, since different societies will monitor and enforce parental power with different levels of enthusiasm and effectiveness). Children are vulnerable to the decisions and choice-making of their primary caretakers, and, initially, wholly dependent on them for their well-being. An adult supervising a child has the power of life or death; and this is not, at least when the child is young, reciprocated. But, more importantly, and less spectacularly, they have the power to make the child's lives miserable or enjoyable (within limits, at least at the enjoyable end).

The second difference between this and most other fiduciary relationships concerns the paternalistic aspect. The parent–child relationship routinely involves coercing the

[3] Frederick Schoemann (1980) puts the interest in intimacy central but fails to recognize the distinctive features of the intimacy specific to parent–child relationships. An account that shares some of the features of ours can be found in MacLeod (2002).

child to act against her own will, or manipulating her will so that it accords with her interests. So, for example, we might lock away the bleach so that she cannot get at it, even though she has displayed great interest in it, or prevent her from having a third helping of ice cream, on the grounds that neither the bleach nor the ice cream will serve her interests. We might persistently serve whole-grain pasta in the face of her frequent (and accurate) complaints that it is tasteless, in order to habituate her to frequent intake of whole grains. We might engineer her social life in order to diminish the significance of a destructive friendship. Although in relationships with other adults we are obliged to take their interests into account, we do not have fiduciary responsibilities of this kind towards them. Indeed, if one saw one's relationship with, say, one's spouse, in this way, one could reasonably be accused of being overbearing, disrespectful, or unloving. In intimate relationships with other adults one might advise and even argue but one does not routinely coerce and manipulate, even in the other's interests. To do so would be to fail as a spouse or friend, just as to refrain from doing so with one's children would be to fail as a parent. And where we do have distinctively fiduciary relationships with other adults—even with ageing parents—coercing or manipulating them may sometimes be required but it is not itself a key part of the job.

A third difference concerns the relationship of the fiduciary (the parent) to the interests of the principal (the child). When the parent–child relationship begins, the child does not have specific beliefs about what is good for her. Later, when she *does* have beliefs, they have been formed in response to the environment structured by the parent and, if the parent has been caring for the child, by someone whose capacities have been shaped by the parent. The parent has a good deal of latitude in shaping the child's emerging values, values that will guide her in her own life. In other fiduciary relationships what the fiduciary should pursue on the principal's behalf is typically fixed by reference to the principal's own beliefs about what is good for her, sometimes expressed directly to the fiduciary, sometimes (as in the case of advanced directives) expressed previously. But the parent does not have and could not have such a standard to guide her. The parent should be guided, rather, by those interests of the child that it is the parent's fiduciary duty to respect and promote. Of course there will be differing accounts of what those interests are but, in our view, one important parental duty is to try to ensure that the child will become an autonomous agent, someone capable of judging, and acting on her judgements about, her own interests. This is a lengthy process, and one that does not just naturally occur but requires active support. It is, for most parents, emotionally as well as practically challenging to prepare a child who has been entirely dependent, and whom the parent loves deeply, to become her own person, capable of effectively challenging the parent and the parent's values; capable, ultimately, of rejecting the adult if she thinks it appropriate. Three natural inclinations are frequently at odds with trying to ensure the child's genuine independence: the inclination to be protective of the loved child, the inclination to promote her well-being according to one's own view of what that would amount to, and the inclination to hold on to her for one's own sake. To overcome these inclinations successfully, when one really loves one's child, is emotionally demanding. Successful parenting is, in this respect, an exercise in maturation because, while the parent has the control that he needs in order to carry out his caring and fiduciary tasks for the child, he simultaneously learns that one *should not* control another person in the way he might like, and learns how not to exercise some of the control he does indeed have. For example, the parent must give the child opportunities for emotional and physical independence, putting the child in situations where she is at risk of failing, but in which the stakes of failure are sufficiently low that the child will be able to bear, and learn from, failure if it happens.

The fiduciary responsibilities of parenthood constitute a distinctive moral burden. But, of course, along with the moral burden come distinctive sources of satisfaction of a much less complicated kind. What children need from parents is not simply the judicious exercise of expertise and authority, of the kind one might hope for from a lawyer or doctor or teacher. What's needed is a *relationship*, and the kind of relationship children need from adults—a parent–child relationship—is also the kind that yields good things to the adults doing the parenting. There is the enjoyment of the love (both the child's for oneself and one's own for the child), but also the enjoyment of the observations the child makes about the world; the pleasure (and sometimes dismay) of seeing the world from the child's perspective; enjoyment of her satisfaction in her successes and of consoling her in her disappointments.

The final difference from other relationships, then, concerns the *quality of the intimacy* of the relationship. The love a parent normally receives from his children, again especially in the early years, is spontaneous and unconditional, and, in particular, outside the rational control of the young child. She shares herself unselfconsciously with the parent, revealing her enthusiasms and aversions, fears and anxieties, in an uncontrolled manner. She trusts the adult in charge until the trust is betrayed, and trust must be betrayed consistently and frequently for it to be completely undermined. Adults do not share themselves with each other in this way: intimacy requires a considerable act of will on the part of adults interacting together. But things are different between parents and children. The parent is bound by his fiduciary responsibilities for the child's emotional development to try to be spontaneous and authentic a good deal of the time, both because the child needs to see this modelled and because the child needs to be in a loving relationship with a real, emotionally available, person. And, of course, the parent will often be inclined to be spontaneously loving. But his fiduciary obligations also often require him to be less than wholly spontaneous and intimate (despite the child's unconditional intimacy with him). The good parent sometimes masks his disappointment with, sometimes his pride in, the child, and often his frustration with other aspects of his life. He may sometimes hide his amusement at some naughtiness of the child, preferring to chide her for the sake of instilling discipline; conversely, he may sometimes control his anger at similar behaviour, substituting inauthentic kindness for the sake of ensuring a better end to the child's day, or because he knows that his angry reaction is, though authentic, inappropriate. He does not inflict on the child, as the child does on him, all of his spontaneous reactions and all of his emotional responses.

These four features combine to make the relationship between parent and child unlike other intimate relationships, and unlike other fiduciary relationships. Children have a weighty interest in the kind of relationship that will meet their needs and promote their vital interests. Given what that involves—given how complex, interesting, and conducive to the adult's own emotional development it is to be the adult in that relationship—adults too have a weighty interest in being in a parenting relationship. The interest is distinctive because what the relationship requires of the adult, and allows the adult to experience, is unique. It cannot be substituted even by other intimate relationships where those are consensual on both sides and in which the parties are symmetrically situated. The relationship as a whole, with its particular intimate character, and the responsibility to play the specific fiduciary role for the person with whom one is intimate in that way, is what adults have an interest in.

The fiduciary aspect to the parental relationship with children has been widely acknowledged since Locke, and is given particular emphasis by so-called "child-centred" justifications of the family.[4] Our claim is adult-centred: many adults have an interest in being in a relationship of this sort. They have a non-fiduciary interest in being in a relationship in which they act as a child's fiduciary. That relationship enables them to exercise and develop capacities the development and exercise of which are crucial to their living fully flourishing lives. The parent comes to learn more about herself, she comes to change as a person, and she experiences pleasures and emotions that otherwise would be unavailable.

We need to tread carefully here. It should be clear that the adult's interest in playing the fiduciary role is not entirely independent of the content of that role. It's because of *what* children need from their parents that adults have such a weighty interest in giving it to them.

Imagine a world in which human children didn't need much more looking after than guinea pigs, or those Tamagotchi toys that were so popular a while back. Imagine that they could fully develop into autonomous, emotionally adjusted adults, and enjoy the intrinsic goods of childhood, with that kind and level of input from adults. We think that, even in that hypothetical world, there would be *some* value to being the person responsible for ensuring that children's interests were met. One would be responsible for the development of a human child, which is a weighty responsibility indeed, and it is good for people, it makes *their* lives go better, to take on that degree of responsibility. So when we say that, in our world, playing the fiduciary role contributes importantly to the flourishing of (most) adults, the sheer fact of being the person responsible for the child is part of the story.

---

[4] Locke (1988) says "parents were, by the Law of Nature, under an obligation to preserve, nourish, and educate the Children they had begotten; [though] not as their own Workmanship, but as the Workmanship of their own Maker, the Almighty to whom they were to be accountable for them" (p. 180, sec. 56). Contemporary theorists who emphasize the fiduciary interest, despite giving otherwise different accounts of the relationship, include Reich (2002: 148–51); Galston (2002: 101–6); Callan (1997: ch. 6); Dwyer (1999); Brennan and Noggle (1997); and Archard (2004).

But only part of it. Properly to see the weight of the adult interest in parenting, we need to keep our eye not on the plain fact of being the fiduciary but on the content of what children need from those who are their fiduciaries. Adults have an interest in being the fiduciary, and parents serving as fiduciaries affects the significance, and hence the value, of so much else that happens in the relationship. But what's really valuable here is not being the fiduciary *per se* but having the kind of relationship that, in fact, is in children's interests. It's that kind of relationship that presents a distinctive challenge, and that kind of challenge that gives adults unique opportunities for flourishing.

Adults can be involved in any number of fiduciary relationships. In our professional lives, as lawyers or social workers or doctors or teachers, we take on duties to serve the interests of our clients or patients or students. In our personal lives, too, we may find ourselves acting as fiduciaries for our ageing parents, for example, if they cease to be able adequately to protect and promote their own interests. If we think about the difference between these other kinds of fiduciary relationships, and the particular case of the parent–child relationship, we can see that some elements in what is special about being a fiduciary for a child concern the fact that what we're talking about here is a *child*. Relevant here is the moral standing of the person for whom one is acting as fiduciary: her possessing the capacity to develop into an autonomous adult, her degree of vulnerability to one's judgements, her involuntary dependence on one, and so on. Failing adequately to discharge one's fiduciary duties to a child would be different from failing to discharge those owed to a client or patient, or even to an ageing parent, even if what was involved in fulfilling the duties were the same. But of course they are not the same. Other elements in what is special about being a fiduciary for a child concern *what* it is that children need from their fiduciaries. They need a special kind of *relationship*—a relationship in which the adult offers love and authority, a complex and emotionally challenging combination of openness and restraint, of spontaneity and self-monitoring, of sharing and withholding. It's that kind of relationship that many adults have an interest in.[5]

To be sure, the fiduciary aspect remains central. Grandparents, uncles and aunts, parents' friends, or nannies, can have close relationships with children, and when they go well those relationships will be conducive to the child's interests and valuable to the adults too. Reading bedtime stories, providing meals, and so on will be contributing to the well-being of both. But there's something distinctively valuable about being the person who not only does those things oneself but has the responsibility to make sure they get done, sometimes by others, and the authority to decide quite how they get

---

[5] It's an interesting question how many parents a child can have consistent with this kind of relationship. Single-parent families clearly qualify, and we see no reason why three or four parents should not share the parenting of a child. More than that and we would start to worry about the dilution of intimacy and authority inherent in "parenting by committee." For discussion see Brennan and Cameron (n.d.).

done. The challenge is different, and the adult who meets that challenge enjoys a special, and especially valuable, kind of human flourishing.[6]

Is our Picture of Parenting Too Rosy?

Our emphasis on the fiduciary aspect of parenting points to something paradoxical
about the widespread desire to be a parent. That is a desire to take on burdens, voluntarily to put oneself in the position of owing things to others that severely limit one's
capacity to pursue other goals.[7] We have tried to explain what adults get out of the
relationship, as it were, in a way that helps to make sense of the paradox, but we suspect
that some readers will find our account of the joys of family life somewhat naïve or
complacent, and suspiciously optimistic in its neglect of the burdens that accompany
parenthood.

   For many, parenthood is indeed a source of deep anxiety and frustration. It is a vital
source of flourishing only if it is carried out in a social environment that renders its
challenges superable. So, for example, poverty and the multiple disadvantages that
accompany it can easily create a micro-environment in which it is very difficult even to
develop, let alone to exercise, the cognitive and emotional skills that successful parenting requires. Meanwhile, children raised in poverty are typically at much higher risk of
very bad outcomes than more advantaged children, so that parents seeking conscien-
tiously to protect their children from such outcomes require greater internal resources
than are needed by the parents of more advantaged children. Adults have a weighty
interest in parenting a child in circumstances that will indeed enable them to realize
the goods we have identified. In another context, we might follow this thought through
to explore the implications for social policy of our account of "family values."[8]

   But parenting a child is not all-consuming. It's true that, done properly, raising
a child severely limits one's opportunities to do other things. Some people choose
not to be parents for precisely that reason. It's true also that raising a child is likely
to be one of the most important things one does with one's life. As Eamonn Callan
(1997: 142) says, "success or failure in the task, as measured by whatever standards
we take to be relevant, is likely to affect profoundly our overall sense of how well or
badly our lives have gone." But although the interest in the fiduciary aspect of the
role is important, parents should not be slaves, entirely and continually subordinat-
ing their own interests to those of their children, or always putting their children first.
We cannot here set out in any detail what rights parents should have with respect to

---

[6] Nannies sometimes experience an almost complete variant of the full package—effectively doing most
of the parenting. In our view, one of the tragedies in that relationship is that its security is vulnerable to the
arbitrary power of the child's official "parents."

[7] That is why Alstott (2004) argues her case for financial support for parents by appeal to the idea that they
should be compensated for their loss of autonomy.

[8] For some thoughts in this direction, see Brighouse and Swift (2008).

the children they parent, but it may be helpful to outline briefly two different ways in which parenting is not like slavery.

On the one hand, parents are not only parents. Quite how much of one's time and energy parenting demands will of course vary with the age and particular characteristics of the child, but generally speaking it is perfectly possible to parent well while performing other roles and pursuing other interests. It is common to talk about the "best interests" of the child, and that may indeed be an appropriate practical criterion for adjudicating custody disputes where things have gone wrong in some way and the child is likely to be at serious risk of serious harm. But it is not plausible to demand that parents always and single-mindedly pursue their child's best interests. Adults who parent also have lives of their own to lead and it is quite appropriate for them sometimes to weigh their own interests, and those of others, against those of their children. Imagine someone who, as well as being a parent, and accepting our view of the fiduciary duties that attend that role, also believes—let us assume rightly—that he has a moral obligation to take part in a political demonstration. Imagine further that he cannot find alternative childcare, so he has to choose between taking his child with him or not going on the demonstration. He accepts that going on the demonstration is not in his child's best interests; those would be better served by their staying home, or going to the zoo instead—the child is not old enough for going on the demonstration plausibly to benefit her in any way at all. As long as going on the demonstration does no harm to the child, bringing or leaving her below some level we might think of as 'good enough', he does indeed have the right to go with her on the demonstration. That is not a right he has qua parent. But it is a right that makes a difference to what he may do with his child.

But we can go further. It is in *children*'s interests that their parents have their own, independent, interests and pursuits, and in children's interests too that their relationship with their parents be one in which their parents are not required always to act with their children's best interests in mind. Someone who was only a parent—someone for whom "parent" was the entire content of their identity—would not be providing the kind of experience that children need, and the parent–child relationship would surely implode in a kind of self-referential black hole. (Of course, that can happen even when the parent *does* have other identities and interests—if he fails to get the balance right—but it looks inevitable if he doesn't.) It is important for children to experience their parents as independent people, with their own lives to lead, not as people whose sole purpose in life is to serve them. So the task of parenting, although indeed extremely demanding, by its very nature allows parents discretionary time and energy: having a life of one's own is, in fact, part of the job description. The point here is not simply that it's good for children if parents get some time off for themselves, or good for children that they have a sense of their parent as having independent interests. The parent's non-parental interests will, and indeed should, manifest themselves, at least sometimes, in the interactions between parent and child. Parents must allow themselves some space, free of self-monitoring, to experience and express to the child their authentic emotions and attitudes. A parent who never said or did anything to or with his child without first asking himself whether it would be in his child's interests would not be spontaneously sharing himself with his child, there would be a lack of genuine intimacy, and he would thus be failing to provide the kind of relationship that *was* in his child's interests. Paradoxically, the kind of parent–child relationship that is good for children is one in which the parent cares about things other than his children, and doesn't spend all his time thinking about, and then trying to deliver, what would be good for them.

Four Clarifications

Four further points of clarification are important. First, we are not saying that there are many adults who cannot flourish *at all* without relationships of the kind we have described. People do indeed go to great lengths in order to raise children, and some consider the inability to do so as a profound blight on their lives, but few who miss out conclude that their lives are thereby worthless. Nonetheless, many regard themselves as having missed out on an experience that would have been necessary for them fully to flourish. Our claim is of that kind—about the contribution parent–child relationships make to a fully flourishing human life.

But, second, this is not true of *all* adults. A significant proportion of people have no desire to raise children, and for many of them the absence of this desire is not an epistemic failing—they are not making a mistake. We are not claiming that all adults need to raise children fully to flourish, and we recognize, further, that there are some for whom parenting would make no contribution to their well-being, and some for whom it would make their lives worse. That the relevant relationship goods contribute importantly to the flourishing of the rights holder does not imply that those goods are good for everybody. In this respect the contribution of this kind of relationship is like that of a romantic sexual relationship. Many people are such that they could not flourish fully without it: it contributes something to their flourishing that nothing else could contribute. Others, however, have no need for it. Similarly, there may be people who do not need to be parents: those who, although they might really enjoy parenting, could indeed flourish fully without it, and those whose lives would actually be diminished by being a parent. In some cases that might be because the person lacks the capacities needed properly to discharge the fiduciary duties (Cassidy, 2006). This does not contradict our general claim about the significance of the relationship.

Third, it may be objected that some parents abandon their children and have little contact. Indeed, even in the nuclear family that emerged after industrialization, fathers have often had very limited time and intimacy with their children. But none of this shows that adults can live well without parenting relationships, for we can ask whether they have really have enjoyed fully flourishing lives. In our terms, such people have not in fact been *parents* at all—they may have helped to create the child, and, in the latter case, they may have provided the financial support necessary for someone else (usually the biological mother) to fulfil the parenting role. (Imagine a society in which the costs of raising children had been fully socialized, so that the citizenry as a whole supported children and those raising them. The job of *parenting* would still exist, and parent–child relationships would be just as important, but it and they would have been separated from financial provision.) The traditional gendered division of labour, in addition to being unjust towards women, has tended to deprive men of something very valuable—a parenting relationship properly understood.

Finally, some parent–child relationships lack some of the features that contribute to the flourishing of the parent, while other kinds of relationship contain some of them. So, for example, the parent of a child with severe cognitive impairments might experience loving intimacy, and the joy in seeing the world reflected through the eyes of someone for whom she acts as fiduciary in some respects, but her fiduciary obligations do not include preparing her child to become an autonomous adult. Maybe some children, perhaps those on the far end of the autism spectrum, cannot be intimate with the parent in the way that we have described as being so important and rewarding. Pet owners take on fiduciary obligations, and some have emotionally rich relationships with their pets, as do many who care for adults with severe cognitive impairments, and for the infirm elderly. So not only does our account of the relationship at stake fail to capture every parent–child relationship, but the contrast between it and other caring relationships is not always as stark as we might have been taken to think. Our conception of the parent–child relationship describes something that many adults have a very strong interest in participating in. Other relationships that resemble it to a greater or lesser degree will yield some of the benefits, but not all. Some of those other relationships will yield benefits for some of the carers that are not made available by our conception of the parent–child relationship.

Alternative Accounts

Our "familial relationship goods" approach to the value of parenting can usefully be contrasted with other approaches. Our analysis is unusual in separating (i) why children should be raised by parents at all and (ii) which children should be parented by which adults. It offers, in Archard and Benatar's terms, an "indirect" justification of any answer to the second question, first justifying the institution of parenthood and then distributing parental roles within that institution (Benatar and Archard, 2010: 18–21). Most answers to the second question, and certainly those prevalent among non-philosophers, offer a direct justification—for example, by appeal to a causal relation between the child and the adult who, it is argued, has a claim to parent her. That kind of answer makes no appeal to the "value" of parenting, at least not in the sense that we have been conceiving it: there is no invocation here of the idea that the parent–child relationship makes a distinctive contribution to human well-being. The thought is more likely to be that the person who brought the child into existence has a right to be its parent, in our sense, with that right not being grounded in any claim about the goods or benefits likely to accrue from that way of justifying or distributing parenthood. From that perspective, whether parenting is valuable, for either parent or child, plays no role in answering either question.

Some of a libertarian persuasion may see the right to parent a child one has procreated as an application of the more general right to own that which one has produced with one's own body. The relationship that matters here is the ownership relationship, which gives the procreative parent certain control rights over the child that in some sense "belongs" to the parent. Another view that also makes biology central points rather to the investment that biological parents, especially gestating mothers, make in "their" children. Bearing the costs and labours of pregnancy gives one a right to parent the child one has worked so hard to produce. This variant on the proprietarian perspective—the idea that the parent "deserves" to parent the child in return for past labours—again gives no special weight to the kind or quality of the relationship between parent and child once it is born nor to the value of that kind of relationship to either parent or child.[9]

But it is also possible to accept the structure of our argument, and our emphasis on "familial relationship goods," while giving it and them different content. For example, some believe that there is a particular value to an adult in having a relationship with a child in which one is able to pass on some aspects of oneself to that child—perhaps one's genes, perhaps one's values, perhaps one's property. The adult interest in parenting, on such a view, does derive from something about the value of a distinctive connection between parent and child; there is something important that one is able to achieve by parenting a child that would not otherwise be available. That may be passing one's deepest religious or cultural commitments on to future generations, extending oneself beyond death, achieving a distinctive kind of connection to posterity, or seeing the fruits of one's labour enjoyed by those whom one loves.

By way of illustration, let us focus on the first of these. Colin MacLeod (2010) identifies a motive of "creative self-extension" which "arises out of the special opportunity... parents have to express their own commitment to ideals and ground-projects by passing them on to their children.... We can see ourselves carried forward in another self we played a significant role in creating" (p. 142). MacLeod seems to endorse this, which brings together two, distinct, motives: one concerning expression of one's own commitment to one's projects and values, the other concerning the carrying forward of one's own self through a creative process. While accepting that parents may often find acting on both motives successfully can be a profound source of "satisfaction," and that in practice parents may often act on these motives, we doubt that they should play any role in grounding adults' claims to parent children.

There are many ways to express commitment to projects and values, and for many projects and values influencing other people to take notice of them, take them

---

[9] For discussion of various views about the (alleged) interest in procreation, see Overall (in this volume). For a view that emphasizes the significance of the gestatory relationship, rather than genetic connection, see Gheaus (2012).

seriously, or adopt them, is part of what it is to be committed to them, or a natural accompaniment to being committed to them. One's children are, like other people, potential adopters. But, as our account of the specificity of the parent–child relationship emphasizes, to parent a child is to have a special kind of power over the emerging values of another human being. The kind of relationship that will deliver the goods we have identified is indeed one in which parents will have some scope to influence their child's emerging values—they will do this as an unintended by-product of the spontaneous sharing of themselves with their children, and they may do it deliberately to the (in our view limited) extent to which the relationship's yielding its benefits requires some degree of shared values between parent and child. The parent's concern to promote her child's well-being may also have implications for the ways in which she may act to shape the child's emerging values—for example, where the parent believes that her child's endorsing a particular project or value will be important for the child's living a successful life.[10] But for a parent to ensure that her child in particular shares some specific value or project *out of commitment to that value or project* rather than *out of commitment to the child and the relationship* strikes us as a case of using the child as a means to the realization of the parent's own goals in a way that has nothing to do with the value of the relationship. On inspection, then, this aspect of MacLeod's claim turns out not to appeal to relationship goods after all.

By contrast, the second thought, that "we can see ourselves carried forward in another self we played a significant role in creating," does seem to put something special about the relationship between parent and child centre stage. The claim that there is something distinctively and importantly valuable, for an adult, about raising a child as an unsubstitutable act of "creative self-extension" does have the same form as our appeal to "familial relationship goods." Though the content differs, the thought is that it is only by raising a child that adults can realize this particular, and weighty, contribution to human flourishing.

Let us explore this alternative specification of the value of parenting by looking at Edgar Page's account, which is the fullest articulation of such a view that we are aware of.[11] Page is concerned to identify a conception of parenthood that is robust and attractive enough plausibly to ground a set of parental rights. That is a project with which we have a good deal of sympathy (and pursue elsewhere).[12] Here we confine ourselves to Page's approach to the value of parenting, with which we disagree strongly. For him:

parents have a positive desire to influence the course of a child's life, to guide the child from infancy to maturity, a desire to mould it, to shape its life, to fix its basic values and broad attitude,

---

[10] For the view that, when it comes to improving their children's lives, it is wrong for parents to be guided by reasons that their children could come reasonably to reject, see Clayton (2006, 2012).

[11] For a more recent (and more subtle) account along these lines, explicitly presented as a critique of our view, see Reshef (2013).

[12] Our theory of familial relationship goods yields a radically different account of parents' rights from that implied by Page's approach. See Brighouse and Swift (2006, 2009, forthcoming).

to lay the foundations of its lifestyle, its priorities, its most general beliefs and convictions, and in general to determine, to whatever degree is reasonable and possible, the kind of person the child will become. It would not be going too far to say that parents have a general propensity to try to send their children forward in their own image, not in every detail, but in broad outline. (Page, 1984: 195)

We do not dispute this as an empirical claim about a "general propensity" on the part of parents. The question is whether the desire of adults to shape a child in this way—to determine the kind of person she will become—should count as an interest weighty enough to constitute a parent-centred justification of the practice of children being raised by parents. If somebody proposed that children should be raised collectively, say in state-run childrearing institutions, there would indeed be compelling objections, appealing to the interests of both children and adults. But would the fact that those collective arrangements denied adults the opportunity creatively to extend themselves via their children be one of them?

As before, it is important carefully to identify the specific claim at stake. Parents inevitably influence, even though they don't "determine," the people their children become. Our own account of the adult interest in the parent–child relationship describes a relationship in which there is plenty of room for such influence to occur, whether as unintended by-product, deliberate concern for shared values, or parental concern for the child's well-being. With respect to the last of these, Page (1984: 196) is admirably clear:

We can normally expect parents to pursue their interest in shaping the child's future with a clear regard for its good. But this does not mean that the parental interest in shaping the child can be reduced to this affection. . . . The propensity of parents to exercise control and guidance over their children, the propensity to determine the development of the child, far from being aimed simply or primarily at the child's good, is the manifestation of a fundamental and unique interest which lies at the heart of human parenthood and at the foundation of parental rights.

To our minds, this clarification, emphasizing the extent to which the motive in question views the child as a vehicle for the realization of the parent's own selfish, and indeed somewhat narcissistic, interests, brings out the latent proprietarianism in the "creative self-extension" account.[13] The child is seen as a canvas on which the parent may objectify herself, or a block of raw marble to be shaped into a future version of herself. But children are entirely separate people from their parents. If collective childrearing arrangements were better for them, the fact that such arrangements would deny adults this particular opportunity for creative self-extension hardly constitutes a ground for insisting that children should be raised by parents.

[13] Cf. Austin (2004: 507–9).

As he develops his view, Page also articulates well what we take to be a common belief about the importance of adults' parenting children they have physically produced. For him:

The parental aim is not simply the creation of a person, but rather the creation of a person in the parents' own image.... One aspect is that in raising their child parents do much to shape the person it will become. This they would do in any case, even if it were not part of their design, but I have argued that parents characteristically have a positive desire to determine the kind of person their child becomes. The other aspect is that natural parents produce from their own bodies the material to be shaped, the organism that is to become a person. (Page, 1984: 200)

For Page, then, the creative dimension of parenting would not adequately be acknowledged by childrearing arrangements that allowed parents to determine the kind of people their children become but allocated children to parents in ways that gave no fundamental importance to any genetic connection between parent and child.

Physically producing the child is itself an essential part of the creative process:

The motive, or the end, of parenthood is surely the creation of the whole person, and this takes within its grasp both the begetting and the raising of the child.... The two parts—begetting and rearing—are clearly complementary to each other and neither is entirely intelligible, as a form of human activity, without the other. (Page, 1984: 199–200)

An obvious problem for such an account, as for all that attach great significance to biological connection, is that it seems to rule out the possibility that adoptive parents can fully realize the value of parenting. Even if they were indeed engaging in the kind of creative self-extension that comes through raising a child, they would inevitably be denied the aspect that comes from having physically created the child they are raising. His response is worth quoting at length:

If all parents were in the position of adoptive parents, i.e. if there were no connection between parenthood and generation, as might be imagined in "science fiction" worlds, parenthood would not have a place of special value in human life, or not the place it now has. Adoptive parenthood is modelled on natural parenthood and the commitment of adoptive parents to the child is parasitic on the special bond characteristic of natural parents. Without this model there would be a question as to the intelligibility of a commitment of adoptive parents to young babies, particularly in conditions which severely test them, and indeed as to the intelligibility of their desire for parenthood. (Would it be comparable to the desire for pets?) For most people, I suspect, adopting a child falls short of being a perfect substitute for natural parenthood, but when they undertake it they can at least borrow from and follow the established patterns and practice and attitudes of parenthood grounded on the physical relation. It is difficult to know what adoptive parenthood would be without this. (Page, 1984: 201)

Where Page believes we can only make sense of adoptive parenthood by thinking of it as parasitic on a parent–child relationship that is grounded in a physical (i.e. biological/natural) connection between parent and child, we have tried to explain the value of parenthood in ways that make no reference to that connection. We accept

that, in a world where there were no connection between parenthood and generation, parenthood would have a different significance from that which it has for most
people today. But we reject the claim that, in such a world, raising a child would be
like keeping a pet. Our aim has been to highlight the specificity of the parent–child
relationship, and to identify the distinctive and weighty contribution it can make to
human well-being. If we are right, there is no reason for adoptive parents to model
themselves on anybody, for what is special about the practice of parenting does not
depend on a biological or natural connection between parent and child.

References

Alstott, A. (2004). *No Exit: What Parents Owe their Children and What Society Owes Parents*.
Oxford and New York: Oxford University Press.

Archard, D. (2004). *Children: Rights and Childhood* (2nd edn). London: Routledge.

Austin, M. (2007). *Conceptions of Parenthood: Ethics and the Family*. Aldershot: Ashgate.

Austin, M. W. (2004). The Failure of Biological Accounts of Parenthood. *Journal of Value Inquiry*, 38, 499–510.

Benatar, D., and Archard, D. (2010). Introduction. In D. Archard, and D. Benatar (eds), *Procreation and Parenthood: The Ethics of Bearing and Rearing Children* (pp. 1–30). Oxford: Oxford University Press.

Brennan, S., and Cameron, B. (2013). How Many Parents can a Child Have? Philosophical Reflections on the Three Parent Case. Draft available at http://works.bepress.com/samanthabrennan/59.

Brennan, S., and Noggle, R. (1997). The Moral Status of Children: Children's Rights, Parents' Rights, and Family Justice. *Social Theory and Practice*, 23(1), 1–25.

Brighouse, H., and Swift, A. (2006). Parents' Rights and the Value of the Family. *Ethics*, 117(1), 80–108.

Brighouse, H., and Swift, A. (2008). Social Justice and Family Policy. In G. Craig, T. Burchhardt, and D. Gordon, *Social Justice and Public Policy* (pp. 139–56). Bristol: Policy Press.

Brighouse, H., and Swift, A. (2009). Legitimate Parental Partiality. *Philosophy and Public Affairs*, 37(1), 43–80.

Brighouse, H., and Swift, A. (forthcoming). *Family Values: The Ethics of Parent–Child Relationships*. Princeton, NJ: Princeton University Press.

Callan, E. (1997). *Creating Citizens*. Oxford: Oxford University Press.

Cassidy, L. (2006). That Many of us Should Not Parent. *Hypatia: A Journal of Feminist Philosophy*, 21(4), 40–57.

Clayton, M. (2006). *Justice and Legitimacy in Upbringing*. Oxford: Oxford University Press.

Clayton, M. (2012). Debate: The Case Against the Comprehensive Enrolment of Children. *Journal of Political Philosophy*, 20(3), 353–64.

Dwyer, J. (1999). *Religious Schools and Children's Rights*. Ithaca, NY: Cornell University Press.

Galston, W. (2002). *Liberal Pluralism*. Cambridge: Cambridge University Press.

Gheaus, A. (2012). The Right to Parent one's Biological Baby. *Journal of Political Philosophy*, 20(4), 432–55.

Locke, J. (1988). *The Second Treatise of Government,* ed. P. Laslett. Cambridge: Cambridge University Press.

MacLeod, C. (2002). Liberal Equality and the Affective Family. In D. Archard, and C. MacLeod, *The Moral and Political Status of Children* (pp. 212–30). Oxford: Oxford University Press.

MacLeod, C. (2010). Parental Responsibilities in an Unjust World. In D. Archard, and D. Benatar, *Procreation and Parenthood: The Ethics of Bearing and Rearing Children* (pp. 128–50). Oxford: Oxford University Press.

Morse, J. R. (1999). No Families, No Freedom: Human Flourishing in a Free Society. *Social Philosophy and Policy*, 16(1), 290–314.

Page, E. (1984). Parental Rights. *Journal of Applied Philosophy*, 1(2), 187–203.

Reich, R. (2002). *Bridging Liberalism and Multiculturalism in American Education.* Chicago, IL: University of Chicago Press.

Reshef, Y. (2013). Rethinking the Value of Families. *Critical Review of International Social and Political Philosophy*, 16(1), 130–50.

Richards, N. (2010). *The Ethics of Parenthood.* Oxford: Oxford University Press.

Schoemann, F. (1980). Rights of Children, Rights of Parents, and the Moral Basis of the Family. *Ethics*, 91, 6–19.

Waldfogel, J. (2006). *What Children Need.* Cambridge, MA: Harvard University Press.

2

The Goods of Childhood and Children's Rights

Samantha Brennan

Introduction

There are basic moral questions about family creation that ought to concern all potential parents. One of the most pressing of these questions concerns the moral status of children and the obligations that stem from their moral status. When one is considering taking on the role of parent and bringing children into one's life (by whatever means) one needs to know what sort of being a child is and what obligations being a parent will entail. The moral obligations that stem from relationships with other adults can look easy by comparison. Most adults contract as equals and agree to the terms of our relationships, whether that relationship is traditional marriage, friendship, workplace colleagues, or something else. However, the moral status of children is a more difficult question and getting it right matters. Children—especially in their early stages—aren't fully fledged rational autonomous persons capable of contract. Instead, they require our protection and our care. Children are not the property of their parents and neither are they merely projects undertaken by parents. It's my view that children have independent moral status as rights bearers and these rights constrain the activity of parenting (Noggle, 1997; Brennan and White, 2008). But what account of children's rights best fits with children's lives? How exactly and in what ways do those rights place obligations on, and constrain the lives of, parents?

I have argued for a developmental account of children's rights, rights that progress from protecting interests to protecting choices (Brennan, 2002). In this chapter I address the trade-offs that occur in protecting children's interests, and in doing so engage with the question, "In what does children's well-being consist?" How much of children's welfare is attached to the child they are now and how much to the future person they become? I argue for the view that parents aren't merely charged with the task of delivering children safely to the threshold of adulthood. Because of the nature of children's rights, parents also have obligations to promote the goods of childhood. Finally, I address parents' role in articulating and adjudicating children's rights and say something both about the obligations children's rights place on their parents and on how those rights constrain the actions of parents. These are the questions this chapter will take up.

The Moral Status of Children

When one is considering forming a family by adding children to the lives of one or more adults, one natural question concerns the moral obligations one will have to the child. How will this relationship change the moral fabric of my life? The aspect of this problem I am most concerned with here involves obligations to children directly: what we owe them because of the sort of thing that they are. That is to say, I am most immediately concerned with the moral status of children. As a philosopher and the parent of three children, I have wondered about the different conceptions we have of children and how these conceptions affect the way we treat the younger members of our societies, especially those who are members of our families. Clearly children are neither like adult persons, fully fledged rights bearers, nor are they property, owned by their parents and subject to their absolute authority.

Perhaps children are neither "property" nor "persons" but something in between. The economist Nancy Folbre worries that many of us now think of children as we do pets, a kind of property that is governed by some, but not many, rules regarding its care. For Folbre (2008) the mistaken children-as-pets assumption lies behind much public policy regarding funding for children. If children are like pets, then those who want them and can afford them, do so. Others are denied access by lack of money and if one's fortunes fail, very little demand can be placed on others for the cost of the supporting children. Pets fall into the general category of thinking of having children as a lifestyle choice, a project in which we decide to engage. Perhaps modern parents do rather think of our children like projects in which we invest a lot of time, energy, and effort. But to the extent that we do think of children like pets or projects, I think we are making a mistake.

Historically, the question of the moral status of children isn't one that has much concerned philosophers. In the past, philosophy has all but ignored children as bearers of moral and political significance. Much of the history of moral and political philosophy has proceeded without consideration of children and their interests. In the history of philosophy, children have either been ignored entirely as part of the family unit which is taken to be the basic unit of society, or if considered at all, treated as property of their parents or the state. Critics who note the absence of children in philosophy are likely to quote from the work of Thomas Hobbes. Hobbes is often maligned for his mushroom metaphor, found in *De Cive*, in which we are asked to view persons as having sprung out of the ground fully formed and completely rational: "Let us return again to the state of nature, and consider men as if but even now sprung out of the earth, and suddenly, like mushrooms, come to full maturity, without all kind of engagement to each other" (Hobbes, 2004: 8.1I). In this metaphor, children and childhood have completely disappeared. We begin our thought experiment about the state and its justification by imagining persons as adults from the start.

But Hobbes does not always make children disappear. Elsewhere, Hobbes argues that parental dominion over children is justified because it would be rational for the child to agree to surrender her autonomy in return for survival. Thus, the child is assimilated to a rational adult decision-maker. In *Leviathan* he writes: "Again, seeing the Infant is first in the power of the Mother, so as she may either nourish, or expose it; if she nourish it, it oweth its life to the Mother; and is therefore obliged to obey her, rather than any other; and by consequence the Dominion over it is hers" (Hobbes, 2011: 182). In some respects Hobbes is better than most historical figures writing about children because he does not think that parental authority is natural. The dominion parents have over children—like that of state authority over citizens—stands in need of justification. But the argument Hobbes offers for parental authority treats children as would-be adults, asking what they would choose were they fully rational and fully informed and able to consent.

One might expect to find that things had changed in the rather large gap between Hobbes and the publication of John Rawls's *A Theory of Justice*. In the course of a very thick book, children are only found in a few places and the subject of justice-based obligations to them is not addressed at all. Rawls does discuss interfamily distribution, justice between families, but not intra-family distribution, justice within families (Brennan and Noggle, 1998). Libertarian political philosophers, such as Jan Narveson and David Gauthier, fare even worse. Limited by the thin conceptual resources of their theories, they cannot place constraints on the treatment of children on the basis of any claims to moral status children might have. Rather, children are either left as part of the terrain of natural sentiment, and not justice, or moral or legal constraints on the treatment of children are argued for on the basis of third-party effects on other adults. It is hard to reconcile the reliance on natural sentiment with the numbers of children who are beaten, abused, and neglected at the hands of family members.

The libertarian third-party effects argument gets better results but it is not a pleasant road to tread. If I beat my children they may turn out badly and pose a threat to you by becoming violent, unstable people. That's the long-term bad effect. Or in the short term the noise of my beating them may disturb your peace and quiet (Narveson, 2001). This approach gets the right answer in some cases (though it does it in the wrong way) but note that it can have nothing to say if what I do is silently, quietly kill my children in the back shed.

Feminist philosophers haven't had much to say to the libertarians, doubting that a moral philosophy which leaves out the disabled, the elderly, along with children would easily be reformed.[1] Feminists were, however, very quick to criticize Rawls for his exclusion of the domestic realm, but here their criticisms mostly focused on the neglect of women and the issue of the division of domestic labour. There is some tension between feminist moral and political philosophers and those whose focus is on the exclusion of children from our theorizing. Not all feminists think that the liberation of women necessarily entails any change in our attitude towards, and our treatment of, children.[2] I do think there is a connection between the exclusion of children and the exclusion of women but I am not going to say much about that here. I will say that the problem of the division of household work between men and women, and the resulting spillover effects in the economic and political realms, is obviously very important and a matter that ought to be of great concern to theorists of justice. However, I don't think it's a particularly challenging problem conceptually. Women and men are equally deserving of respect and the just society will take women into account on equal terms.

But adding children to the list of subjects of justice is not so simple. Insofar as moral and political philosophy tends to idealize the subjects of justice, children cannot simply be added to our theories and remain children. We can imagine idealizing men and women, and asking were everyone perfectly, equal, rational, and autonomous whether they would choose the existing distribution of burdens and benefits. We can use the idealizing conditions to test relationships between men and women but it's not so easy when we try to include children. The problem is that idealizing theories tend to include children by asking what they'd choose for themselves were they adults (and perfectly smart, autonomous, self-interested adults at that).

I think most of us have no difficulty in thinking of children as beings with independent moral status. I am not going to argue for that claim here. The failure of contemporary political philosophy to cope with the inclusion of children ought to count against the general applicability of such views. Children have rights that protect them from harm. The hard question is the shape these rights take.

Rights, Choices, and Interests

On one standard account of rights, rights protect choices. But a choice-protecting conception of rights is not well suited to children. Indeed, so bad is the fit between the choice conception of rights and children's rights that advocates of children's rights have used the existence of children's rights to argue against the choice conception of rights. Likewise, advocates of the choice conception of rights have argued that children cannot have rights since it makes no sense to assign to them rights that protect their choices. To recap their reasons: Children make lousy choosers, both in terms of their ability and in terms of the content of the choices they'd make. They also lack a stable

---

[1] For an exception, see Okin (1989).
[2] For an argument against equal rights for children, by a feminist, see Purdy (1992).

identity and as developing moral agents would always be choosing for future versions of themselves that are radically different from the selves they are now. On the strongest version of this worry, it would be like assigning me rights to make choices that affect your future.

But the choice conception isn't the only conception of rights that's available. Proponents of the interest conception of rights argue that rights, properly understood, protect our well-being. The protection of well-being as the content of a right's claim is a good fit with rights for children. I argue elsewhere that the interest conception won't work for adults and so if we wanted rights for both children and adults we would need two completely different conceptions of rights. That cannot be correct, though, as the move from childhood to adulthood isn't a binary thing. We aren't children one day, and adults the next. So I've argued for a developmentally sensitive account of rights in which we begin with rights that protect our well-being and end with rights that protect our choices. In the middle stages we have both sorts of rights and an important task of parents and the state lies in articulating and protecting the rights of children (Brennan, 2002).

So if children have rights that protect their interests, what sort of interests do children have? The problem is that even those people who have done ground-breaking work on children's rights have argued for the better treatment of *children* on the basis of the *adults* they'll become. Consider a classic paper on children's rights by Joel Feinberg. In "A Child's Right to an Open Future," Feinberg describes rights children have as children and rights adults have as adults and then the rights adults and children share (1980). But his main argument for children's rights is on the basis of the autonomous capacity to choose that children have the potential to develop. Also for Feinberg, the most important right a child has, the right to an open future, is a right that benefits the adult the child becomes, not the child as he or she is now. So children's rights theorists include children but very much on the basis of concern for the adult rights bearers they'll come to be. As parents it's our job, according to Feinberg, to bring our children to adulthood with a robust range of life choices available to them. We educate them, nurture them, and help them develop but we have to do so in a way that respects their own autonomy. That sounds good, and it's a much-needed correction to the children-as-property model I mentioned earlier, or the contract model that can't account for children, but it leaves childhood out of the picture. Our main focus as parents, Feinberg thinks, ought to be producing autonomous adults.

Claudia Mills criticizes Feinberg's view in her paper "The Child's Right to an Open Future," arguing that the open options programme of childrearing denies the child the richness of experiencing a single religion, language, tradition, or musical instrument in depth (2003). Mills's main arguments target the smorgasbord approach that Feinberg is advocating, claiming that its implementation is often shallow, superficial, frenetic, and exhausting. However, Mills also notes that it's not very child-centric and she thinks that Feinberg's approach to parenting fails to appreciate childhood. On her view, we ought to respect the present child for who she or he is now. Paying attention to children will also guard against the danger posed by parenting that is too focused on instilling a particular good into the lives of one's children as well. A more child-centric approach speaks in favour of options but not for the reason that motivates Feinberg. Writes Mills (2003: 509): "The chief benefit of a wealth of options here as elsewhere lies not in their providing for a child's open future, but in their opening up to her a more rich and diverse life right now. The open options view prioritizes the future over the present; the alternative view I would defend focuses on a celebration of the present instead."

Feinberg is not the only rights theorist to conceive of children's rights primarily in terms of rights that protect the future choosers they may become. Ironically, much of the literature on children's rights seems caught up in the concern we have for autonomous adults. Tom Campbell attributes this to the prevalence of the choice (or as he calls it, "power") theory of rights. If we think of rights as serving the purpose of protecting choices, then it seems natural to conceive of children's rights in terms of rights held in trust for the adults they'll become. Writes Campbell (2002: 20):

It is these rights of the child as future adult which tend to be identified as the distinctive rights of children. This is, however, a manifestation of the dominance of the power theory of rights which tends to interpret the significance of children either in terms of their emerging adult-like capacities, or because of their position as rational choosers of the future.

Campbell argues that this misses the mark in two ways. First, he worries that children's rights that are justified in terms of the child's development into an autonomous adult are only properly rights of the child if the adult person and the child she or he was are one and the same person. Some theories of personal identity might cast doubt on this claim. Second, such rights pay insufficient attention to children as children. According to Campbell (2002: 20), "This approach neglects the significance of the experiences of children in their present situations, their present happiness, and current concerns."

I agree entirely with Campbell's analysis of the mistake that rights theorists have made. With Campbell I think it is an error to subordinate the "current interests of the child" to the "training needs of the future person." However, while Campbell attributes the error to the dominant liberal ideology and the choice theory of rights, I think the basis for the mistake is more widespread. If Campbell were right about the basis for the mistake, one would expect that other sorts of moral and political theories would cope better with the inclusion of children. In particular, one might hope that consequentialist approaches to ethics—that is, those approaches according to which the only relevant moral factors can be found in an action's results—included children on their own terms.

Children's Well-Being and the Goods of Childhood

If children's rights protect their interests, then we need to know about children's interests. What's good for children? Again, the history of philosophy isn't much help.

However, more recently, attempts have been made by moral and political philosophers to include children and there is now a growing body of literature in moral and political philosophy that examines the rights and interests of children and the responsibilities we have toward them. See, for example, recent work by David Archard, Adam Swift, Harry Brighouse, Colin Macleod, and Robert Noggle. However, while we no longer neglect children, our philosophical broadmindedness has not expanded to include childhood. Insofar as philosophers now pay attention to children, we tend to do so in terms of the adults they will become. But this future-directed thinking may be a mistake if there are goods intrinsic to childhood and if they carry any weight in our evaluation of well-being over the life span. Are there intrinsic goods of childhood? How are we to weigh the goods of childhood against the goods of adult life?

A point of clarification is needed before we begin. There are two senses of "intrinsic" that might be relevant to our question. In the first case, the contrast is between intrinsic and instrumental goods. One might hold a monistic theory of the good according to which pleasure is the only good and all other goods are valuable only insofar as they contribute to pleasure. If knowledge, for example, contributes to pleasure, then it will have instrumental value. Thus, we might be asking, when we ask about the intrinsic goods of childhood, whether they are good in their own right, or merely because they contribute to some other good that itself has intrinsic value. However, while I am pluralist about the good and I do think that there are multiple goods that are valuable intrinsically, that's not the distinction of interest to me here. The contrast I'm interested in is between the goods of childhood and the goods of adult life. One might hold that only the goods of adult life really count and that childhood goods only count when they contribute to adult goods. This is the view that childhood goods have merely instrumental value and it's the position I mean to take on in this chapter. Thus, when I ask whether there are intrinsic goods of childhood, what I am asking is whether there are goods the value of which doesn't follow from their contribution to the goods of adult life.

The tendency to focus on future goods is not just restricted to rights theorists or to deontological approaches to moral philosophy. The mistake can be found in consequentialist approaches to ethics as well. This is especially troubling because of the widespread application of consequentialist moral reasoning in cases that concern children. Many philosophers, parents, and policy-makers who would not normally be tempted by consequentialist considerations in ethics find them compelling when it comes to the treatment of children. (I am reminded here of Robert Nozick's slogan from *Anarchy, State, and Utopia*: "Utilitarianism for animals, Kantianism for people": 1974: 41.) When we inquire whether a particular practice or policy is good for children, our usual entry into that problem is in terms of its long-term effects. For example, when philosophers ask about the morality of corporal punishment or turn our minds to the content of adequate education, we ask how people benefit as adults from this form of discipline or from this kind of education. Contrast this with the way that we treat other adults. When I was chair of my department I might have wondered whether the research productivity or teaching efforts of colleagues might be improved by imposing a system of beatings for bad behaviour, but no one thinks the most interesting or relevant question is whether such a plan would work. Whether or not it works is beside the point since such treatment is ruled out by moral constraints regardless of how good the end result might be. It may be that, given their limited abilities to reason and act on reasons, children merit a kind of paternalism that other adults do not.

But it's not just consequentialism that is the focus of my objection here. In fact, in this chapter it's my goal to be neutral between consequentialist and deontological approaches to ethics. The account of children's rights I've offered in detail elsewhere contains both consequentialist and deontological elements (Brennan, 2002). The deontological element is that children have rights and the consequentialist element is that those rights protect their interests. I'm interested in examining consequentialism's account of the good even though I think it's flawed as a complete story about right action in the case of both children and adults. The focus of my criticisms is the predominant application of consequentialist reasoning in which goods that occur in adult life count for more than what goes on in childhood. Those arguing for corporal punishment don't seem to weigh childhood suffering against the goods of adult life. It's as if the only relevant question is whether corporal punishment works in the long run, not whether it's justified overall.[3]

Another example of a public policy area that frequently takes the approach of thinking only about the long-term effects on children, asking how they turn out, rather than caring about how they are treated, is education. Daniel Weinstock has criticized approaches to education that justify injustices that occur to children in schools on the basis of goods that these children receive later in life, as adults. He believes there are goods that are intrinsic to childhood, the loss of which cannot be made good by anything that happens later in life. Any view that fails to recognize these goods and treats childhood only as a way station to adulthood is deeply mistaken, on Weinstock's account.[4]

Indeed many, or most, children's rights are justified in terms of the adult persons that the children may become and the goods those adults' lives may contain. For the most part, I think this makes sense. But I also think there is a danger in focusing too much on the future and neglecting the goods of childhood. This is especially true if some of the goods of childhood are valuable in their own right, and even more so if some of those goods are incommensurable with the goods of adult life. I think there is a connection between a focus on children-as-future-adults and a negative conception of

---

[3] Even if consequentialism's account of the good was repaired to take account of children properly, i.e. to include the goods of childhood, I don't think consequentialism offers the correct account of how children ought to be treated, for the same reason that consequentialism is flawed as a moral theory for adults. I'm interested in consequentialism's account of the good because the interest theory of rights also needs an account of the good and faces the same issues as consequentialism.

[4] Weinstock has communicated this view in personal correspondence with me.

childhood, that is a conception of childhood according to which nothing of value can be found there.

So if the easy problem of children and moral philosophy is the decision to include children, then the hard problem is this: Just how do we include them? On what terms? To ask the question in Tamir Schapiro's terms (1999), "What is a child?" In this section I examine competing conceptions of childhood, contrasting a negative conception of childhood with a positive conception. Gareth Matthews describes the negative conception of childhood as Aristotelian in origin. He writes:

> According to what we have called the "Aristotelian conception," childhood is an essentially prospective state. Given this, what is good for a child will tend to be understood as something that will contribute to its good in adulthood. Moreover, the goods of childhood will be, on the whole, derivative from the goods of adulthood. (Matthews, 2010: sec. 6)

One way to test one's intuitions about whether one thinks the negative or the prospective conception is correct is to ask whether one would, if one could, simply give children a pill to have them grow up.[5] Is childhood merely something to be gotten over, grown out of as quickly as possible? Of course, children give us pleasure, but what I am asking is whether childhood is good separate from its instrumental benefits, including the pleasure children give to adults. So, to put the question differently, would it be good for the person who is the child to forgo childhood? On the negative conception, childhood is seen simply as a time of deficits. Compared to adults children are smaller, weaker, less rational, less autonomous, and so on, and the process of growing up consists largely in acquiring the skills and abilities and physical form of a human adult. On the negative conception, children are merely inferior adults and childhood is a state to be survived. Because children are weaker we owe them protections, on this view, but we owe them that protection in virtue of the person they will become.[6]

Childhood itself on this view is seen as prep school for adult life—filled with study, exercise, and lessons in socialization—so that the child can emerge well prepared for adult life. Childhood is the period in which raw material is drilled into shape. Note that this view is often associated with rugged individualism and personal success, but it need not be. Some people who criticize the instrumental conception of childhood do so because they worry we are moulding children into the round pegs for the round holes required by consumer society. Thus we give them bank accounts and teach them about finances, and work with them to plan careers, at increasingly tender ages.

---

[5] I owe this way of putting the problem to Daniel Weinstock.

[6] The negative conception of childhood has in recent years been criticized by neuroscientists. Against the blank slate/empty container view of childhood, psychologist Alison Gopnik offers a very different image of babies as empathetic, logical, and "the best learners in the universe" (2009). Gopnik writes that recent research in human cognition has revealed that babies have a rich array of neural pathways. On her account of infancy and childhood, growing up appears to be, in part, a process of narrowing: prioritizing certain pathways and letting others atrophy. "A baby's full-tilt ability to take in the world, think and imagine," says Gopnik (2010), is like "being in love in Paris for the first time after you've had three double espressos" (para. 3).

When children complain that they are miserable sitting at their desks for hours a day at school, we think "Just wait until you start work." But this connection between the instrumental conception of childhood and the child-as-future adult worker and consumer is not a necessary one. The negative conception can come in both "nice" and "nasty" versions. On the "nice" version we might, for example, care about connections, love, and deep personal relationships, but it will be the adult's ability to conduct such relationships by which childhood is judged a success or failure. The important point is that the only value childhood friendships will have is as a training ground for adult friendships.

There are more laissez-faire versions of the view too, in which the transition from child to adult is seen as natural and does not require any special tutoring or preparation. Childhood, on this view, is not like prep school. Instead, it is like a long sleepy summer vacation during which nothing of any particular importance happens. Much of my own childhood was more like this. I am sure many of us can recall being told to go outside and play and come back hours later. I enjoyed a freedom to explore woods and water in Newfoundland that I think very few children who live there today get to experience. In *Children at Play: An American History*, Howard Chudacoff (2007) describes the period of time in which I grew up as a particularly idyllic time for children and play. He writes that children won a burst of autonomy when there were still woods and parks at hand to roam, yet fewer chores to keep children busy at home. The next obstacles to children's free play, according to Chudacoff, were parental fear and suburban development. Chudacoff cites findings from a recent survey showing that, after school and on the weekends, kids on average spend only one half-hour a week in unstructured play outside, compared to 14.5 hours playing inside, and 12 more hours watching television.

It might be that these hours of rest and play have positive value for children but arguably children had this freedom to play in this era not because people thought childhood play had intrinsic value, but rather because childhood is largely a prospective state. On this model childhood is about waiting for adulthood, rather than frantically preparing for it, but the idea remains the same. Not much that happens in childhood counts. What matters most is the adult one becomes.

I think the negative conception of childhood is intuitively familiar. Many of us may even think of our own lives that way, or look this way at the lives of our children. Certainly it is an alarming trend I see among parents of my generation for whom the "children as property" model has given way to "children as personal projects" model.

Nancy Folbre has another suggestion, though I think it's really a variant of the projects model. In *The Invisible Heart: Economics and Family Values* she has a chapter called "Children as Pets" which more clearly captures the economic implications of thinking of children as personal projects of the parents. She opens the chapter this way:

I think I know why so many people seem to think that parents, especially mothers, should pay most of the costs of raising children. These people think of children as pets. Parents acquire them because they provide companionship and love…. those who care for them are the ones who get the fun out of them; therefore, they should pay the costs. (Folbre, 2001: 109)

The reason I don't think the pets model quite works though is that modern middle-class parents have expectations for their children. They invest considerable energy, time, and thought into their children's lives, in ways that seem to me to be new. It's not a simple matter of enjoying their company. Rather, today's parents have plans and schedules, read books, and have expectations of achieving results.

Consider the phenomenon of the "helicopter parents" of university age "children." Dubbed "helicopter parents" because of their habit of hovering around campus, at my university they make an average of six visits to campus a year. The phenomenon is so common that "helicopter parent" now merits an entry in Wikipedia. It defines them this way: "a person who pays extremely close attention to his or her child or children, particularly at educational institutions. They rush to prevent any harm or failure from befalling them or letting them learn from their own mistakes, sometimes even contrary to the children's wishes. They are so named because…. they hover closely overhead, rarely out of reach whether their children need them or not." An extension of the term, "Black Hawks," has been coined for those who cross the line from a mere excess of zeal to unethical behaviour such as writing their children's college admission essays (the reference is to the US military helicopter of the same name). I associate helicopter parents with the negative conception of childhood and the idea of children as projects, because of the thought that one's goal as a parent is to ensure that we produce a successful adult.

But when do children become adults? In addition to disagreements about the goals of parenting, it might be that we also disagree about the boundaries of childhood; that is, where childhood ends and adulthood begins. Increasingly parents are financially responsible for children into their twenties and thirties and the average of moving out is now in the late twenties. A *New York Times* columnist goes so far as to say our times have given rise to a new life stage between adolescence and adulthood, which he dubs "the Odyssey Years." During the Odyssey Years, "20-somethings go to school and take breaks from school. They live with friends and they live at home. They fall in and out of love. They try one career and then try another." One interesting philosophical question, which I will save for another time, is what moral obligations parents have to children in their twenties and beyond (Brooks, 2007).

But the negative conception of childhood is not just a reflection on popular culture. It has its philosophical defenders. Recent advocates of negative conceptions of childhood are (*a*) Tamir Schapiro and (*b*) Michael Slote. Schapiro's version is one rooted in a Kantian account of human agency. It's the account of childhood that is connected to a deontological approach to ethics. Slote's conception is connected to views about human well-being and the aggregation of goods across a life span. His approach is a consequentialist one. And so I continue to look at this problem as one shared by consequentialist and deontological moral theorists.

a. Tamir Schapiro

In "What is a Child?" Schapiro characterizes childhood as a predicament. She writes, "Thus the condition of childhood is one in which the agent is not yet in a position to speak in her own voice because there is no voice which counts as hers" (Schapiro, 1999: 729). Roughly speaking, on the Kantian view, childhood is problematic because while adults have characters, children lack stable characters. The undeveloped agent is unable to have a rational plan of life. Now on Kant's view, it is all or nothing and what Schapiro adds is a developmental account. Adolescents typically try on various characters through their identifications with pop stars, peer groups, sports teams, and so on. Young children work at developing a sense of self through the activity of play. Writes Schapiro (1999: 732):

> It may make sense to see play as a strategy—perhaps the strategy—for working through the predicament of childhood. By engaging in play, children more or less deliberately "try on" selves to be and worlds to be in. This is because the only way a child can "have" a self is by trying one on. It is only by adopting one or another persona that children are able to act the part of full agents, to feel what it must be like to speak in their own voices and to inhabit their own worlds.... I do not mean to suggest that this is the only thing children are doing when they play. Children also play purely for the sake of stimulation and diversion, as do animals and adults. But play serves an essential function in children's lives which it does not serve in the lives of either animals or adults. Play is children's form of work, for their job is to become themselves.

We can agree with Schapiro about the role of play in the lives of children while disagreeing with her about the necessary importance of rational agency, or with the idea that the source of value in human life stems from rationality. There is nothing in Schapiro's view that rules out there being value in childhood, but her description of childhood as a puzzle and as a predicament that must be solved suggests that the sooner this puzzle can be solved, the better. Returning to our earlier thought experiment about the pill that bypasses childhood, there would be no reasons on a Kantian account not to take it.

b. Michael Slote

In his book *Goods and Virtues*, Slote argues against the view that all the goods of life, at whatever stage they occur, count equally. Instead, he thinks, the goods of childhood count for less. He writes: "within a very wide range, the facts of childhood simply don't enter with any great weight into our estimation of the (relative) goodness of total lives" (Slote, 1983: 14). Slote believes "an unhappy schoolboy career" followed by "happy mature years" is such that the latter "wipes the slate clean." Thus on Slote's view, success following failure wipes out the disvalue of earlier failure but not vice versa. His view is not merely the endorsement of temporal asymmetry where future goods (and bads) count for more and past goods (and bads) count for less. Rather, he thinks that they count for less because of their specific temporal location, in childhood. This is true, thinks Slote, even though there are goods specific to the stage of childhood. On Slote's view, these goods just do not count for very much. As an aside I think Slote is wrong about what the goods of childhood are—he counts badges, school awards, sports trophies as among the goods—but more importantly I think he is wrong to say that, whatever the goods of childhood are, they automatically count for less.

Another possible explanation for this intuition about the insignificance of childhood goods comes from our preference for certain kinds of life stories. We prefer lives in which things start out bad and get better over lives that start out good and then go downhill. We like authors Augusten Burroughs and David Sedaris in large part for having survived their childhoods and grown up to be funny writers with happy lives. But this won't count in favour of Slote's intuitions about the goods of various life stages. I won't say much about this here but Slote has the same view about old age, namely that it has its own goods, and that they count for less than goods that occur during one's adult years. While I may care very much about the beautiful Christmas tree ornaments I make in the seniors' craft room, their beauty doesn't add much value to my life. The same holds true for winning the over-70 cards tournament and so on.

There are two different questions here and I think it would be helpful to distinguish them. The first question is whether the goods of childhood are different in any way than the goods of adult life. Are there childhood-specific goods? The second question is whether we weight goods that occur in childhood the same as goods that occur in adult life. I think we can understand Slote's mistake by seeing the connection between these two questions. If the goods of childhood were the sort of things that Slote takes them to be, then he would be right that they count for less. I just do not think that winning cards tournaments, for most people, figures much into our calculation of how well a particular life went, regardless of when the tournament took place. It may be that children value some relatively trivial things (to update Slote, we might note the concern of my sons for how their battles and conquests are going in Skyrim, or a teenager's fixation with how many friends he has on Facebook) but we needn't take what's valuable in childhood directly from what it is that children seem to value. Further, I am not sure that understanding what's valuable just in terms of what people value is the right way to proceed in the case of understanding the goods of adult life, and it is even less plausible in the case of children. What seems to me to be right about Slote's view is his idea that there are childhood goods. I disagree about what those are and I also disagree about how much weight they get in evaluating well-being over a life.

We tend to assume that our goal as educators, parents, legislators is to produce the best possible adult we can from the material that is the child. But I think that this gets it wrong. As we noted earlier, it is not obviously the case that any parenting method that produces good adults would be justified. Suppose that all of our current social scientific research was wrong and that adults who were beaten as children were happier, more successful, etc., than those who were not. This does not mean that beating children would be required, or even allowed. There are a range of views here. One extreme view might discount all childhood goods in favour of adult goods. A more moderate view is one that counts childhood goods equally. But note that a maximizing across the life span view might get it wrong if the goods of childhood were incommensurable. It seems to me that we owe obligations to the child qua child, not just in terms of the future person she'll become. I have phrased this in terms of negative obligations but one might think that we also have positive obligations to the child as a child regardless of the impact that fulfilling these obligations will have on the adult person she'll become. Suppose that reading to children, or playing with them, did not in fact make a difference to their future reading or relationship skills. It might still be that parents have a moral obligation to read to and play with their children because it's good for them as children. Consider the case of the parents of developmentally disabled children, children who may never learn to read. There may well be a duty to read to such children because of the pleasure it gives them now and the connection it forges between parent and child.

What are the Goods of Childhood?

So if I am right that we cannot read the goods of childhood directly off children's desires, then what approach ought we to use to determine what the goods of childhood are? Let me offer the following list of childhood goods in the spirit of various objective lists of human goods that have been put forward in recent years.

- Unstructured, imaginative play
- Relationships with other children and with adults
- Opportunities to meaningfully contribute to household and community
- Time spent outdoors and in the natural world
- Physical affection
- Physical activity and sport
- Bodily pleasure
- Music and art
- Emotional well-being
- Physical well-being and health

One might ask about the proposed list of childhood goods, in what sense are they goods of childhood if they are attainable outside of childhood? How are they different from human goods? I'm not claiming here that the intrinsic goods of childhood are distinct to childhood. Rather, my claim is that these goods matter intrinsically—that is, for their sake—and they can't be discounted completely in favour of goods that occur during adulthood as much writing on childhood suggests. Obviously most of life's goods are continuous, though they may be realized in different ways across the life span. For example, friendship is a good that clearly matters across the life span even though the friendships that occur in childhood are very different than adult friendships.

One might wonder whether there are any goods that one can only attain in childhood. My answer is that I am not certain. It seems to me that there are life-stage-specific experiences that are good. Some of these are biologically determined, as in child-bearing, which can only occur during certain life stages. Others are significant because they are firsts, such as first love and first crush and first job. These usually occur in the early stages of one's life but they need not. There is a certain kind of play that seems unique to childhood and which it's hard to make up later in life if missed out on.

Despite the rise of play therapy for adults and discussions of making peace with one's inner child, there seems to be something unique about play as a good of childhood. There also seems to be something distinctive about childhood friendships and relationships. Friends play a different sort of role in childhood than they do in later life and people report feeling an attachment to childhood friends out of all proportion to the sorts of shared interests and beliefs that usually form the basis of adult friendships. There is also a sense of time as endless, as having one's whole life stretched out ahead (think too of endless summer vacations) that one never has again in life. Likewise, there is the sense that all doors are open and that anything is possible. This sense fades as one leaves childhood. Finally, there is a kind of absolute trust in others, possible in childhood but then never again. I am not committed to the claim that there are goods only attainable in childhood. Really, my interest is in showing that there are childhood goods with a value that goes beyond their instrumental value. And I am interested in what moral philosophy would look like if we include children as children, including the goods of childhood.[7]

The Role of Parents in Weighing Interests and Choice, Current, and Future Goods

So far I've concluded that children have rights, and that the best account of children's rights is a developmental one in which rights start out protecting interests and progress to protecting choices as children mature into adults. I have also argued that children's interests include the goods of childhood, not just the goods they will come to have as adults. Now I want to address parents' roles in articulating and adjudicating children's rights. I also want to say something both about the obligations children's rights place on their parents and on how those rights constrain the actions of parents.

At the most basic level, parents share with all other persons the obligation not to infringe on children's rights. But insofar as parents act as advocates for the child and his or her rights, the obligations of parents extends to interpreting and evaluating the rights of their children. There are at least two aspects to this task. First, as children move from having rights that protect interests to rights that protect choices, parents

---

[7] I owe three of these suggestions—timelessness, possibility, and absolute trust—to Larry Temkin, with whom I discussed these ideas while we were both visitors at the Australian National University.

need to be closely involved in managing that transition. This means seeing when children are ready to make choices and educating children about what it means to be a good chooser. Parents might, for example, let children "try out" certain choices as part of the process of parenting. But even when the rights that children have are clearly rights that protect their interests, I think we have seen that there are trade-offs to be made. Parents need to make trade-offs between what's good for the child now, as a child, and what's good for the child, in the future, as the adult he or she will become. As parents we need to know—whether they are unique to childhood or not—how to weight goods that occur during childhood. I am dismissive of the negative conception of childhood and the view that childhood goods count for nothing. I think it's a deeply mistaken and pervasive view of parenting that our main obligations consist in shepherding our children safely to the threshold of adulthood, well prepared for autonomous adult lives. Such a view neglects the importance of the goods both parents and children get from childhood itself. However, it doesn't follow from the view that childhood matters to children's rights that I think life-stage egalitarianism is the right view either. It may be that some sort of discounting of childhood goods makes most sense of our intuitions about the distribution of life's goods. I am not committed to the claim that the goods of childhood count equally. A variety of factors may go into decisions about how to weight goods across the life span and in the case of children those decisions are usually made by parents. An important question facing parents is how do we, as agents who act on behalf of children, balance things that are good for the child-as-child with the things that are good for the child-as-future-adult? It's my hope that seeing the obligations that parents have towards children in this way demonstrates what a complex and creative task parenting can be. While parenting becomes more demanding than the old-fashioned "deliver them safely to the threshold of their adult years" approach, I also think the ways in which it's more demanding show how engaging an endeavour parenthood can be.

Acknowledgements

I owe a great many people, places, and research communities thanks for their help with this chapter. The bulk of it was written while I was a Visiting Faculty Fellow in Philosophy, Research School of Social Sciences, at the Australian National University in 2011 and while I was the Taylor Fellow in Philosophy at the University of Otago, New Zealand in 2012. Thanks also to the audience at the Rights and Realities: Children and the Australian State conference, held at ANU in November 2011 where the chapter was presented as "Children's Rights and Children's Lives." I would also like to thank the SSHRC-funded research group "Children, Family, and the State" (Colin Macleod (UVic), Daniel Weinstock (U de Montréal), PI, and Shauna Van Praagh) of which I was a part. In particular, I would like to thank Larry Temkin and Daniel Weinstock for their inspiration for the ideas in this chapter and their help in developing them.

References

Anonymous. (2012). Helicopter Parent. *Wikipedia*. Retrieved Aug. 2012 from http://en.wikipedia.org/wiki/Helicopter_parent.

Brennan, S. (2002). Children's Choices or Children's Interests: Which do their Rights Protect? In C. Macleod and D. Archard (eds), *The Moral and Political Status of Children* (pp. 53–69). New York: Oxford University Press.

Brennan, S., and Noggle, R. (1998). John Rawls' Children. In S. Turner and G. Matthews (eds), *The Philosopher Child: Critical Essays in the Western Tradition* (pp. 203–32). Rochester, NY: University of Rochester Press.

Brennan, S., and White, A. (2008). Responsibility and Children's Rights: The Case for Restricting Parental Smoking. In S. Brennan and R. Noggle (eds), *Taking Responsibility for Children* (pp. 97–111). Waterloo, ON: Wilfrid Laurier University Press.

Brooks, Andrew. (2007). The Odyssey Years. *New York Times*, 9 Oct. Retrieved June 2013 from http://www.nytimes.com/2007/10/09/opinion/09brooks.html?_r=0.

Campbell, T. (1992). The Rights of the Minor: As Person, as Child, as Juvenile, as Future Adult. *International Journal of Law, Policy and the Family*, 6(1), 1–23.

Chudacoff, H. (2007). *Children at Play: An American History*. New York: NYU Press.

Feinberg, J. (1980). A Child's Right to an Open Future. In W. Aiken and H. LaFollette (eds), *Whose Child? Parental Rights, Parental Authority and State Power* (pp. 124–53). Totowa, NJ: Littlefield, Adams, & Co.

Folbre, N. (2001). *The Invisible Heart: Economics and Family Values*. New York: New Press.

Folbre, N. (2008). *Valuing Children: Rethinking the Economics of the Family*. Cambridge, MA: Harvard University Press.

Gopnik, A. (2009). *The Philosophical Baby*. New York: Farrar, Straus & Giroux.

Gopnik, A. (2010). Alison Gopnik: Psychology and Philosophy (Berkeley). *University of California Research Profiles*. Retrieved June 2013 from: http://research.universityofcalifornia.edu/profiles/2012/05/alison-gopnik.html.

Hobbes, T. (2004). *De Cive*. Whitefish, MT: Kessinger Publishing.

Hobbes, T. (2011). *Leviathan*, ed. A. P. Martinich and B. Battiste. Peterborough, ON: Broadview Press.

Matthews, G. (2010). The Philosophy of Childhood. *Stanford Encyclopedia of Philosophy*. Retrieved June 2013 from http://plato.stanford.edu/entries/childhood.

Mills, C. (2003). Child's Right to an Open Future. *Journal of Social Philosophy*, 34(4), 499–509.

Narveson, J. (2001). *The Libertarian Idea*. Peterborough, ON: Broadview Press.

Noggle, R. (1997). The Moral Status of Children: Children's Rights, Parents' Rights, and Family Justice. *Social Theory and Practice*, 23(1), 1–26.

Nozick, R. (1974). *Anarchy, State, and Utopia*. New York: Basic Books.

Okin, S. M. (1989). *Justice, Gender, and the Family*. New York: Basic Books.

Purdy, L. M. (1992). *In their Best Interest? The Case Against Equal Rights for Children*. Ithaca, NY, and London: Cornell University Press.

Schapiro, T. (1999). What is a Child? *Ethics*, 109(4), 715–38.

Slote, M. (1983). *Goods and Virtues*. New York: Oxford University Press.

PART II

Bionormativity: Philosophical and Empirical Perspectives

3

A Critique of the Bionormative Concept of the Family

Charlotte Witt

The last good-bye I ever said to my grandmother (on my Dad's side) was when I was ten years old and it was September 8th. She died September 12th. My grandmother, Jeannette Okrent, grew up and lived in New York. In Brooklyn to be exact. She was the typical Jewish mother; full of life, opinion, and care for those in her family. In a way I never felt like I would ever have to say good-bye to her..... Death was never imminent in my mind. Especially not about my grandmother. As long as I can remember my family and I always visited her in New York every summer. The city became a second home to me, and to this day I still feel at home only whenever I'm there.

(from "The Last Good-bye" by Anna Witt)

Introduction

Everyone knows that children like Anna, who was adopted from Vietnam in infancy, are family members in the full sense. Indeed, everyone knows that families with children can originate in several ways. They are formed through adoption (both formal and informal) and ART (assisted reproductive technologies) as well as via sexual intercourse. However, everybody knows something else as well, which is that families with children who are not genetically related to both their parents are not the gold standard or Platonic form of the family, even though it is hard to pinpoint exactly what is wrong with them. I call this the bionormative conception of the family. The two sides of what everyone knows work together to maintain the superiority of families with children genetically related to their parents by affirming both that all families are the same, and that some families are different (and lacking in some respect).

The bionormative conception of the family is so deeply embedded in our culture that we find it in unlikely places like the adoption world and ART communities, and even in individual families with children who are not genetically related to their parents. It is also highly resistant to change, despite the increasing pace of change in family formations. And, even though there is a large and growing literature on the ethical challenges surrounding the institutions and practices of both adoption and ART, there is relatively little that focuses on the idea central to the bionormative conception of the family, namely that families formed via biological reproduction (in which there is a genetic relationship between parents and children) are, *for that reason*, superior to families formed in other ways. The relative lack of explicit defence of this position might well be the consequence of two factors; namely, the widespread acceptance of the bionormative conception of the family, coupled with the psychological tendency to infer from what often happens to what ought to happen. Whatever the explanation, I think it is important to challenge the bionormative view of the family because it has negative social and political implications for the standing of families that originate in other ways. If there are no good reasons to accept the bionormative view of the family, then we can begin to undermine its widespread, tacit acceptance. Or so I hope.

In this chapter I criticize two arguments that support the bionormative conception of the family. The first, which I call the argument from family resemblances, claims that there is something morally important missing from the lives of children in families formed in non-standard ways, who are not genetically related to their parents. In "Family History" and other essays J. David Velleman (2005, 2008) argues that the biological family plays an essential role in the healthy psychological development of children. The second argument, which I call the argument from human sociality, claims that the biocentred family plays a necessary role in the fashioning of human sociality itself. In "Blood is Thicker than Water, *nicht wahr?*", Robert Wilson (2008a) argues that the biocentred family is a key element in the development of human sociality. Each of these arguments makes claims about what seem to be important normative tasks of families, namely that they provide an environment for the healthy psychological development of children and that they help children to develop an ability to participate successfully in the human social world. Although I will be critical of these claims, I appreciate the opportunity they provide us to consider possible reasons for elevating some family formations above others rather than simply echoing and reinforcing a widespread cultural prejudice.

It is important to note at the outset, however, that neither of these philosophers is directly arguing for a bionormative conception of the family. Of the two, J. David Velleman comes closer. Although his actual target is the practice of anonymous gamete vending, his critique of this practice—in fact—applies to all families that are not formed via genetic reproduction. Indeed, Velleman takes the inadequacies of families formed through adoption as a premise in his argument against the ethics of using anonymous gamete vending to form families. However, he thinks that, in the case of adoption, the second-best status of their families is the best one can do for already existing children. The idea that adoption, and the situation of adoptees, can serve as a template for the experiences of those born via anonymous gamete vending is quite widespread, as can be seen in the arguments in the *Pratten* (2011) decision. In this decision the plaintiffs referenced the psychological harm caused by closed adoption records to argue against anonymity in gamete vending. Velleman's family resemblances argument targets a feature common to many families with children who are not genetically related to one or both of their parents and so it supports a bionormative concept of the family.[1]

Robert Wilson's (2008a) central project is a critical assessment of the way in which kinship studies has been conceptualized within anthropology. Wilson's particular focus is Schneider's critique of traditional kinship studies, which holds that the notion of kinship involves an ethnocentric projection by anthropologists from Western cultures to the cultures they study. Moreover, the particular concept of kinship that has been projected onto non-Western cultures is bio-essentialist in that it attaches particular importance to bio-genealogical and reproductive relations. In the course of Wilson's rich and interesting discussion of developments in anthropological kinship theories, an argument for the centrality of biocentred kinship relations for the development of human sociality emerges. My focus is on this relatively small part of Wilson's much larger project.

Velleman's Argument from Family Resemblances

In a series of essays, J. David Velleman (2008) addresses the question of how we should understand our responsibilities to future persons. Velleman argues that our responsibilities towards future generations is not best understood in terms of balancing out future harms and benefits that might accrue to future persons. In particular, the morality of bringing a person into existence cannot be measured in terms of benefits and harms because the procreative act can neither harm nor benefit the person (Velleman, 2008: 247). So how can we assess its morality? Velleman (2008: 249–50) endorses the Aristotelian idea that human flourishing requires the development of characteristic human abilities and capacities. But the child will not develop towards human flourishing inevitably; she requires support of various kinds. So parents have an obligation to support and facilitate the development of future children so that they can attain human flourishing. In this context Velleman argues that the practice of anonymous gamete vending is harmful to the resultant children and should be abandoned because it precludes or damages development towards human flourishing. Velleman's central and original argument in support of this claim is the argument from family resemblances, namely that we need direct acquaintance with biological relations in order

---

[1] Some families with children formed via adoption maintain fully open relationships with members of the children's birth families and the children in these families would presumably not lack direct acquaintance with biological family members.

to develop an adequate sense of self. Hence, children who live in families created by anonymous gamete vending are, by that fact alone, deliberately denied the materials necessary for development towards human flourishing. The same is true of children living in families formed through adoption of course, but for an already existing child this might be the best option available, preferable to institutional life or life with people who cannot provide adequate care (Velleman, 2008: 256). Since the argument from family resemblances is the centrepiece of Velleman's case for the bionormative family, we will begin with it.

In "Family History" Velleman reflects upon the significance that knowledge of his familial history—his ancestry—has had on his self-understanding. But Velleman (2005: 358) is particularly interested in the significance of his biological or genetic inheritance and how it factors into the significance of his ancestry:

That I am the great-grandson of Russian Jewish immigrants, that I also enjoy the fruits of their strivings—this much I know with certainty. I also know that I inherited not just the fruits but the striving too. What I don't know is how to understand that latter piece of my inheritance. Was it passed down entirely through my mother's upbringing by her father, and my upbringing by her? Or is the push in my personality a genetic endowment, from great-grandparents who twice pushed on?

After several autobiographical vignettes, Velleman argues that the practice of anonymous gamete vending is morally wrong because it deliberately creates a child who will lack direct knowledge of one (or both) of his or her biological parents and hence will lack access to his or her biological ancestry. But what is the precise harm here?[2] Velleman (2005: 365, 368) thinks that self-understanding, knowing what one is like, is usually accomplished by seeing resemblances between oneself and one's biological relatives:

In coming to know and define themselves, most people rely on their acquaintance with people who are like themselves by virtue of being their biological relatives.... If I want to see myself as another, however, I don't have to imagine myself as seen through other people's eyes: I just have to look at my father, my mother and my brothers, who show me by way of family resemblance what I am like.

In other words it is by virtue of seeing similarities between myself and other family members that I come to understand myself. So far, so good. But Velleman actually has a stronger view in two respects. He holds that directly observing familial similarities is *necessary* to form an adequate self-image (not just one way among many others), and that the enlightening similarities must hold between a self and his or her *biological* kin: "Not knowing any biological relatives must be like wandering in a world without reflective surfaces, permanently self-blind" (Velleman, 2005: 368). Velleman (2005: 366) also makes weaker claims: "I think that forming a useful

---

[2] It is important to note that Velleman does not argue that there is anything intrinsically wrong with the care provided in families not formed via biological reproduction. So, if there is a wrong here, it must be located elsewhere.

family-resemblance concept of myself would be *very difficult* were I not acquainted with people to whom I bear a literal family resemblance" and "knowing one's relatives and especially one's parents provides a kind of self-knowledge that is *of irreplaceable value* in the life-task of identity formation" (2005: 357). Clearly for Velleman a literal family resemblance refers to a resemblance to biological relatives. And, even though *very difficult* is weaker than *impossible,* the clear sense of the three passages taken together is that, without direct acquaintance with biological relatives, a child would lose the central and irreplaceable resource for identity formation. That is the precise harm to children in families not formed via biological reproduction; it is avoidable in the case of anonymous gamete vending and unavoidable in the case of adoption. Since forming an adequate sense of self plausibly is required for a child to develop towards human flourishing, Velleman's argument from family resemblances supports a bionormative view of the family.

Is Velleman correct that, without direct acquaintance with biological relatives, a child would lose the central and irreplaceable resource for identity formation, and hence be thwarted in her progress towards human flourishing? Consider the epigraph to this chapter written by my daughter. My daughter Anna shares with her grandmother Jeannette the attribute of "only feeling at home in New York" whereas Mark, Jeannette's son and Anna's father, dislikes New York and never wanted to live there. We might say Anna resembles Jeannette in her attitude towards New York, but Mark does not. However, it is Mark who stands in the correct causal relationship to Jeannette and Anna who does not. Anna is our adopted daughter and Mark is Jeannette's biological son. It is just false that a family resemblance must be grounded in a biological kinship relation or that a biological kinship relation is sufficient to ground a similarity between two individuals. It is also noteworthy that the resemblance is one that Anna employs in her self-description; it is an element in her self-understanding. As a New-York-o-phile Anna sees herself as like her grandmother.

However, perhaps all that Velleman wants to claim is that biological kinship results in similarities among individuals in some respects. This seems plausible. After all, we know that a number of the traits of individuals are heritable (via the gene) through biological reproduction. But the weaker claim does not provide what is needed to make Velleman's argument in support of the biocentric concept of the family. Let's recall how the argument goes. Children who lack direct acquaintance with their biological relatives will for that reason lack the necessary ingredients to form an adequate self-image because the necessary ingredients are gathered from direct acquaintance with biological relatives who look like them or resemble them in personality, etc.

It is worth noting two preliminary points. First, many of us lack direct acquaintance with some or most of our biological relatives due to death, divorce, immigration, and other events. So, there is reason to doubt the degree to which even the family with children formed via traditional biological reproduction provides the raw materials for self-understanding as Velleman describes them. Second, although the statistics vary, a significant number of presumed biological fathers are not—in fact—the biological parent of their child. This presents Velleman with a dilemma. It seems that Velleman cannot say that it makes no difference whether the father is the biological parent of the child or not, because he holds that biological kin alone can play the appropriate role in a child's self-understanding. But it also seems very implausible to think that the fact that the child's father is not biologically related to her would make a difference in the child's psychological development.

But there is a more serious difficulty facing the weaker argument. There is a big gap between the premise and the conclusion that an individual's self-understanding will be seriously compromised without access to biological relatives. This is because the weaker claim loosens the connection between a biological relationship between individuals and the existence of observable similarities among them. It allows for there being no similarities between biologically related individuals on the one hand, and for the existence of similarities among non-biologically related individuals on the other. Recall the example of Anna and her grandmother. So, the problem with the weaker premise is that it does not establish the desired conclusion, which is that children need direct acquaintance with their biological relatives in order to form a psychologically adequate self-image. And without that conclusion Velleman's argument provides us with no reason to think that biocentred families are the gold standard or the Platonic form of the family.

Velleman (2005: 365) bolsters the family resemblance argument by connecting it to the philosophical concept of family resemblances coined by Wittgenstein:

Philosophers should not have to be reminded that living things tend to resemble their biological relatives. After all the philosophical term for indefinable similarities is "family resemblance." Though much has been written by philosophers about family resemblance in this technical sense, little has been written about literal resemblance within families, which is after all the paradigm case of technical family resemblance.

Of course Velleman is referring to Wittgenstein's theory of family resemblance concepts, according to which a concept is unified by an overlapping and open-ended series of resemblances, which are not further analyzable. And Velleman assumes that the relevant resemblances hold among biological family members, who are paradigmatic exemplars of this kind of resemblance relation. Hence Velleman uses Wittgenstein's metaphor of family resemblance to reinforce the connection he draws between biological kinship and resemblances among family members.

But it is not at all clear that Wittgenstein's family relations concept supports Velleman's family resemblance argument. In an important article Hans Sluga (2006: 15) points out that Wittgenstein's metaphor of family resemblance conflates two very different kinds of concepts: kinship concepts and similarity concepts. A kinship concept is one that maps real causal connections between individuals whether those connections are established by biology, culture, or law. The concept of nationality is an example of a kinship term. An individual can become a German biologically by being the offspring of a German or legally by immigration to Germany. Looking a lot like a

German does not make you a German and conversely many Germans do not look like the stereotypical German. Family, like nationality, is a kinship term. "The conclusion I draw then is that 'family' in the human sense is a kinship and not a similarity term" (Sluga, 2006: 16). The concept of resemblance, pertinent to the notion of family resemblance, is a similarity term and not a kinship term. Similarity terms admit of degrees and they do not require a causal connection among individuals.

Where does Sluga's distinction between kinship terms and similarity terms leave Velleman's appeal to Wittgenstein's metaphor of family resemblances in support of a bionormative conception of family? Pretty clearly, if Sluga is correct, then there is no conceptual connection tying biological ancestry relations to resemblances as the metaphor of family resemblances might suggest. If a case for the biocentric conception of the family presupposes a conceptual connection between biological ancestry and similarity of appearance, we have good reason to reject it. Further, the example of Anna, Mark, and Jeannette shows that being a biological relative is neither necessary nor sufficient to establish a likeness between individuals. And, as we have seen, the weaker connection does not establish the conclusion that Velleman's position requires.

A third strand of Velleman's (2008: 255) argument makes an appeal to what he calls "universal common sense":

I claim that a life estranged from its ancestry is already truncated in this way. This claim is no less than universal common sense—though it is also no more, I readily admit. I cannot derive it from moral principles; I can at best offer some reflections on why we should trust rather than override common sense in this instance.

Of course, as Velleman notes, nothing of ethical significance follows from universal common sense. Moreover, some of the evidence Velleman (2008: 256) mentions is itself questionable, like the idea that it would be hard to understand great works of literature by Homer, Sophocles, and even the story of Moses in the Bible, without accepting his view about the necessity of direct acquaintance with biological kin for the development of a flourishing human life. Other evidence Velleman mentions, like the heritability of psychological characteristics based on twin studies, is controversial and, perhaps more importantly, its relevance to the argument from family resemblances is not entirely clear. At most it would establish that a child might be more likely to resemble her biological kin in some respects. But, as I explained earlier, it does not follow from this that the child could not find other sources of resemblance (and difference) in those around her sufficient to develop a sense of self as required for human flourishing.

Finally, Velleman points to the phenomenon of adoptees searching for their birth families, and the more recent movement by vendor-conceived children to search, as evidence that might be interpreted in support of the family resemblance argument.[3] According to a literature study (Muller and Perry, 2001) cited by Velleman (2005), up

---

[3] Velleman (2008: 261) allows that the connection he draws between the phenomenon of adoptee (and vendor-conceived) search and the argument from family resemblances is speculative.

to 50 per cent of adoptees will search for their birth parents at some point in their lives. Velleman says that "we can only speculate why" and then provides his speculation, which is a version of the family resemblance argument. However, the literature study Velleman cites provides a different interpretation of the reasons for search, one that is not speculative but supported by the studies under review. For the estimated 50 per cent of adoptees who will undertake a search during their lifetimes, Muller and Perry describe a wide range of motives for search including: identity-related motives, the need for factual or medical information, the need to fill a gap in an individual's history, and just plain curiosity.

In addition to extensive description of studies that record considerable variation in motivation for search and in degree of intensity, Muller and Perry propose three theoretical frameworks for explaining adoptee search. The social interactionist and the normality models, some combination of which the authors find supported by the evidence, and the psychopathological model, a model not supported by the evidence. According to the normativity model, search is an expression of normal psychosocial development for an adoptee and the social interactionist model emphasizes that adoptees' searching may be a result of our culture's definition of kinship in terms of blood ties. Although it is not a perfect match the psychopathological model (sometimes called "genealogical bewilderment") is the closest to Velleman's view of how to understand adoptee searches. It is striking that this is the model that the study authors *reject* based on the evidence; "the vast majority of searchers appear to be rather well-adjusted" (Muller and Perry, 2001: 31). Furthermore, Velleman simply rejects the social interactionist model, which the survey authors find partially supported by the evidence. He comments: "But maybe they (adoptees who search) are simply confused, because they live in a culture that is itself confused about the importance of biological ties" (Velleman, 2005: 360). In sum, the study Velleman cites to support "what everyone knows" about adoptees, biological ancestry, and searching turns out not to support "universal common sense." Another study concludes: "The vast majority of children who are adopted are well within the normal range of adjustment and show behavioural patterns that are similar to their nonadopted peers" (Brand and Brinich, 1999: 30).[4]

Velleman develops the family resemblance argument in order to defend the bionormative conception of the family and to criticize what he calls "the new ideology of the family." According to Velleman (2005: 360), the new ideology of the family was "developed for people who want to have children but lack the biological means to 'have' them in the usual sense... It [the new ideology] says that these children will have families in the only sense that matters, or at least in a sense that is good enough." But, as we have seen, Velleman's argument from family resemblances fails to establish its conclusion. And since he presents no other compelling evidence in support of the bionormative concept of the family, we have no reason to think that these families are the

---

[4] Brand and Brinich (1999) is cited in Blake, Richards, and Golombok (in this volume) which contains a nuanced and rich discussion of families formed via adoption and ART.

gold standard or Platonic form of the family. And so we have no reason to accept the bionormative concept of the family.

The Argument from Human Sociality

Recently Robert Wilson has defended a biocentred kinship concept rather than the pluralist approach to kinship relations advocated by some, perhaps most, anthropologists. One might think that which kinship concept to use is simply an empirical question, one that could be answered by gathering facts about how various cultures define kinship relations. In short, one might think that this is a question for anthropologists to answer. But, as I mentioned, Wilson is critical of a dominant shift in the way in which anthropologists conceptualize kinship, and so they are not a reliable resource for determining the character of kinship relations. Although Wilson situates his defence of a biocentred concept of kinship against the backdrop of the debate in anthropology concerning the issue of ethnocentric projection in the understanding of kinship relations, his direct argument for a biocentred kinship concept centres on the distinctive features of human sociality as he understands it.[5] In other words Wilson thinks that there are certain features of human social life that favour or require a biocentred family concept. Let's look at his argument, which requires a bit of preliminary conceptual stage-setting.

According to Wilson, human sociality is distinctive because of its *externally mediated, cognitively driven normativity.*[6] Human normativity is "group-focused," which connects with a central feature of human sociality, namely that we distinguish different sorts of people, and these different sorts of people occupy different normative positions in our social world:

People sort one another (and themselves) into many different categories; by their height and weight, their eye, hair, and skin color, their sex and sexual behaviour, their income level and type of employment, their personality and beliefs, their tastes in recreation and entertainment, their ancestry, religion and ethnicity, their astrological sign and year of birth, and their marital and parental status. (Wilson, 2008a: 40)

This sorting process has three important features.[7] First, it has normative resonance since how we interact with another person is affected by what kind of individual they are. We treat colleagues differently from students; we treat family differently from

---

[5] Wilson may also think that a biocentred notion of kinship is established through his critique of the turn taken in kinship studies under the leadership of Schneider and others (private communication).

[6] Wilson is a proponent of the extended mind view of cognition, which is what he means by "externally mediated." This is the idea, roughly, that our thinking is not confined within the bounds of an individual mind but can extend far beyond—to a social context as in the case of language use or to a piece of technology like the computer I am typing this on. This feature of human sociality is not central to the case Wilson makes for a biocentred kinship concept and so I will not have any more to say about it.

[7] According to Wilson, other animals that live in social groups may satisfy these conditions individually, but it is distinctive of human sociality that it satisfies all three of them.

strangers. Second, we tend to sort others as either of our kind or not, as like us or not, as one of us or not. Finally, the categories and concepts that we use to classify sorts of people are constrained by our bodily and social location. While it is in theory possible for beings like us to categorize people in an entirely third-person way, in fact we categorize others in our social world from a first-person perspective and that perspective is situated in a particular body and in a particular cultural location.

The existence of one important cognitive mechanism of categorization, the "like me" detector, has been established in developmental psychology. The "like me" detector is in play in the imitative behaviour of infants, and has been documented in the research of the developmental psychologist Andrew Metzger. Wilson postulates a sister cognitive mechanism, the "like us" detector, which underlies our ability to sort other people into kinds. "Thus, the postulated 'like us' mechanism tells those who possess it not simply whether another is the same as or different from oneself, as does Meltzoff's 'like me' detector, but by doing so with respect to social group membership carries information about the corresponding social groups" (Wilson, 2008a: 44).

Finally, Wilson introduces the psychological concept of "cognitive pull" to explain the importance of the "like us" detectors. The cognitive pull of a concept refers to its "disposition or tendency to impact on cognitive or behavioral functioning of the organism who has it" (Wilson, 2008b: 3). Concepts have different degrees of cognitive pull:

In saying that concepts and categories have differential pull I am saying that concepts and categories differ in the grip they have on us cognitively, emotionally, and behaviourally, and I take this piece of common sense knowledge to have been substantiated by a large body of psychological work on concepts and categories. (Wilson, 2008a: 46; citation to Murphy, 2002, given in the original)

As one might expect, "like us" detectors also have a high level of cognitive pull in that they have a significant impact on our understanding (of ourselves and others) and on our behaviour towards others.

We have almost arrived at the issue of why biocentric kinship is an important ingredient or necessary condition of human sociality and hence why the biocentred kinship concept is preferable to kinship pluralism.[8] According to Wilson, kinship is a powerful and universal "like us" concept with a high degree of cognitive pull. Recognizing another as kin has a powerful effect on how we understand and treat that other person:

Concepts of kinship *do* inform us about whom is like us, and they are acknowledged by all as having a high level of cognitive pull. Precisely in which respects kin are like us, and precisely what forms this cognitive pull take can vary, but I know of no societies in which these connections are absent or weak, and I suspect there are none. (Wilson, 2008a: 47)

---

[8] Wilson positions his view as opposed to both relativism or pluralism about kinship, in which biocentred kinship is simply one view among many, and biological reductionism about kinship relations in which human kinship and non-human primate kinship relations are on a continuum.

While none of Wilson's claims are uncontroversial, I am not interested in exploring them further here because they are only preparatory steps in the argument. One could accept all of them and remain a pluralist about kinship relations. However, it is important to underline that Wilson is arguing that the concept of kinship—a powerful "like us" detector—plays a significant role in our self-understanding and our understanding of the sorts of people who inhabit our social world. While kinship is not the only "like us" detector available to us, it is fair to say that without it an individual would be lacking an important cognitive and normative resource. Indeed, according to Wilson, it is a resource central to what is distinctive about human sociality.

The crucial step in Wilson's argument is the claim that it is a particular concept of kinship, the biocentric concept, that has a special role in human sociality and its *externally mediated, cognitively driven normativity*. Wilson (2008a: 47) thinks that "a biocentred concept of kinship is among those concepts with a high level of cognitive pull, not just for those in Western cultures, but universally." How does he support this claim?

Wilson (2008a: 48) argues that "biology imposes constraints on any concept of kinship" and that no concept of kinship can be biologically innocent. To explain what he means Wilson (2008a: 48) lists the following primary biosocial facts:

1. Babies are born from the body of a female person.
2. They are not self-sufficient until years after their birth, and require nurturance from a care-provider for sustenance, protection, and development for an extended period of time.
3. An individual carried in the womb during pregnancy is born and then develops over time into an adult person.
4. Biparental sex between a male and a female is needed for the conception of a foetus.
5. Both males and females are capable of having sex with different partners at different times.
6. Multiple individuals can be born to the same parents, to the same mother only, or to the same father only.
7. Lone care providers and their dependants are subject to vulnerabilities that are minimized by a system that delivers multiple care providers.
8. Biological substances are transmitted intergenerationally from parents in the creation and upbringing of children.
9. The reciprocal filiative relations "parent of" and "child of" are iterative, and their iteration generates at least some ancestral-descendant chains that go beyond these basic filiative relations.

These biosocial facts are "widespread, but not exceptionless" according to Wilson; for example, #4 is no longer a fact that obtains in Western culture because of advances in ART. Taken together, however, they work as constraints upon possible concepts of kinship, which as a consequence cannot be biologically "innocent." Notice that Wilson's list avoids both the facts of reproduction developed in Western biological science, which he calls "secondary biosocial facts," and institutional facts about how individual societies organize reproduction. For example, in some cultures marriage is a social institution that regulates sexual behaviour. Secondary biosocial facts cannot provide adequate constraints on the notion of kinship because they are the product of a particular scientific regime and not available to all cultures. But we find kinship in *all* cultures. Institutional facts concern how reproduction is organized in particular cultures; they are too variable to account for the unity of the concept of kinship. We find *kinship* in all cultures. Hence, only the primary biosocial facts can provide a core meaning to the concept of kinship.[9]

However, I think that Wilson's list of biosocial facts is compatible with a pluralistic concept of kinship. Consider for a moment a culture in which many of the facts on Wilson's list hold. In this culture the concept of kinship would not be biologically innocent because it satisfies many of the primary biosocial facts on Wilson's list. But this society could also allow for multiple concepts of kinship. Some kinship groups in this society would reflect #4 and #8 but others would not. Assuming that this culture might exist (since it does exist), we find a counter-example to the idea that this list will underwrite a single concept of kinship. Moreover, there is also tension in the list itself, which throws into question whether or not this list comprises a coherent whole that can underwrite a unified notion of kinship. Recall that Wilson (2008a: 52) rejects institutional facts as a basis for kinship is because "they do not form the kind of cluster that primary biosocial facts do. The regularities and norms that they express will often compete or pull in opposite directions." Now consider #1 and #2 again:

1. Babies are born from the body of a female person.
2. They are not self-sufficient until years after their birth, and require nurturance from a care-provider for sustenance, protection, and development for an extended period of time.

It is a persistent cultural trope that these two biosocial facts can and do pull in different directions. To see this we just need to consider cultural archetypes like Oedipus or Moses where the female whose body births an offspring and the provider of care are different persons. These two biosocial facts can pull in different directions and can underwrite two different concepts of kinship.

Biosocial facts 5 and 6 also provide occasion for normative tension:

5. Both males and females are capable of having sex with different partners at different times.

---

[9] Wilson's view of kinship draws on the homeostatic property cluster theory of kinds which holds that kinds are defined by clusters of properties, only a subset of which need to be instantiated for membership in the kind. The clustering of properties is established by natural and social mechanisms so that the clustering is systematic and not accidental. Wilson's list of biosocial facts record biological and social relations that might be co-instantiated by kinship relations in a given society. I think that the homeostatic property cluster theory of kinds is compatible with kinship pluralism.

6. Multiple individuals can be born to the same parents, to the same mother only, or to the same father only.

Consider facts 5 and 6 in relation to the issue of determining the paternity of an off-spring. In our culture the determination of paternity is institutional; the legal system recognizes a male to be the father of the children born to his wife (during the period of the marriage). That is one way of conceiving of kinship, but fact 5 points in a different direction and could underwrite a different kinship concept. Clearly this tension, which is internal to Wilson's list, generates normative conflict.

Let me summarize. Wilson thinks that there exists a set of biosocial facts that must constrain any concept of kinship. Wilson also thinks that these biosocial facts underwrite a single, unified biocentric concept of kinship and rules out kinship pluralism. I disagree because I think that these facts plausibly underwrite kinship pluralism. I have argued two points:

1. You can generate different concepts of kinship depending on which biosocial facts are included and so the list is compatible with multiple kinship concepts, each of which is constrained by the list and hence is not biologically innocent.
2. There is internal tension among the facts themselves that underwrite multiple kinship concepts, each of which is constrained by the list and is not biologically innocent.

Kinship pluralism is compatible with Wilson's biosocial facts, and indeed is required by them. Although Wilson may be right that the requirements of human sociality include the concept of kinship as a powerful "like us" indicator, he has not established that only the biocentric concept of kinship can do the job. It turns out not to be the case that the biocentric concept of kinship (or the family) is required in order for human sociality to develop.

Conclusion

Both Velleman and Wilson focus on the crucial role of likeness or similarity in forging a sense of self or for developing a sense of belonging to a culture (or to a larger social group). These are, of course, important aspects of human development. And they are two sides of the same coin as each develops in tandem with the other. It is fascinating that both philosophers find familial relationships or kinship relations to be central to these developmental tasks even though their emphasis falls in slightly different places.

Velleman is interested in an individual's autobiography, or in how an individual human being charts her understanding of herself. What landmarks will she use? What story will she have available to her in this endeavour? Velleman's description of his family history and the way in which it informs his self-understanding is genuine and moving. But so is the story my daughter Anna tells of her connection to Jeannette Okrent, who was both her grandmother and someone different from her in race, religion, and ancestry. Nonetheless, Anna draws inspiration and self-understanding from her similarity to Grandma Jeannette. If family resemblances are central to our self-understanding and that is an important ingredient in human flourishing as Velleman believes, then Anna is all set. And, contra Velleman, the argument from family resemblances gives us no reason to consider families with children who are genetically related to their parents to be the gold standard and Platonic form of the family.

Interestingly enough, Wilson is also concerned with the role of likenesses, his "like us" detectors, in enabling the development of characteristically human sociality. Wilson thinks, reasonably enough, that kinship relations provide a basis for activating the detectors, as do other kinds of relationships like friendships. Wilson also thinks, reasonably enough, that concepts of kinship have a high degree of cognitive pull, meaning that they have a strong grip on us cognitively and emotionally. Arguably, kinship concepts are central to our understanding of our social milieu and our place in it. What Wilson has failed to show is that a particular kind of kinship concept, a biocentred kinship concept, is required to do the job. So, his argument from human sociality fails to support the bionormative family concept and gives us no reason to consider biocentred families (or biocentred kinship relations) to be the Platonic form of family life.

In this chapter I have focused fairly narrowly on the question of whether or not families formed via biological reproduction with children genetically related to their parents are, for that reason alone, superior to families formed in other ways. It might seem that I have achieved very little since I only reject two philosophical arguments that support the bionormative family concept. What about all the other reasons and arguments out there? The surprising fact is that there aren't any other persuasive arguments out there.[10] There is simply what everyone knows about families.

References

Brand, A. E., and Brinich, P. M. (1999). Behavior Problems and Mental Health Contacts in Adopted, Foster, and Nonadopted Children. *Journal of Child Psychology and Psychiatry*, 40(8), 1221–9.

Buller, D. J. (2005). *Adapting Minds: Evolutionary Psychology and the Persistent Quest for Human Nature*. Cambridge, MA: MIT Press.

Haslanger, S., and Witt, C. (eds) (2005). *Adoption Matters: Philosophical and Feminist Essays*. Ithaca, NY: Cornell University Press.

[10] Evolutionary psychologists describe the "Cinderella Effect" that "substitute" parents will tend to care less profoundly for children than "natural" parents and they develop the hypothesis that children with "substitute" parents will be more subject to abuse and mistreatment than children living with their genetic parents. However, this hypothesis is not supported by the evidence: "the rates of physical abuse of children living with two unrelated adoptive parents were significantly lower than the rates of physical abuse of children living with both genetic parents" (Buller, 1995: 380).

Muller, U., and Perry, B. (2001). Adopted Persons' Search for and Contact with their Birth Parents: I. Who Searches and Why? *Adoption Quarterly*, 4(3), 5–34.

Pratten v British Columbia (Attorney General), 2011 BCSC 656. (2011). Retrieved July 2013 from http://www.courts.gov.bc.ca/jdb-txt/SC/11/06/2011BCSC0656cor1.htm.

Sluga, H. (2006). Family Resemblance. *Grazer Philosophische Studien*, 71(1), 1–21.

Velleman, J. D. (2005). Family History. *Philosophical Papers*, 34(3), 357–78.

Velleman, J. D. (2008). Persons in Prospect. *Philosophy and Public Affairs*, 36(3), 221–88.

Wilson, R. (2008a). Blood is Thicker than Water, *nicht wahr?* Unpublished manuscript, Department of Philosophy, University of Alberta, Edmonton, AB, Canada. On file with author.

Wilson, R. (2008b). What is so Special about Kinship? Unpublished manuscript. Department of Philosophy, University of Alberta, Edmonton, AB, Canada. Retrieved July 2013 from http://www.artsrn.ualberta.ca/raw/5WhatsSpecialmay08.pdf.

Witt, C. (2005). Family Resemblances: Adoption, Personal Identity and Genetic Essentialism. In S. Haslanger and C. Witt (eds), *Adoption Matters* (pp. 135–45). Ithaca, NY: Cornell University Press.

4

The Families of Assisted Reproduction and Adoption

Lucy Blake, Martin Richards, and Susan Golombok

Introduction

The theme of this book is the ethics of choosing between having children through adoption or by the use of assisted reproductive technologies (ARTs). This chapter focuses on the empirical evidence about the consequences of adoption and assisted reproduction for maternal health, parenting, and child development. After considering the social and regulatory contexts for choices between adoption and using ARTs, we will proceed to a discussion of the risks and hazards associated with ARTs for mothers and children, and their use by married or cohabiting heterosexual couples who for reasons of sub-fertility may use ARTs. We then consider their use by those without a reproductive partner: single women, and lesbian and gay couples. Next, we discuss the children and families of adoption. We conclude with a brief comment about the fact that our empirical research does not support claims in favour of the bionormative view of the family, according to which families formed via assisted reproduction with children who are genetically related to both parents are superior to families formed in other ways (i.e. adoption). (See Witt, in this volume.)

Contexts of Choosing between Adoption and ARTs

In the UK those seeking adoption face hurdles of "suitability." Public policy is focused on the interests and welfare of the adopted child, rather than satisfying the needs of couples unable to conceive themselves, as was the case until the 1960s. The number of children available for adoption is limited: in 2011 in England, 2,450 children were placed for adoption, with an average age at adoption of 3 years and 10 months (Adoption UK, 2012). This may be contrasted with the wide use of ARTs by both infertile couples and those without reproductive partners, with about 2 per cent of all children born in the UK conceived with the use of IVF. In 2011, 45,264 women had a total of 57,652 cycles of IVF or Intra Cytoplasmic Sperm Injection (ICSI) and 1,985 women (about half of whom had a female partner) had cycles of donor[1] insemination. There were about 1,500 cycles involving egg donation (Human Fertilisation and Embryology Authority, 2012).

Children available for adoption (in the UK) have generally been removed from their parents by state intervention, usually on the grounds of abuse or neglect, and typically will also have spent time in foster care before adoption. Thus, by the time of adoption children may well have had troubled and changing social relationships which may adversely affect their development and well-being. In addition, some 300 children per year are adopted from overseas (Adoption UK, 2012; Selman, 2002). This is strictly regulated and would-be adopters are assessed as eligible and suitable, as home adopters are, but in addition, adopters must show that a child cannot be cared for in a safe environment in their own country and that the adoption would be in the child's best interests. In the UK, as in some other countries, overseas adoption is relatively rare. This seems to be partly a result of the barriers to such adoption, including the high costs, and despite the existence of many orphans in some other countries—and the arguments about a duty to adopt (Rulli, in this volume).

Surveys indicate that many see adoption as a last resort, perhaps only to be contemplated after all the possibilities of assisted reproduction have been tried unsuccessfully (Van Den Akker, 2001). Most see ARTs as offering a much better option because there are fewer barriers to access, and, not least, because for most this includes the possibility of baby making with their own gametes. There are criteria for access, based on medical and social factors for free "treatment" in the UK National Health Service, but not for private clinics. However, the latter are expensive, charging some of the highest fees in Europe. About three-quarters of all IVF in the UK is carried out in the commercial sector where IVF typically may cost more than £5,000, and the use of associated techniques and treatments can add considerably to this sum.

Since the late 1930s when artificial insemination by donor began to become available in the UK, the list of procedures available to assist reproduction has grown, especially since 1978 when the first IVF baby was born. These now include IVF, egg donation, embryo cryopreservation (freezing that allows storage for later use or donation), and most recently, egg cryopreservation. ICSI was introduced in the early 1990s (Palermo et al., 1992). This technique, necessarily also involving IVF, is the injection of a single

---

[1] In using the terms donation and donors we are aware that in some countries these terms are misleading as the provision of eggs, sperm, or embryos for reproductive use is in the context of a market (see e.g. Almeling's (2011) account of the market for sex cells in the USA). However, in the UK (and more widely in Europe) donation is regulated and payment is prohibited. Some of those who donate eggs, sperm, or embryos receive a sum to cover their expenses (£35 for men, £750 for women, with a provision to claim an excess). In the UK egg "sharing" schemes a woman may receive IVF at a reduced cost and her "surplus" eggs are provided for other women who cover her treatment costs. In this chapter, which is written from a European perspective, we will use the term donation as this accurately describes the process and is the term used in the social science and ethics literatures, but we will avoid this when we are describing procedures where gametes etc. are provided on a commercial basis.

sperm (perhaps immature sperm removed directly from the testes) into an egg *in vitro*. It can be used to overcome many male fertility problems and it has to a significant extent replaced the use of insemination with sperm from others. Recently, its use has become more general as it is believed to be more effective for conception and so may increase the success rate of IVF. This is a crucial issue because, although IVF success rates have increased a little over the past decades, still less than a third of all IVF treatment cycles result in the birth of babies.

ARTs can also be used in combination with surrogacy to allow those who cannot carry a pregnancy to have children who may be conceived with their own gametes, or those from donor(s). Thus, gay couples will often use an egg provider and another woman to carry the pregnancy to have children. This may appeal to those prospective parents who want and value having a biological connection with their baby (Lev, 2006). However, surrogacy can be a prohibitively costly procedure for these prospective parents,[2] who will often need to pay expenses to the surrogate mother and to an egg donor in addition to agency fees (Golombok and Tasker, 2010).

Regulation and access to ARTs varies widely geographically, especially with regard to the use of gametes from others (Richards *et al.*, 2012). While in some countries ARTs are provided exclusively through commercial clinics, in other countries, especially for married couples, this may be subsidized by the state or available through health services. There is now a global market for ARTs, with growing numbers of prospective parents travelling internationally for services for reasons of accessibility, cost, or to find services unavailable for reasons of regulation in their own countries (e.g. Culley *et al.*, 2011; Pennings and Broadbent, 2012). So, for example, women from the UK may travel to Spain where costs of IVF and eggs are lower, and they are able to use sex selection by PGD or anonymous sperm provision which are both prohibited in the UK.

In choosing between adoption and assisted reproduction as routes to family building, the context in which choices are made will vary considerably from country to country. Public policies, service provision, and cultural attitudes will influence access to both. However, as we shall consider in the next section, the use of ARTs entails some risks to the health of mothers and to the health and well-being of children conceived through the use of these technologies.

The Risks and Hazards of ARTs

As compared with family building through adoption (or sexual reproduction), there are risks to the health of the mother and children associated with the use of ARTs. These include the risks associated with the stimulation of egg production and the surgical

---

[2] Even if they travel to low-wage economies where fees and expenses paid to the women who carry the pregnancy may be much lower than in the UK or USA (Audi and Chang, 2010).

hazards of egg removal during IVF, ICSI, and egg provision (Kennedy, 2005). Most significant here is ovarian hyperstimulation syndrome which may follow the use of hormones to trigger egg release. It occurs in about 5 per cent of those undergoing IVF and is severe in 1 per cent of cycles. Very rarely it is life threatening and fatalities have been reported (Delvigre and Rosenberg, 2003). The other notable risk for mothers is ectopic pregnancy which has about twice the normal rate and so affects about 1–3 per cent of all pregnancies resulting from embryo transfer. This is usually treated by removing the fallopian tube or the embryo from the tube.

Multiple births continue to be the major risk for those using fertility treatments. Ovarian stimulation, with or without intrauterine insemination, and IVF cycles in which more than one embryo are transferred, have led to what has been described as an iatrogenic epidemic of multiple births, with frequencies of multiple births in different countries reported to range from 25 per cent to nearly 50 per cent (Fauser et al., 2005). While in Europe (and some other countries) rates may be falling, these remained above 20 per cent in 2005 (Nyobe-Andersen et al., 2008). While twins may be a wanted instant family by some sub-fertile couples, all multiple births carry significant medical and social risks for mothers, children, and their families. The mortality and morbidity of pregnancies are high for triplets and there are significant risks for twins, in addition to the long-term medical, educational, and social consequences of prematurity and low birth weight (Mwaniki et al., 2012; Nyobe-Andersen et al., 2008; Pinborg et al., 2004; Salgal and Doyle, 2008). Recent research suggests that, even for moderate/late-preterm and early-term babies, health outcomes in childhood are significantly worse than those for full-term babies (Boyle et al., 2012). While a few have argued that hazards of twins are minimal if measured against the alternative of failure to conceive (Gleicher and Barad, 2009), there would seem to be wide agreement that high rates of multiple births are a significant public health issue which also places a heavy burden on obstetric, paediatric, and educational services and that regulatory efforts to reduce their number should continue. Multiple births following ARTs use are largely preventable. In the UK, the regulatory body has a policy to reduce these through limits on multiple embryo implantation and encouragement of single embryo transfer.

However, while much of the increased risk for mothers and children associated with the use of ARTs is related to multiple births, not all is and there are also raised risks of interuterine and subsequent perinatal complications with singleton births (Helmerhorst et al., 2004; Halliday, 2007). While some of this may reflect the underlying health factors of sub-fertility and increased maternal age, some may result directly from the use of ARTs (Sutcliffe and Ludwig, 2007). These risks include pre-eclampsia of pregnancy (odds ratio 1:55) and the risk of having a low-birth-weight baby born small for gestational age (Jackson et al., 2002). Children conceived with ARTs are more likely to use hospital services through childhood, in part because they have an increased risk of cerebral palsy (Ericson et al., 2002). Male children conceived after

ICSI have greater needs for urological surgery. These boys could be at high risk of urogenital defects which are known to be associated with male sub-fertility (Foresta et al., 2005) or transmitted infertility (Halliday, 2012). Though the condition is very rare, there is also evidence of raised risks of Beckwith-Wiedmann Syndrome (an epigenetic imprinting disorder following IVF). Though long-term follow-up data on IVF children are generally reassuring, it remains possible that alterations in epigenetic genomic imprinting caused by manipulations of the embryo might have other unrecognized health implications for those who were conceived through IVF (Arnor and Halliday, 2008; Halliday et al., 2004).

Sutcliffe and Ludwig (2007: 357) in their major review recommend that continuing monitoring of outcomes of ARTs should be obligatory, and they also note a concern about "the extent to which the interests of potential children might be subsumed by the rights of putative parents and the growing technological imperative of assisted reproduction."

In addition to the medical risks outlined, concerns have been expressed that the experience of undergoing fertility treatment may have a negative impact on parents' psychological well-being. Fertility treatment has been identified as being a stressful experience for women in particular (Greil, 1997; Holter et al., 2006), although the arduous elements of IVF treatment (such as daily injections and waiting to see if fertilization is successful) have been recognized as being emotional and physically burdensome for both partners (Eugster and Vingerhoets, 1999). In those families in which parents have conceived using gametes from others, it was feared that the involvement of a third party in reproduction might disturb the marital relationship of the mother and the father (Department of Health and Social Security, 1984).

Another concern about family functioning in families created using collaborative reproductive technologies is the issue of whether to tell the child about the nature of their conception or keep a family secret. In the 1980s and 1990s, the vast majority of parents using collaborative reproduction had not told their child about their mode of conception and did not plan to do so (e.g. Golombok et al., 1996). Early research on adoption found that those children who were denied knowledge of their origins were at risk of becoming confused about their identity and developing emotional problems (Sants, 1964), and that discovering the nature of their origins later in life or through a third party often led to feelings of resentment and distress (Triseliotis, 1973). These outcomes have been considered as just as relevant to families in which children are unaware that they are donor-conceived (Evan B. Donaldson Adoption Institute, 2009). Although studies reveal that disclosure is becoming more common, a substantial number of parents continue not to tell (Golombok et al., 2011b). Although disclosure is currently recommended by the Human Fertilisation and Embryology Authority in the UK, and the American Society of Reproductive Medicine in the US, the issue of whether disclosure is in the best interests of children continues to be debated (Appleby et al., 2012).

Family Functioning and ARTs

We now explore the evidence that is available regarding family functioning in families that parents conceived using: (1) IVF and ICSI; (2) gamete and embryo donation; and (3) surrogacy. Some parents who choose to conceive using gametes from others do so outside a clinic (e.g. a couple may use sperm provided by a friend). In such arrangements, intending parent/s may be exposing themselves to health risks and uncertainties regarding legal parenting rights and responsibilities. Little is known of these families and therefore this chapter will focus on those who conceive in a clinic.

1. Family Functioning in Heterosexual Two-Parent ART Families: IVF and ICSI families

Pregnancy has been found to be a stressful and anxious time for mothers who conceive using IVF (Eugster and Vingerhoets, 1999; van Balen *et al.*, 1996) which is perhaps unsurprising, given that IVF pregnancies are known to be at higher risk for adverse outcomes. Once the child has been born, differences between natural conception and IVF mothers seem to disappear. Studies conducted when children were in infancy, childhood, and adolescence have generally found few differences in the psychological well-being of IVF and ICSI parents compared to those parents who conceived naturally (Barnes *et al.*, 2004; Colpin and Bossaert, 2008; Colpin and Soenen, 2002). In a similar vein, a review of over 700 journal articles concluded that IVF treatment does not evoke long-term emotional problems for parents (Verhaak *et al.*, 2007). Although undergoing fertility treatment has been found to be stressful, couples undergoing IVF treatment report having close and supportive relationships both during and immediately after treatment (Holter *et al.*, 2006).

In terms of the quality of the mother–child relationship, some aspects of parenting in IVF families have been found to be superior to that of natural conception families. Compared to families in which parents conceived naturally, IVF mothers have been found to rate their young children more positively (Sydsjö *et al.*, 2002), and are rated by teachers as being more affectionate (Hahn and Dipietro, 2001). ICSI parents have likewise been found to show higher levels of commitment to parenting when compared to those who conceived naturally (Barnes *et al.*, 2004). In terms of the father–child relationship, a history of infertility, especially when IVF is involved, has been found to lead to a stronger involvement of fathers in family life (van Balen *et al.*, 1996). These findings may reflect parents' commitment, gratitude, and pleasure in finally having a family, having endured the physically and emotionally draining experience of infertility, ICSI, and IVF.

In addition to findings of more positive parenting, there is some evidence to suggest that IVF mothers are overprotective of their children. In a study conducted in the UK in the early 1990s, mothers' reports showed IVF mothers to be more protective of their infants than mothers who conceived naturally (Weaver *et al.*, 1993). The infants in this study were the first IVF babies to be born in a particular geographical region, therefore mothers may have been concerned about the wider social reaction to their children. A more recent study obtained observational data of mother–child interactions and concluded that IVF mothers were no more overprotective or intrusive than mothers who conceived naturally (McMahon and Gibson, 2002).

What do we know of children's well-being in IVF and ICSI families? Two studies conducted in the 1990s were indicative of negative outcomes for IVF offspring. First, a study conducted in Israel found that children conceived by IVF and embryo transfer scored lower on teacher-ratings of socio-emotional well-being compared to children who were naturally conceived, and IVF children scored higher on self-report measures of anxiety, aggression, and depression (Levy-Shiff et al., 1998). The authors suggested that these findings could be explained by the older age of IVF parents and/or their parenting style (e.g. IVF parents may be more likely to be overprotective). However, the unmatched nature of the groups, and the culture and context in which the study was conducted, may also explain these findings. Secondly, a French study found that IVF infants had more feeding difficulties and sleep disorders compared to control groups of infants who were conceived naturally, although these group differences disappeared over time (Raoul-Duval et al., 1994).

These two early studies aside, the majority of studies that have utilized standardized measures of children's psychological well-being have found no differences between offspring conceived by IVF and children who were conceived naturally, in both childhood (Colpin and Soenen, 2002; Golombok et al., 1996, 2002; Montgomery et al., 1999) and adolescence (Colpin and Bossaert, 2008; Golombok et al., 2009; Murray et al., 2006). Likewise, the majority of studies have found no difference between ICSI offspring and their naturally conceived counterparts in terms of psychological well-being (Barnes et al., 2004; Knoester et al., 2007), or the development of behavioural problems (Middelburg et al., 2008). However, studies exploring the psychological well-being of ICSI offspring in adolescence and beyond are required before conclusions about the long-term consequences of ICSI on children's psychological well-being can be drawn.

2.   Gamete and Embryo Donation Families

Of studies that have compared the psychological well-being of parents who conceived using donated gametes and those parents who conceived naturally or adopted, few differences have emerged (Golombok et al., 1996, 2002, 2004, 2005, 2006b; Murray et al., 2006). Similarly, comparisons between embryo donation families and adoptive families have revealed few differences between groups (MacCallum and Golombok, 2007; MacCallum and Keeley, 2008).

As for the stability of the marital relationship, an early study in the United States found that the rate of parental separation was low compared to population norms (Amuzu et al., 1990). Likewise, over the five phases of a longitudinal study of assisted reproduction families in the UK, few differences were found between gamete donation families and natural conception families in terms of mothers' or fathers' marital stability or quality (Blake *et al.*, 2012). However, some evidence of instability was found in a study in New Zealand of forty-four families who were revisited fourteen years after undergoing donor insemination; 46 per cent of couples had either divorced or separated (Daniels *et al.*, 2009), although this rate was not compared to national norms.

A number of studies have found evidence of more positive parenting in gamete donation families. For example, in the first phase of a longitudinal study of families created by assisted reproduction in the UK when children were one year old, mothers in gamete donation families were found to show higher levels of warmth towards their children and experience greater enjoyment in motherhood compared to those who conceived naturally (Golombok *et al.*, 2004). Findings of more positive parenting also emerged when the children were two (Golombok *et al.*, 2005) and three years old (Golombok *et al.*, 2006b).

There is some evidence to suggest that mothers in donor insemination egg donation and embryo donation families show greater levels of emotional overinvolvement with their child compared to mothers in natural conception, adoptive, and IVF families (Golombok *et al.*, 1996, 2002; MacCallum and Golombok, 2007; MacCallum and Keeley, 2008). Maternal overinvolvement in ART families is likely to be related to parents' experiences of infertility and assisted conception in general, as the parents in these families have waited a long time to have a family and therefore may be more likely to perceive their child as particularly special or precious.

With respect to children's psychological well-being, three studies conducted in the 1980s and 1990s indicated that there was some (albeit limited) evidence of increased problems in children's emotional and behavioural development in donor insemination families. First, a French study found that donor-conceived children showed signs of disturbed eating and sleeping patterns compared to naturally conceived children (Manuel *et al.*, 1990). Secondly, in a study of fifty-two donor insemination families, fourteen were reported to show hyperactive behaviour (Clayton and Kovacs, 1982) but a follow-up study of these families, in which comparison groups of adoptive and natural conception families were included, found no differences between family types in child well-being (Kovacs *et al.*, 1993). Thirdly, in the Dutch sample from the European Study of Assisted Reproduction Families, children in donor insemination families showed a higher incidence of emotional and behavioural problems compared to their naturally conceived counterparts, as reported by both mothers and teachers (Brewaeys *et al.,* 1997a). However, this finding was not replicated in the larger sample of families from Italy, Spain, and the United Kingdom (Golombok *et al.*, 1996).

Unlike the early studies outlined, the majority of studies conducted over the past two decades have found children in gamete donation families to be well adjusted and no different to their naturally conceived counterparts (Golombok *et al.*, 1996, 2002, 2004, 2006b, 2005, 2011b; Kovacs *et al.*, 1993). Likewise, children in embryo donation families have been found to be no different from children in control groups of IVF and adoptive families, and to be at no greater risk of developing psychological problems (MacCallum and Golombok, 2007; MacCallum and Keeley, 2008).

What do we know of family functioning in families in which parents conceived using gametes from a friend or family member? Two studies, conducted in the UK (Jadva et al., 2011) and Canada (Yee et al., 2011), have found mothers to be satisfied with the donor's level of involvement in the child's life, and that the donor maintained their social role in the family (e.g. that of the child's aunt in the case of sister-to-sister donation). Relationships between the egg donor (sister, sister-in-law, friend, or niece) and members of the family (the child, mother, and father) were generally found to be positive. Another finding emerging from these studies was that disclosure was uncommon. Few children were aware that they were conceived using donated eggs, and thus that the donor was a family member or friend of the family.

3. Surrogacy Families

Little research has focused on family functioning and child well-being in families in which the child was born using a surrogate and this remains the most controversial of the collaborative reproductive practices (Braverman et al., 2012). A longitudinal study of families created by ARTs in the UK included a sample of forty-two families who had used a surrogate. The families were assessed at five time-points, when the children were between the ages of one and ten years old (Golombok et al., 2004, 2006a, 2005, 2011a). As with findings relating to families in which parents conceived using IVF, ICSI, and gamete and embryo donation, more positive parenting was found for mothers and fathers in surrogacy families compared to parents who conceived naturally. For example, levels of parental warmth were higher for mothers in surrogacy families when the children were three years old (Golombok et al., 2006b) and fathers showed lower levels of parenting stress when children were aged two (Golombok et al., 2006a).

In contrast to parents in families using donated gametes, those who use surrogacy are more likely to be open with the child about their use of assisted reproduction. At the age of ten, almost all parents had told their child about their surrogacy birth (91 per cent), and of those parents who had used a genetic surrogate, 58 per cent had told the child about the use of the surrogate's egg and 32 per cent planned to do so in the future. The children in surrogacy families were found to be psychologically healthy and no different to their naturally conceived peers at all five time-points (see also Braverman et al., 2012).

4. Single, Lesbian, and Gay Parent ART Families

Access to ARTs for single, lesbian, and gay prospective parents varies widely in different cultures and countries. In the UK, recent changes to legislation have made ARTs more accessible to single, lesbian, and gay prospective parents. Where practitioners were once required to consider the welfare of the child and the child's "need for a father" (Human Fertilisation and Embryology Act, 1990), which led some clinics to deny lesbian couples and single women fertility treatment, practitioners are now encouraged to consider instead the child's need for "supportive parenting" (Human Fertilisation and Embryology Act, 2008). Since April 2010, gay men are enabled by the same Act to become the legal parents of a child born to them through a surrogacy arrangement and a fertility agency specializing in surrogacy services for gay men has now appeared in the UK (British Surrogacy Centre, 2012).

A number of studies over the past two decades have compared family functioning in lesbian mother families who conceived using donor insemination compared to two-parent heterosexual parent families who conceived with donor insemination or conceived naturally. In general, few differences emerged between the different family types. However, compared to fathers in heterosexual families, social mothers in donor insemination families have been found to be more sensitive and effective in responding to their child's needs (Flaks *et al.*, 1995), to have a greater quality of interaction with their child (Brewaeys *et al.*, 1997b), and to be more involved with and more concerned about their child (Bos *et al.*, 2007; Vanfraussen *et al.*, 2003). A number of early studies found no differences in well-being between children in lesbian mother families and in heterosexual families (Brewaeys *et al.*, 1997a; Flaks *et al.*, 1995). Recent studies, benefiting from greater sample sizes, reached the same conclusion (Bos *et al.*, 2007; Bos and Sandfort, 2010; Gartrell *et al.*, 2005). In a small study in Belgium, teachers, but not mothers, grated children from lesbian families as showing more attention problems (Vanfraussen *et al.*, 2002). No studies to date have examined family functioning in gay father families in which the child was conceived using a surrogate.

As for families of single women using donor insemination (commonly referred to as single mothers by choice or solo mothers), we know relatively little. In a comparison of family functioning with those of heterosexual couples in which children were two years old, solo mothers were found to experience greater joy in their child and perceive their child as less clingy (Murray and Golombok, 2005a), although findings were more mixed at an earlier phase of the study when children were one year old (Murray and Golombok, 2005b). Of the few studies that have examined children's psychological well-being, children have been found to be functioning well as rated by both parents and the child's teachers (Chan *et al.*, 1998a) and to have fewer emotional and behavioural problems than children in families of heterosexual parents (Murray and Golombok, 2005a). Studies conducted when the children in these families are in late childhood and beyond that draw on larger samples are required in order to address questions regarding the long-term consequences for children being raised in these families.

Adoptive Families

Adoption has been defined as the legal placement of children who have been abandoned, relinquished, or orphaned with an adoptive family (Juffer and van IJzendoorn, 2009). Although the definition of adoption seems relatively simple, the term is used to refer to numerous family situations. Parents may adopt a child of a different race or ethnicity to themselves, or a child to whom they are related (in the case of intra-family adoptions), and/or a child with disabilities. Adoptive families also differ in composition (adoptive parents may be single or in a heterosexual, lesbian, or gay relationship) and differ on important dimensions such as the child's age at adoption, the nature of the child's early experiences, and the child's level of contact with the birth family.

As with families created using collaborative reproduction, adoptive parents face the task of telling the child about their origins. Knowing what and how much information to share with the child may be a particularly difficult task. It is generally recommended that the disclosure process begins in the early preschool years, in which a simple story-like description of adoption becomes increasingly complex as the child's level of understanding becomes increasingly sophisticated (Wrobel et al., 2003). Adoption of children in infancy is now relatively rare, therefore children may have memories of their birth families or foster families, which parents need to take into account in the process of disclosure. Children's understanding and feelings about adoption have been found to shift from naïve and positive attitudes in toddlerhood to feelings of ambivalence, sadness, and even anger in middle childhood (Brodzinsky et al., 1984). As children develop a more sophisticated understanding that being adopted involves gaining a family, they also begin to understand that they have lost one as well, and this experience of loss may lead to a sense of ambivalence about being adopted and the emergence of psychological problems (Brodzinsky et al., 1984).

Another task is that of responding to children's potential curiosity about their birth family, and in some cases, supporting their child in negotiating and maintaining a relationship with the birth family. Adoptive arrangements range from those that are confidential (in which there is no exchange of information between the birth family and the adoptive family), to open adoptions, in which the adoptive family has identifying information about the birth family and they may have meetings with one another, ranging from the exchange of letters to face-to-face meetings. In a study of open adoption arrangements and adolescent psychological well-being, Grotevant (2007) concluded that fears about open arrangements were unfounded; children did not become confused or struggle with divided loyalties between their birth families and adoptive families, and openness did not interfere with adoptive parents' sense of entitlement. However, the authors advised against a single approach, concluding that no type of arrangement should be considered inherently better than another, as adoption is a dynamic process in which the needs and desires of the different kinship members may change over time (Grotevant, 2007; Grotevant et al., 2005).

A distinction has been made between structural openness in adoption arrangements (i.e. how much contact/communication there is) and communication openness (i.e. the nature of communication about adoption between and within the adoptive family network) (Brodzinsky, 2005). Communication between parent and child that is open, direct, and non-defensive has been found to be beneficial to children's development (Brodzinsky, 2005; Kirk, 1959; Wrobel et al., 2003).

Heterosexual Two-Parent Adoptive Families

The majority of studies examining the psychological well-being of adoptees in families of heterosexual couples conclude that a minority of adopted children fare worse in terms of behavioural and mental health compared to their naturally conceived peers (Palacios and Brodzinsky, 2010). In the early 1960s, an American psychiatrist reported that children who had been adopted were 100 times more likely to present a range of serious emotional problems at his clinical practice (Schechter, 1960). Although early studies such as this were methodologically flawed (e.g. making generalizations from small clinical samples), subsequent studies also found that adopted children were overrepresented in both outpatient and inpatient clinical settings (Brodzinsky et al., 1998). Similar findings emerged from studies utilizing large nationally representative survey data (Brand and Brinich, 1999; Miller et al., 2000a) and studies using meta-analysis (Juffer and van IJzendoorn, 2005, 2007). Adopted children have been found to experience problems that are typically externalizing in nature (e.g. attention deficit hyperactivity disorder, conduct problems) (Brodzinsky et al., 1998), and most likely to occur during middle childhood and early adolescence (Bohman and Sigvardsson, 1990; Hoopes, 1982; Maughan and Pickles, 1990).

Why are adopted children more likely to experience psychological problems than their non-adopted peers? Adopted children are more likely to face multiple risk factors that their peers do not. For example, adopted children often experience adversity in their pre-natal and/or pre-adoption environments which may involve neglect, abuse, malnourishment, and understimulation. These early experiences of adversity may result in developmental delays and can have a negative impact on the quality of the attachment relationship that infants and young children form with their birth parents and subsequently their adoptive parents (Haugaard and Hazan, 2003). A longitudinal study of internationally adopted children revealed that adversity in early childhood prior to adoption was associated with increased levels of psychiatric problems in adulthood, especially where earlier maltreatment was severe (van Der Vegt et al., 2009). Another important factor when considering outcomes for adopted children is their age at adoption (Brodzinsky and Pinderhughes, 2002). Those adopted at a later age are more likely to have experienced neglect or abuse (Dozier and Rutter, 2008), are more likely to feel that they do not belong in their adoptive family, and are less likely to feel loved by their adoptive parents (Howe, 2001).

Although the finding that adopted children are overrepresented in clinics is relatively robust, adoptive parents have been found to be more likely to use mental health services than non-adoptive parents when emotional or behavioural problems are at a relatively low level (Miller et al., 2000b). This tendency has been attributed to adoptive parents' familiarity and knowledge of mental health services that are available to them having undergone the adoption process (Brodzinsky and Pinderhughes, 2002). Also important to note is that the differences that have been identified between adopted and non-adopted families have typically been found to be small or moderate in size (Juffer and van IJzendoorn, 2005). The vast majority of children who are adopted are well within the normal range of well-being and show behavioural patterns that are similar to their non-adopted peers (Brand and Brinich, 1999).

In summary, the majority of studies that have compared adopted and non-adopted children have found that a small yet notable group of adopted children experience significant behavioural and/or mental health problems. But what about studies that have compared the development of adopted children to more appropriate comparison groups of children living in adverse social conditions, who experience the type of physical, social, and economic disadvantage that often characterizes the early life of adoptees? Studies that have compared the development of adopted children to children residing in long-term foster care, institutional environments, or those living with biological parents in disadvantaged backgrounds have typically found that outcomes favour adopted children (Brodzinsky and Pinderhughes, 2002). Thus adoption enhances the life chances of children who have been exposed to adverse early experiences.

The vast majority of children who are adopted experience a transition from a birth family of low socio economic status and an unstable home environment to an adoptive family of a higher socio economic status and a more stable, nurturing home environment. This shift is particularly dramatic in the context of international adoption, in which children in countries experiencing crisis (such as war, social upheaval, and poverty) are placed with socio economically advantaged parents in industrialized countries. Researchers have explored to what extent adopted children can overcome the consequences of early deprivation and childhood experiences of adversity. In a longitudinal study of children who were adopted after having lived in conditions of severe deprivation in Romanian institutions, Rutter *et al.* (2009) concluded that, following deprivation lasting up to three and a half years, there can be (and usually is) great improvement in children's functioning following placement into a well-functioning family environment. Improvement initially occurs at a rapid pace, but then continues for years following initial placement. In a recent review of studies exploring children's recovery following early adversity, Palacios and Brodzinsky (2010) concluded that the effects of adopted children's early experiences of adversity do not simply disappear, but that many children can and do experience a remarkable recovery.

Family Functioning in Lesbian and Gay Parent Adoptive Families

The possibility of lesbian and gay individuals and couples adopting a child varies in different countries and cultures. In the UK, single, lesbian, and gay prospective parents are able to adopt (The Adoption and Children Act, 2002), whereas in the USA access to adoption by single, lesbian, and gay prospective parents varies state by state (Gates *et al.*, 2007).

Research on lesbian and gay adoptive families is growing. In a comparison of adoptive families of gay/lesbian parents and those of heterosexual parents, the quality of attachment between adolescents and their adoptive parents was found to be unrelated to parents' sexual orientation (Erich *et al.*, 2009). Likewise, a comparison of three types of adoptive families (lesbian/gay couples, heterosexual couples, and the adoption of children with special needs) concluded that there were no negative effects on family functioning associated with the sexual orientation of adoptive parents (Leung *et al.*, 2005). In a study of trans racial adoptive families of lesbian mothers, gay fathers, and heterosexual couples, parents and children were found to be psychologically healthy, regardless of their parents' sexual orientation (Farr and Patterson, 2009).

Conclusions

Most of those in the UK who are unable to conceive through their own efforts turn to fertility clinics and only consider adoption when they have exhausted all the possibilities that ARTs have to offer. One underlying reason for this is the widespread preference to have children conceived with their own gametes, even when this may require the use of probably expensive and perhaps risky ARTs. However, the preference for ARTs may also reflect the perceived difficulties in achieving an adoptive family.

Although research has generally concluded that ART families, in terms of their dynamics and the social and emotional development of children, are very much like families where there was "natural" conception, the use of ARTs, especially IVF and ICSI, does carry some raised risks to health for mothers and children. In studies of family functioning and child well-being, there is a tendency to confine these to ART families with healthy singleton children, to avoid the potentially confounding effects of a multiple birth. However, as we have discussed earlier, IVF and ICSI families have high rates of multiple births and raised risks of children being born early, at low birth weight and with other perinatal problems. Thus these studies do not illustrate the well-documented negative psychosocial and emotional effects of prematurity and other perinatal problems and their consequences for family functioning (e.g. Cronin *et al.*, 1995; Moore *et al.*, 2006).

As for children who have been placed in adoptive families, although a minority are more likely to have behaviour problems and experience other difficulties, the majority have been found to do well. Most children are put up for adoption because they have suffered abuse or neglect in their birth families and the evidence suggests that their life chances will be improved by adoption despite the difficulties that are sometimes seen in adoptive families.

The evidence we have discussed strongly suggests that the families of ARTs, including those children who have been conceived with gametes from others, like the majority of adoptive families, function well and their children grow up very much like those in families built through the usual processes of sexual reproduction.

And finally, we turn to the bionormative conception of the family—the view that families formed via biological reproduction where there is a genetic relationship between the parents and the children are superior to families formed in other ways. As we have seen in this chapter, there seem to be no significant differences between families formed using donor gametes and those created via "natural" conception. Thus, if bionormativity is taken to favour the latter, then concerns such as Velleman's (2008; see also Witt, in this volume) about donor-conceived children and their families are unfounded, at least according to the empirical evidence discussed here. Concerns about adoptive families are similarly unwarranted. While some differences between adoptive and non-adoptive children have been noted here, as we have discussed, they seem attributable to the children's pre-adoption experiences, rather than to the fact they were being reared in a non-biological family. Indeed, the supposition would be that such differences would not exist for adoptive children who had less troubled early years or who had been adopted at birth. Similarly, some of the problems we have described that are associated with the use of IVF stem from the employment of the technologies rather than the non-biological nature of some of the families that use assisted reproduction.

Acknowledgements

We would like to acknowledge the support of a Wellcome Trust Biomedical Ethics Enhancement Award.

References

Adoption UK (2012). Adoption UK. Retrieved July 23, 2013 from www.adoptionuk.org.

Almeling, R. (2011). *Sex Cells: The Medical Market for Eggs and Sperm*. Berkeley, CA: University of California Press.

Amuzu, B., Laxova, R., and Shapiro, S. S. (1990). Pregnancy Outcome, Health of Children, and Family Adjustment After Donor Insemination. *Obstetrics and Gynecology*, 75(6), 899–905.

Appleby, J., Blake, L., and Freeman, T. (2012). Is Disclosure in the Best Interests of Children Conceived by Donation? In M. Richards, G. Pennings, and J. B. Appleby (eds), *Reproductive Donation: Practice, Policy and Bioethics* (pp. 231–49). Cambridge: Cambridge University Press.

Arnor, D. J., and Halliday, J. (2008). A Review of Known Imprinting Syndromes and their Association with Assisted Reproduction Technologies. *Human Reproduction*, 23, 2826–34.

Barnes, J., Sutcliffe, A. G., Kristoffersen, I., Loft, A., Wennerholm, U. B., Tarlatzis, B. C., Kantaris, X., *et al.* (2004). The Influence of Assisted Reproduction on Family Functioning and Children's Socio-Emotional Development: Results from a European Study. *Human Reproduction*, 19(6), 1480–7.

Blake, L., Casey, P., Jadva, V., and Golombok, S. (2012). Marital Stability and Quality in Families Created by Assisted Reproductive Technologies: A Follow-Up Study. *Reproductive BioMedicine Online*, 25(7), 678–83.

Blake, L., Casey, P., Jadva, V., and Golombok, S. (2013). "I was Quite Amazed": Donor Conception and Parent–Child Relationships from the Child's Perspective. *Children and Society.* doi:10.1111/chso.12014.

Bohman, M., and Sigvardsson, S. (1990). Outcome in Adoption: Lessons from Longitudinal Studies. In D. M. Brodzinsky and M. Schechter (eds), *The Psychology of Adoption* (pp. 93–106). New York: Oxford University Press.

Bos, H., and Sandfort, T. G. M. (2010). Children's Gender Identity in Lesbian and Heterosexual Two-Parent Families. *Sex Roles*, 62, 114–26.

Bos, H. M. W., Van Balen, F., and Van Den Boom, D. C. (2007). Child Adjustment and Parenting in Planned Lesbian-Parent Families. *American Journal of Orthopsychiatry*, 77(1), 38–48.

Boyle, E. M., Poulsen, G., Field, D. J., Kurinczuk, J. J., Wolke, D., Alfirevic, Z., and Quigley, M. A. (2012). Effects of Gestational Age at Birth on Health Outcomes at 3 and 5 Years of Age: Population Based Cohort Study. *British Medical Journal*, 344, 896–900.

Brand, A. E., and Brinich, P. M. (1999). Behavior Problems and Mental Health Contacts in Adopted, Foster, and Nonadopted Children. *Journal of Child Psychology and Psychiatry*, 40(8), 1221–9.

Braverman, A., Casey, P., and Jadva, V. (2012). Reproduction through Surrogacy: The UK and US Experience. In M. Richards, G. Pennings, and J. Appleby (eds), *Reproductive Donation: Practice, Policy and Bioethics* (pp. 289–307). Cambridge: Cambridge University Press.

Brewaeys, A., Golombok, S., Naaktgeboren, N., de Bruyn, J. K., and van Hall, E. V. (1997a). Donor Insemination: Dutch Parents' Opinions about Confidentiality and Donor Anonymity and the Emotional Adjustment of their Children. *Human Reproduction*, 12(7), 1591–7.

Brewaeys, A., Ponjaert, I., Van Hall, E. V., and Golombok, S. (1997b). Donor Insemination: Child Development and Family Functioning in Lesbian Mother Families. *Human Reproduction*, 12(6), 1349–59.

British Surrogacy Centre (2012). British Surrogacy Centre. Retrieved from http://www.british-surrogacycentre.com.

Brodzinsky, D. M. (2005). Reconceptualizing Openness in Adoption: Implications for Theory, Research, and Practice. In D. M. Brodzinsky and J. Palacios (eds), *Psychological Issues in Adoption: Research and Practice* (pp. 117–44). Westport, CT: Praeger.

Brodzinsky, D. M. (2006). Family Structural Openness and Communication Openness as Predictors in the Adjustment of Adopted Children. *Adoption Quarterly*, 9(4), 1–18.

Brodzinsky, D. M., and Pinderhughes, E. (2002). Parenting and Child Development in Adoptive Families. In M. H. Bornstein (ed.), *Handbook of Parenting* (vol. 1, pp. 279–311). Mahwah, NJ: Lawrence Erlbaum.

Brodzinsky, D. M., Singer, L. M., and Braff, A. M. (1984). Children's Understanding of Adoption. *Child Development*, 55(3), 869–78.

Brodzinsky, D. M., Smith, D. W., and Brodzinsky, A. B. (1998). *Children's Adjustment to Adoption: Developmental and Clinical Issues*. Thousand Oaks, CA: Sage.

Chan, R. W., Raboy, B., and Patterson, C. J. (1998a). Psychosocial Adjustment among Children Conceived via Donor Insemination by Lesbian and Heterosexual Mothers. *Child Development*, 69(2), 443–57.

Clayton, C. E., and Kovacs, G. T. (1982). AID Offspring: Initial Follow-Up Study of 50 Couples. *Medical Journal of Australia*, 1(8), 338–9.

Colpin, H., and Bossaert, G. (2008). Adolescents Conceived by IVF: Parenting and Psychosocial Adjustment. *Human Reproduction*, 23(12), 2724–30.

Colpin, H., and Soenen, S. (2002). Parenting and Psychosocial Development of IVF Children: A Follow-Up Study. *Human Reproduction*, 17(4), 1116–1123.

Cronin, C. M., Shapiro, C. R., Casiro, O. G., and Cheang, M. S. (1995). The Impact of Very Low-Birth-Weight Infants on the Family is Long Lasting. *Archives of Pediatrics and Adololescent Medicine*, 149(2), 151–8.

Culley, L., Hudson, N., Rapport, F., Blyth, E., Norton, W., and Pacey, A. A. (2011). Crossing Borders for Fertility Treatment: Motivations, Destinations and Outcomes of UK Fertility Travellers. *Access*, 26(9), 2373–81.

Daniels, K. R., Gillett, W., and Grace, V. (2009). Parental Information Sharing with Donor Insemination Conceived Offspring: A Follow-Up Study. *Human Reproduction*, 24(5), 1099–1105.

Delvigre, A., and Rosenberg, S. (2003). Review of Clinical Course and Treatment of Ovarian Hyperstimulation Syndrome. *Human Reproduction Update*, 9, 77–96.

Department of Health and Social Security (1984). *Report of the Committee of Inquiry into Human Fertilisation and Embryology (Warnock Report)*. Cmnd. 9314. London: HMSO.

Dozier, M., and Rutter, M. (2008). Challenges to the Development Of Attachment Relationships Faced by Young People in Foster and Adoptive Care. In J. Cassidy and J. Shaver (eds), *Handbook of Attachment: Theory, Research and Clinical Applications* (2nd edn, pp. 1083–95). New York: Guilford Press.

Erich, S., Hall, S. K., Kanenberg, H., and Case, K. (2009). Early and Late Stage Adolescence: Adopted Adolescents' Attachment to their Heterosexual and Lesbian/Gay Parents Early and Late Stage Adolescence. *Adoption Quarterly*, 12(3–4), 152–70.

Ericson, A., Nygren, K. G., Olausson, P. O., and Källén, B. (2002). Hospital Care Utilization of Infants Born After IVF. *Human Reproduction*, 17(4), 929–32.

Eugster, A., and Vingerhoets, A. J. (1999). Psychological Aspects of In Vitro Fertilization: A Review. *Social Science and Medicine*, 48(5), 575–89.

Evan B. Donaldson Adoption Institute (2009). *Old Lessons for a New World: Applying Adoption Research and Experience to Assisted Reproductive Technology. A Policy and Practice Perspective*. New York: Evan B. Donaldson Adoption Institute.

Farr, R. H., and Patterson, C. J. (2009). Transracial Adoption by Lesbian, Gay, and Heterosexual Couples: Who Completes Transracial Adoptions and with What Results? *Adoption Quarterly*, 12(3), 187–204.

Fauser, B. C. J. M., Devroey, P., and Macklan, N. S. (2005). Multiple Births Resulting from Ovarian Stimulation for Sub-Fertility Treatment. *Lancet*, 365(9473), 1807–16.

Flaks, D. K., Ficher, I., Masterpasqua, F., and Joseph, G. (1995). Lesbians Choosing Motherhood: A Comparative Study of Lesbian and Heterosexual Parents and their Children. *British Journal of Developmental Psychology*, 31(1), 105–14.

Foresta, C., Garolla, A., Bartoloni, L., Bettella, A., and Ferlin, A. (2005). Genetic Abnormalities among Severely Oligospermic Men who are Candidates for Intracytoplasmic Sperm Injection. *Journal of Clinical Endocrinology and Metabolism*, 90(1), 152–6.

Gartrell, N., Deck, A., Rodas, C., Peyser, H., and Banks, A. (2005). The National Lesbian Family Study: Interviews with the 10-Year-Old Children. *American Journal of Orthopsychiatry*, 75(4), 518–24.

Gates, G. J., Badgett, M. V. L., Macomber, J. E., and Chambers, K. (2007). *Adoption and Foster Care by Gay and Lesbian Parents in the United States*. Los Angeles, CA: UCLA.

Gleicher, N., and Barad, D. (2009). Twin Pregnancy, Contrary to Consensus, is a Desirable Outcome in Infertility. *Fertility and Sterility*, 91(6), 2426–31.

Golombok, S., and Tasker, F. (2010). Gay Fathers. In M. E. Lamb (ed.), *The Role of the Father in Child Development* (5th edn). New York: Wiley.

Golombok, S., Brewaeys, A., Cook, R., Giavazzi, M. T., Guerra, D., Mantovani, A., Hall, E., *et al.* (1996). The European Study of Assisted Reproduction Families: Family Functioning and Child Development. *Human Reproduction*, 11(10), 2324–31.

Golombok, S., Brewaeys, A., Giavazzi, M. T., Guerra, D., MacCallum, F., and Rust, J. (2002). The European Study of Assisted Reproduction Families: The Transition to Adolescence. *Human Reproduction*, 17(3), 830–40.

Golombok, S., Jadva, V., Lycett, E., Murray, C., and MacCallum, F. (2005). Families Created by Gamete Donation: Follow-Up at Age 2. *Human Reproduction*, 20(1), 286–93.

Golombok, S., Lycett, E., MacCallum, F., Jadva, V., Murray, C., Rust, J., Abdalla, H., Jenkins, J., and Margara, R. (2004). Parenting Infants Conceived by Gamete Donation. *Journal of Family Psychology*, 18, 443–52.

Golombok, S., MacCallum, F., Murray, C., Lycett, E., and Jadva, V. (2006a). Surrogacy Families: Parental Functioning, Parent–Child Relationships and Children's Psychological Development at Age 2. *Journal of Child Psychology and Psychiatry*, 47(2), 213–22.

Golombok, S., Murray, C., Jadva, V., Lycett, E., MacCallum, F., and Rust, J. (2006b). Non-Genetic and Non-Gestational Parenthood: Consequences for Parent–Child Relationships and the Psychological Well-Being of Mothers, Fathers and Children at Age 3. *Human Reproduction*, 21(7), 1918–24.

Golombok, S., Owen, L., Blake, L., Murray, C., and Jadva, V. (2009). Parent–Child Relationships and the Psychological Well-Being of 18-Year-Old Adolescents Conceived by In Vitro Fertilisation. *Human Fertility*, 12(2), 63–72.

Golombok, S., Readings, J., Blake, L., Casey, P., Marks, A., and Jadva, V. (2011a). Families Created through Surrogacy: Mother–Child Relationships and Children's Psychological Adjustment at Age 7. *Developmental Psychology*, 47(6), 1579–88.

Golombok, S., Readings, J., Blake, L., Casey, P., Mellish, L., Marks, A., and Jadva, V. (2011b). Children Conceived by Gamete Donation: Psychological Adjustment and Mother–Child Relationships at Age 7. *Journal of Family Psychology*, 25(2), 230–9.

Greil, A. (1997). Infertility and Psychological Distress: A Critical Review of the Literature. *Social Science and Medicine*, 45(11), 1679–1704.

Grotevant, H. D. (2007). Openness in Adoption: Re-Thinking "Family" in the US. In M. C. Inhorn (ed.), *Reproductive Techniques: Gender, Technology, and Biopolitics in the New Millennium*. New York: Bergshaw Books.

Grotevant, H. D., Perry, Y. V., and McRoy, R. G. (2005). Openness in Adoption: Outcomes for Adolescents within their Adoptive Kinship Networks. In D. M. Brodzinsky and J. Palacios (eds), *Psychological Issues in Adoption: Research and Practice* (pp. 167–85). Westport, CT: Praeger.

Hahn, C. S., and Dipietro, J. A. (2001). In Vitro Fertilization and the Family: Quality of Parenting, Family Functioning, and Child Psychosocial Adjustment. *British Journal of Developmental Psychology*, 37(1), 37–48.

Halliday, J., Oke, K., and Breheny, S. (2004). Beckworth Wiedemann Syndrome and IVF: A Case Control Study. *American Journal Human Genetics*, 75(3), 526–8.

Halliday, J. (2007). Outcomes of IVF Conceptions: Are they Different? Best Practice and Research. *Clinical Obstetrics and Gynaecology*, 21(1), 67–81.

Halliday, J. (2012). Outcomes for Offspring of Men Having ICSI for Male Factor Infertility. *Asian Journal of Andrology*, 14(1), 116–20.

Haugaard, J. J., and Hazan, C. (2003). Adoption as a Natural Experiment. *Development and Psychopathology*, 15(4), 909–26.

Helmerhorst, F. M., Perquin, D. M., Donker, D., and Keirse, M. J. N. C. (2004). Perinatal Outcome of Singletons and Twins After Assisted Conception: A Systematic Review of Controlled Studies. *British Medical Journal*, 328(7434), 261.

Holter, H., Anderheim, L., Bergh, C., and Möller, A. (2006). First IVF Treatment—Short-Term Impact on Psychological Well-Being and the Marital Relationship. *Human Reproduction*, 21(12), 3295–3302.

Hoopes, J. L. (1982). *Prediction in Child Development: A Longitudinal Study of Adoptive and Nonadoptive Families*. New York: Child Welfare League of America.

Howe, D. (2001). Age at Placement, Adoption Experience and Adult Adopted People's Contact with their Adoptive and Birth Mothers: An Attachment Perspective. *Attachment and Human Development*, 3(2), 222–37.

Human Fertilisation and Embryology Act 1990, c. 37. (1990). Retrieved July 2013 from http://www.legislation.gov.uk/ukpga/1990/37.

Human Fertilisation and Embryology Act 2008, c. 22. (2008). Retrieved July 2013 from http://www.legislation.gov.uk/ukpga/2008/22.

Human Fertilisation and Embryology Authority (2012). Latest UK IVF Figures—2010 and 2011. Retrieved July 2013 from http://www.hfea.gov.uk/ivf-figures-2006.html.

Jackson, R. A., Gibson, K. A., Wu, Y. W., and Croughan, M. S. (2002). Perinatal Outcomes in Singletons Following In Vitro Fertilization: A Meta-Analysis. *Obstetrics and Gynecology*, 103(3), 551–63.

Jadva, V., Casey, P., Readings, J., Blake, L., and Golombok, S. (2011). A Longitudinal Study of Recipients' Views and Experiences Of Intra-Family Egg Donation. *Human Reproduction*, 26(10), 2777–82.

Juffer, F., and van IJzendoorn, M. H. (2005). Behavior Problems and Mental Health Referrals of International Adoptees. *Journal of the American Medical Association*, 293(20), 2501–15.

Juffer, F., and van IJzendoorn, M. H. (2007). Adoptees Do Not Lack Self-Esteem: A Meta-Analysis of Studies on Self-Esteem of Transracial, International, and Domestic Adoptees. *Psychological Bulletin*, 133(6), 1067–83.

Juffer, F., and van IJzendoorn, M. H. (2009). International Adoption Comes of Age: Development of International Adoptees from a Longitudinal and Meta-Analytic Perspective. In G. M. Wrobel and E. Neil (eds), *International Advances in Adoption Research for Practice* (pp. 169–92). New York: Wiley.

Kennedy, R. (2005). *Risks and Complications of Assisted Conception: Factsheet*. London: British Fertility Society.

Kirk, D. (1959). A Dilemma of Adoptive Parenthood: Incongruous Role Obligations. *Marriage and Family Living*, 21(4), 316–28.

Knoester, M., Helmerhorst, F. M., van der Westerlaken, L., Walther, F. J., and Veen, S. (2007). Matched Follow-Up Study of 5 8-Year-Old ICSI Singletons: Child Behaviour, Parenting Stress and Child (Health-Related) Quality of Life. *Human Reproduction*, 22(12), 3098–3107.

Kovacs, G. T., Mushin, D., Kane, H., and Baker, H. W. G. (1993). A Controlled Study of the Psycho-Social Development of Children Conceived Following Insemination with Donor Semen. *Human Reproduction*, 8(5), 788–90.

Leung, P. T., Erich, S., and Kanenberg, H. (2005). A Comparison of Family Functioning in Gay/ Lesbian, Heterosexual and Special Needs Adoptions. *Children and Youth Services Review*, 27(9), 1031–44.

Lev, A. I. (2006). Gay Dads: Choosing Surrogacy. *Lesbian and Gay Psychology Review*, 7(1), 73–7.

Levy-Shiff, R., Vakil, E., Dimitrovsky, L., Abramovitz, M., Shahar, N., Har-Even, D., Gross, S., *et al.* (1998). Medical, Cognitive, and Behavioural Outcomes in School-Age Children Conceived by In-Vitro Fertilization. *Journal of Clinical Child Psychology*, 27(3), 320–9.

MacCallum, F., and Golombok, S. (2007). Embryo Donation Families: Mothers' Decisions Regarding Disclosure of Donor Conception. *Human Reproduction*, 22(11), 2888–95.

MacCallum, F., and Keeley, S. (2008). Embryo Donation Families: A Follow-Up in Middle Childhood. *Journal of Family Psychology*, 22(6), 799–808.

McMahon, C., and Gibson, F. L. (2002). A Special Path to Parenthood: Parent-Child Relationships in Families Giving Birth to Singleton Infants through IVF. *Reproductive Biomedicine Online*, 5(2), 179–86.

Manuel, C., Facy, F., Choquet, M., and Grandjean, H. (1990). Les Risques psychologiques de la conception par IAD pour l'enfant. *Neuropsychiatrie de l'enfance et de l'adolescence*, 38, 642–58.

Maughan, P. J., and Pickles, A. (1990). Adopted and Illegitimate Children Growing Up. In L. Robins and M. Rutter (eds), *Straight and Deviant Pathways from Childhood to Adulthood* (pp. 36–61). New York: Cambridge University Press.

Middelburg, K. J., Heineman, M. J., and Bos, A. F. (2008). Neuromotor, Cognitive, Language and Behavioural Outcome in Children Born Following IVF or ICSI: A Systematic Review. *Human Reproduction*, 14(3), 219–231.

Miller, B. C., Fan, X., Christensen, M., Grotevant, H. D., and van Dulmen, M. (2000a). Comparisons of Adopted and Nonadopted Adolescents in a Large, Nationally Representative Sample. *Child Development*, 71(5), 1458–73.

Miller, B. C., Fan, X., Grotevant, H. D., Christensen, M., Coyl, D., and van Dulmen, M. (2000b). Adopted Adolescents' Overrepresentation in Mental Health Counseling: Adoptees' Problems or Parents' Lower Threshold for Referral? *Journal of the American Academy of Child Psychiatry*, 39(12), 1504–11.

Montgomery, T. R., Aiello, R. D., and Adelman, R. D. (1999). The Psychological Status at School Age of Children Conceived by In Vitro Fertilization. *Human Reproduction*, 14(8), 2162–5.

Moore, M., Gerry T, H., Klein, N., Minich, N., and Hack, M. (2006). Longitudinal Changes in Family Outcomes of Very Low Birth Weight. *Journal of Pediatric Psychology*, 31(10), 1024–35.

Murray, C., and Golombok, S. (2005a). Solo Mothers and their Donor Insemination Infants: Follow-Up at Age 2 Years. *Human Reproduction*, 20(6), 1655–60.

Murray, C., and Golombok, S. (2005b). Going it Alone: Solo Mothers and their Infants Conceived by Donor Insemination. *American Journal of Orthopsychiatry*, 75(2), 242–53.

Murray, C., MacCallum, F., and Golombok, S. (2006). Egg Donation Parents and their Children: Follow-Up at Age 12 Years. *Fertility and Sterility*, 85(3), 610–18.

Mwaniki, M. K., Atieno, M., Lawn, J. E., and Newton, C. R. J. C. (2012). Long-Term Neurodevelopmental Outcomes with Intrauterine and Neonatal Insults: A Systematic Review. *Lancet*, 379(9814), 445–52.

Nyobe-Andersen, A., Goossens, V., Ferraretti, A. P., Bhattacharya, S., Felberbaum, R., Mouzon, J. D., and Nygren, K. G. (2008). Assisted Reproductive Technology in Europe, 2004: Results Generated from European Registers by ESHRE. *Human Reproduction*, 23(4), 756–71.

Palacios, J., and Brodzinsky, D. M. (2010). Review: Adoption Research. Trends, Topics, Outcomes. *International Journal of Behavioral Development*, 34(3), 270–84.

Palermo, G., Joris, H., Devroey, P., and Van Steirteghem, A. C. (1992). Pregnancies After Intracytoplasmic Injection of a Single Spermatozoon into an Oocyte. *The Lancet*, 340(8810), 17–18.

Pinborg, A., Loft, A., Schmidt, L., Greisen, G., Rasmussen, S., and Andersen, A. N. (2004). Neurological Sequelae in Twins Born After Assisted Conception: Controlled National Cohort Study. *British Medical Journal*, 329(7461), 311–17.

Raoul-Duval, A., Bertrand-Servais, M., Letur-Konirsch, H., and Frydman, R. (1994). Psychological Follow-Up of Children Born After In-Vitro Fertilization. *Human Reproduction*, 9(6), 1097–1101.

Richards, M., Pennings, G., and Appleby, J. (eds) (2012). *Reproductive Donation: Practices, Policy and Bioethics*. Cambridge: Cambridge University Press.

Rutter, M., Beckett, C., Castle, J. E. C., Kreppner, J., and Mehta, M. (2009). Effects of Profound Early Institutional Deprivation: An Overview of Findings from a UK Longitudinal Study of Romanian Adoptees. In G. M. Wrobel and E. Neil (eds), *International Advances in Adoption Research for Practice* (pp. 147–67). New York: Wiley.

Salgal, S., and Doyle, L. W. (2008). An Overview of Mortality and Sequelae of Preterm Birth from Infancy to Adulthood. *The Lancet*, 371(9608), 261–9.

Sants, H. J. (1964). Genealogical Bewilderment in Children with Substitute Parents. *British Journal of Medical Psychology*, 37(2), 133–41.

Schechter, M. D. (1960). Observations on Adopted Children. *Archives of General Psychiatry*, 3(1), 1–32.

Selman, P. (2002). Intercountry Adoption in the New Millennium: The "Quiet Migration" Revisited. *Population Research and Policy Review*, 21(3), 205–25.

Sutcliffe, A. G., and Ludwig, M. (2007). Outcome of Assisted Reproduction. *The Lancet*, 370(9584), 351–9.

Sydsjö, G., Wadsby, M., Kjellberg, S., and Sydsjö, A. (2002). Relationships and Parenthood in Couples After Assisted Reproduction and in Spontaneous Primiparous Couples: A Prospective Long-Term Follow-Up Study. *Human Reproduction*, 17(12), 3242–50.

The Adoption and Children Act 2002, c. 38. (2002). Retrieved July 2013 from http://www.legislation.gov.uk/ukpga/2002/38.

Triseliotis, J. (1973). *In Search of Origins: The Experience of Adopted People*. London: Routledge.

van Balen, F., Naaktgeboren, N., and Trimbos-Kemper, T. C. M. (1996). In-Vitro Fertilization: The Experience of Treatment, Pregnancy and Delivery. *Human Reproduction*, 11(1), 95–8.

Van Den Akker, O. B. A. (2001). Adoption in the Age of Assisted Reproduction. *Journal of Reproductive and Infant Psychology*, 19(2), 147–59.

van Der Vegt, E. J. M., van Der Ende, J., Ferdinand, R. F., Verhulst, F. C., and Tiemeier, H. (2009). Early Childhood Adversities and Trajectories of Psychiatric Problems in Adoptees: Evidence for Long Lasting Effects. *Journal of Abnormal Child Psychology*, 37(2), 239–49.

Vanfraussen, K., Ponjaert-Kristoffersen, I., and Brewaeys, A. (2002). What does it Mean for Youngsters to Grow Up in a Lesbian Family Created by Means of Donor Insemination? *Journal of Reproductive and Infant Psychology*, 20(4), 237–52.

Vanfraussen, K., Ponjaert-Kristoffersen, I., and Brewaeys, A. (2003). Family Functioning in Lesbian Families Created by Donor Insemination. *American Journal of Orthopsychiatry*, 73(1), 78–90.

Velleman, J. D. (2008). Persons in Prospect. *Philosophy and Public Affairs*, 36, 221–88.

Verhaak, C. M., Smeenk, J. M. J., Evers, A. W. M., Kremer, J. A. M., Kraaimaat, F. W., and Braat, D. D. M. (2007). Women's Emotional Adjustment to IVF: A Systematic Review of 25 Years of Research. *Human Reproduction Update*, 13(1), 27–36.

Weaver, S. M., Clifford, E., Gordon, A. G., Hay, D. M., and Robinson, J. (1993). A Follow-Up Study of "Successful" IVF/GIFT Couples: Social-Emotional Well-Being and Adjustment to Parenthood. *Journal of Psychosomatic Obstetrics and Gynecology*, 14(suppl.), 5–16.

Wrobel, G. M., Kohler, J. K., Grotevant, H. D., and McRoy, R. G. (2003). The Family Adoption Communication (FAC) Model: Identifying Pathways of Adoption Related Communication. *Adoption Quarterly*, 7(2), 53–84.

Yee, S., Blyth, E., and Tsang, A. K. T. (2011). Oocyte Donors' Experiences of Altruistic Known Donation: A Qualitative Study. *Journal of Reproductive and Infant Psychology*, 29(4), 404–15.

PART III

Becoming a Parent: Personal Choices

5

What is the Value of Procreation?

Christine Overall

What is the Value of Procreation as a Means of Family-Making?[1]

There may be several ways to interpret this question. I shall construe it as asking about the *justification* of procreation, and in particular, whether there are good reasons, moral or pragmatic, for prospective parents to prefer the creation of genetically related children over the adoption of children who are not genetically related to them.

The paradigm case of procreation is the situation in which one or two prospective parents create a child using their own gametes, and one of them gestates him. The paradigm case of adoption is the situation in which one or two people take parental responsibility for a child for whom they have not provided the gametes and whom neither of them has gestated. There are more complex cases, such as "procreative adoption" (Crawford, in this volume), the use of others' embryos, and contract motherhood. Here I focus on the paradigm cases of procreation and adoption, but I believe my arguments apply beyond the paradigms.

Without procreation the human species would no longer exist. Advanced as it is, technology does not yet permit the reproduction of human beings without the work of human beings, especially female human beings. That much is obvious. But many other reasons have been presented, or merely assumed, to justify procreation. In this chapter I analyze and evaluate them. I show that while there can be good reasons to choose procreation, not all such reasons are plausible and none are universally persuasive. Moreover, procreation should not be regarded as being morally or pragmatically superior to adoption as a means of family-making. Most of the reasons ordinarily given for choosing procreation also apply to adoption.

[1] Parts of this chapter are based on material from my book, *Why Have Children? The Ethical Debate* (2012) and are used here with permission from MIT Press. In the book I discuss at length the general question of whether and how choosing to procreate may be morally justified. Of course, within the scope of just one chapter, it is not possible to pursue all of the arguments to the extent that they warrant.

Bearing Children as Intrinsically Worthwhile and a Source of Moral Superiority

People commonly believe that having a child may be, in part, an expression of gratitude that one exists, oneself. They say they give the "gift of life" because others gave it to them. This viewpoint appears to make procreation a source of moral superiority over adoption because procreation passes on the gift of life, whereas adoption merely involves caring for a life already created.

There are several problems with the "gift" metaphor as a basis for the supposed superiority of procreation. First, the metaphor assumes what it is supposed to show: It assumes that coming into existence is a benefit that must then be passed on to others. Whether coming into existence is a benefit is philosophically controversial (e.g. Benatar, 2006). Second, we don't usually assume that we should pass on a gift that we received; often, doing so would be highly inappropriate, since what is a gift for one person could be irrelevant or a burden to someone else. Third, the meaning of gift-giving is precisely that it need provoke no further action; a gift is something that has "no strings attached," and there is no moral reason to pass it on.

J. David Velleman argues that life is not a gift at all, because the gift has "no intended recipient. It is a 'gift' that is launched into the void, where some as-yet nonexistent person may snag it. Such untargeted benefits do not fit our usual concept of gift-giving" (Velleman 2005: 372 n. 7). That's not quite right, for sometimes "untargeted" benefits can also be gifts. Consider monetary donations to charities. These gifts have no specific intended recipients; they are "launched into the void" in the sense that one does not know, and is likely never to know, who are the specific people who are helped. Instead, what makes procreation an odd "gift" is the fact that the recipient does not yet exist. In this one case the "gift" *creates* the recipient, and there is no particular being on whom the "gift" of existence is bestowed. As David Benatar (2006: 129–30) puts it, since procreation is not a matter of bringing "the benefit of life to some pitiful non-being suspended in the metaphysical void and thereby denied the joys of life," children are never brought into being for their own sake.

Rosalind Hursthouse argues in a different way for the allegation that procreation is a source of moral superiority. She says that men and women (and especially women who have borne a child) are not "morally equal," because women are better: "[I]n bearing children, Mrs. Average does something morally significant and worthwhile which Mr. Average does not match, whereby they are not morally on a par" (Hursthouse, 1987: 299). Hursthouse (1987: 300) compares pregnancy and labour to going "into battle," and says that they require "courage, fortitude and endurance". She questions why women are not "praised and admired" for all of this. After all, she says, "A man who goes through something like what women go through in childbearing (such as a painful illness or operation) with the same unthinking courage, fortitude and endurance is counted as particularly admirable" (Hursthouse, 1987: 304).

Thus, Hursthouse emphasizes the burdensome aspects of pregnancy and delivery, and sees child-bearing as (usually) a heroic endeavour. She romanticizes women for their sheer capacity to gestate and give birth to children. She writes, for example, that "women are, in one respect, born superior to men—superior in the straightforward sense that they are born with a capacity to do something worthwhile, *viz.* bear children (and no corresponding incapacity, such as being unable to think logically) which men lack" (Hursthouse, 1987: 298). She even goes as far as to suggest, at least tentatively, that anyone who has not "borne children well" might have to say, "I haven't done anything with my life really" (Hursthouse, 1987: 318).

It's rather pleasant to read the work of a philosopher who so powerfully values the work of child-bearing. Nonetheless, in her enthusiasm for women's procreative capacities, Hursthouse goes too far. It is implausible and even sexist to suppose that merely possessing a biological capacity makes women superior to men. Some women never exercise that capacity; being the possessor of a uterus does not make them better human beings. Yet using one's capacity to gestate and give birth is also not automatically value-conferring. If it were, then the more children one had, the better and more moral a person would be. Mere numbers of offspring do not make multiparous women morally superior to those who are childless or who have only one; the mother of five is not more advanced than the mother of two. The fact is, depending on their medical condition and the available medical resources, some women are heroic in pregnancy and delivery and some are not. Most simply get through it, and most are fairly gracious about it. Although pregnancy and delivery are significant, even more important is how one raises the child, for that endeavour requires at least an eighteen-year moral commitment.

To claim that persons who have not "borne children well" have not done anything with their lives is preposterous, given all the other valuable, even heroic activities that many human beings engage in. The idea is especially dangerous for women. It could be used to get women to give up other activities on the grounds that all that matters is having a baby. To suppose that every single childless person has "not done anything" with her or his life is to debase our human history of creativity and achievement.

Hursthouse also describes child-bearing as "intrinsically worthwhile." The reason for this notion is not, she says, that human beings in general are intrinsically valuable. Rather, the idea that child-bearing is intrinsically worthwhile expresses important ideals with respect to "the value of love, of family life, of our proper emotional development through a natural life-cycle and what counts as enrichments of this emotional development" (Hursthouse, 1987: 311). Hence, it is more accurate to think of creating a child as intrinsically worthwhile to oneself (Hursthouse, 1987: 312).

But if the supposed intrinsic value of child-bearing is actually a function of the other values Hursthouse lists, then it is not in fact worthwhile for its own sake. Love, an enhanced family life, and greater emotional development are indeed important. But bringing a child into one's life is not a guarantee of attaining them. And when they are attainable in childrearing, they can be achieved as much through adoption as through procreation.

The Intrinsic Value of Human Life and Human Beings

Another possible justification for procreation is the claim that sheer human life has intrinsic value. Since procreation creates new human lives, procreation would therefore be highly important. Indeed, some writers have claimed that, because of the value of human life itself, we have a moral responsibility to bring as many human beings into existence as possible. Sahin Aksoy, for example, says that human existence "is essential and prerequisite to everything good or bad." He adds, "[E]very life is worth living, even if it is worse than some other lives, if the only alternative is non-existence" (Aksoy, 2004: 382). Aksoy makes the error identified by Benatar: He believes that human beings have an ethereal existence prior to conception, waiting for their potential parents to call them into the material world. He writes, "[L]ife and existence is [*sic*] always better than non-existence" and "[t]herefore, it is irrational and immoral to 'sentence' someone to non-existence while you have the chance to bring them into life and existence" (Aksoy, 2004: 383).

Moreover, there are good reasons to question whether procreation can derive justification from the hypothesis of the intrinsic value of human life. First, we might doubt whether there are adequate grounds for supposing that human life and human beings have intrinsic value. It is unclear what property or properties would be the basis for intrinsic value. If it is sheer humanness, then a fertilized egg has the same inherent value as a 10-year-old child, and merely creating fertilized eggs would be a valuable action—a conclusion that is implausible. If it is some other property associated with humanness, then we would have to say that certain other beings are intrinsically valuable and entitled to the same protections as human beings, and that procreating them is also valuable. As Thomas Young (2001: 189) writes,

The familiar problem is that whatever trait(s) one selects—such as language or self-consciousness—it, or they[,] will be present in some non-human animals as well (or be pitched so high that many humans are excluded!); thus, it is extremely difficult to locate a non-arbitrary characteristic which most members of our species possess and all non-human animals lack.

And even if some or all human beings do have intrinsic value (according to some not-yet-obvious, non-arbitrary criterion that also bestows such value on some non-human entities), it might not be enough to establish the justification of human procreation. For if intrinsic value is possessed by several different forms of life, the procreation of human beings could result in a net loss of value, since the more human beings are created, the more other life forms are threatened, and the less likely it is that large numbers of members of other life forms will be created. That is, more procreation of (allegedly intrinsically valuable) human beings seems to lead to less procreation of (allegedly equally intrinsically valuable) members of certain other species. If procreation derives instrumental value from the alleged intrinsic value of living beings, then *human* procreation would decline in value as more human beings are produced.

Reflecting on the purported intrinsic value of different forms of life and whether it shows that extinction is inevitably bad, James Lenman suggests that claims about intrinsic value have no implications for the value of procreation. He uses an analogy with white rhinoceroses. Lenman suggests that even if we assume, for the sake of argument, that white rhinos are intrinsically valuable, it does not follow that the procreation of more white rhinos is good, or that they should be spread all over the planet. No one laments the fact that there are no white rhinos in Scotland, for example. Nor does it imply that there should be more of them spread out through time. "If it is unclear how it would make things better to stretch out, synchronically, in a single generation, the numbers of white rhinos, it is unclear why it should make things better to stretch them out diachronically by having more generations." We need not regret that there are no white rhinos in Scotland, and we also need not regret that there may be no white rhinos a million years from now (Lenman, 2004: 138–9).

In other words, the fact (if it is a fact) that *y* is intrinsically valuable does not imply that there should be more of *y*, whether simultaneously or serially. Even if entities of certain kinds are intrinsically valuable, there nonetheless is an infinite number of such intrinsically valuable entities that have never come into existence and never will. "But it is hard to make much sense of the thought that this [is] a bad thing—either for those individuals themselves or otherwise" (Lenman, 2004: 139). Therefore, the fact (if it is a fact) that human beings are intrinsically valuable does not make procreation better than adoption.

Creating Happy People

Another possible justification for procreation lies in the human capacity for happiness: Procreation is good because it creates happy people. John Leslie (1996: 181, 178), for example, writes that if you did *not* see a world of very happy people, with only a small number unhappy, as "remarkably good,"

> then you'd have fairly strong grounds for thinking it right to annihilate the human race in some quick and painless fashion.
>
> ...Just as a planet of utterly miserable people could be worse than nothing, so also a planet of happy people could be better than nothing. If a philosopher had a chance to create the first planet simply by lifting a finger, then prima facie the finger oughtn't to be lifted.... Similarly with the second planet. Assuming that creating it wouldn't produce harm elsewhere, it ought to be created....

Leslie therefore writes of there being a "moral need," wherever possible, to replace miserable people with happy people. He asks "what our duty would be in a situation where absolutely nothing could be done to help the miserable, no matter how hard we tried" (Leslie, 1996: 182), and he draws an analogy to "miserable" individuals who are born ill or impaired and cannot be helped. His implication is that since such individuals cannot be changed, one must instead go on and create new ones who will not be ill or impaired and hence (according to him) will be happy.

If the creation of happy people generates a "moral need" to procreate, then the justification of procreation appears to rest upon its consequences: the greater the total happiness of the population, the more justified is procreation. Taken to its extreme limit, this consequentialist justification for procreation famously results, on a global scale, in what philosopher Derek Parfit calls the "Repugnant Conclusion": "Compared with the existence of very many people—say, ten billion—all of whom have a very high quality of life, there must be some much larger number of people whose existence, if other things are equal, would be better, even though these people would have lives that are barely worth living" (Parfit, 2004: 10). In other words, if the goal is simply to produce as much good as possible then the procreation of as many people as possible would be justified, even if their individual lives are horrible, because the total amount of good would thereby be maximized. Such a conclusion, resulting in lives that are at best minimally tolerable, is not morally acceptable.

In response to this unpalatable implication, Leslie might insist that his point is that procreation is justified by the creation of genuine happiness, not merely some bare minimum balance of good over bad. Even so, a version of the Repugnant Conclusion would still apply: On Leslie's construal, procreation would be good and morally justified, provided it results in happy people. The prospect of increasing happiness would thus give procreation a moral advantage over adoption. Indeed, Leslie's viewpoint makes procreation so important that human beings should continue creating human beings and spreading them throughout the galaxy, provided they are happy. Once the science and technology are developed, he postulates that humanity could have "a very strong duty" to expand right across our galaxy (Leslie, 1996: 183).

But these consequentialist claims entirely overlook an essential value: women's autonomy. A woman with one happy child, whose second child, if created, would also be happy, does not have a moral "duty" to have that child. The happiness of the potential child does not trump her liberty to determine whether her body is used for procreation. Torbjörn Tännsjö (2004: 233), who thinks we should simply accept the Repugnant Conclusion, exclaims, "Such a want of generosity, if we do not welcome such a creature [a newborn infant]!" But if a non-existent individual is never conceived, no one has been wronged. Generosity does not require us to create the baby in the first place. Although some children end up having very good lives, we do not have an obligation to non-existent beings to bring them into existence. In fact, if we care about increasing happiness, then adopting a parentless child would seem to be a much more direct way to do it. After all, the adoptive child already exists and has material, social, and emotional needs, whereas the potential child who will be brought into existence if a woman decides to procreate has no needs whatsoever.

Moreover, even if many people have happy lives, not everyone agrees that it is even possible for procreation to create a net balance of benefit over harm. Thomas Young

(2001: 185–6) argues that, if we regard having children as morally permissible, let alone desirable, then we must say the same thing about "eco-gluttony;" that is, increasing one's own consumption to a level equal to adding another human being, who might live to eighty, to the American population. Eco-gluttony is clearly wrong. He concludes that because "having even just one child in an affluent household usually produces environmental impacts comparable to an intuitively unacceptable level of consumption, resource depletion, and waste," human procreation is morally wrong in most cases (Young, 2001: 183). Even if one has just two children, those children will use huge amounts of resources during their lives, and they will then probably go on to have children of their own, compounding the problem. "[T]wo more children…in a world with over six billion people is insignificant; yet most agree that the cumulative effect of a number of people acting that way is, and will continue to be, disastrous for species diversity, ecosystem preservation, and future generations" (Young, 2001: 185). The implication is that, in a prosperous nation like the United States, having any children at all is likely to be morally wrong.

However, given the importance of child-bearing to many people's life-projects and sense of self, an obligation not to have any children at all would be an enormous sacrifice, one that is too much to expect of persons who want to procreate. Moreover, people are not likely to adhere to such an obligation, not only because it would be so difficult (given how much some people care about procreation), but because others would surely violate it, thereby lowering their own motivation and drastically increasing resentment. It would also be hard to undertake such an obligation knowing that, once the population was sufficiently reduced, people in the future might no longer have to adhere to it. Nonetheless, Young's argument for the negative results of procreation is hard to reject entirely: It is undeniable, on a crowded planet, that every new person, especially in the wealthy nations, adds substantially to resource demands and waste production. For that reason, arguably, people in the developed nations have a responsibility to limit their procreation.[2] At the same time, the same degree of liability does not automatically attach to adoption, because it involves not the creation of a new human being but simply taking over the care of an existing one. (However, if the child is adopted from a poor country, becoming part of a wealthier one will inevitably cause a substantial increase in the child's environmental footprint.[3])

Benatar also believes that procreation produces no net balance of benefit over harm. In fact, he argues that "coming into existence is *always* a serious harm." If he is right then there is a prima facie reason never to procreate and procreation has no positive value. The main argument behind this claim is straightforward:

Although the good things in one's life make it go better than it otherwise would have gone, one could not have been deprived by their absence if one had not existed. Those who never exist

---

[2] Elsewhere, I argue that the Chinese state-mandated limit of one genetically related child per family is unfair and draconian, but that a self-assumed responsibility to have no more than one genetically related child per person is reasonable (Overall, 2012).

[3] I owe this point to Queen's Ph.D. student Josephine Nielsen.

cannot be deprived. However, by coming into existence one does suffer quite serious harms that could not have befallen one had one not come into existence. (Benatar, 2006: 1)

Benatar's argument relies on what he thinks is a key asymmetry between the absence of good and the absence of bad:

The absence of bad things, such as pain, is good even if there is nobody to enjoy that good, whereas the absence of good things, such as pleasure, is bad only if there is somebody who is deprived of these good things. The implication of this is that the avoidance of the bad by never existing is a real advantage over existence, whereas the loss of certain goods by not existing is not a real disadvantage over never existing. (Benatar, 2006: 14)

What Benatar advocates is fundamentally a small-c conservative approach to existence. Imagine that each of us somehow had a choice about coming into existence on this earth. In effect, Benatar's advice to the would-be earthling is, "You're gonna get hurt, so don't risk it." But a high degree of risk aversion is not the only plausible approach to human existence. There may be some, probably many, people who say that life's experiences are worthwhile provided at least that there are enough good ones and the bad ones are not too numerous. My point here is that not only do many of us in retrospect say we are glad to be in existence; in addition, if we notice that Benatar's highly risk-averse outlook is not incontrovertible, then we might conclude that we would have chosen to come into existence ourselves if that option had, *per impossible*, been offered to us. Such observations count against the claim that it is better never to have been.

Of course, in reality we do not make the decision whether or not to come into existence. Instead, others make that decision for us. It is risky. The question whether or not one is justified in imposing the risks of existence on *other* potential people is not the same as the question whether or not one should take on those risks oneself; that is, it is distinct from the question whether or not, for each of us, it is better never to have been. But surely Benatar cannot say that it is unacceptable to commit another person to risking any suffering at all. After all, adults commit their minor children to risks of suffering all the time. Every time a parent transports her child in a car, visits a park with the child, or signs the child up for sports, the child will run the risk of suffering if something goes wrong. It would be bizarre to take the view that parents are wrong in presenting these opportunities to their children, for the correct assumption is that the likelihood of benefits far outweighs the risk of harms.

Benatar (2006: 59) also deprecates what he calls "the unduly rosy picture most people have about the quality of their own lives". There's something far-fetched about the idea that I, along with everyone who is happy to be alive, could be badly mistaken about the quality of my life, yet Benatar (2006: 64) insists that it is so. He cites psychological studies indicating that people usually remember positive rather than negative experiences, and also states, "We tend to have an exaggerated view of how good things will be." Moreover, "Many studies have consistently shown that self-assessments of well-being are markedly skewed toward the positive end of the spectrum" (Benatar,

2006: 65). Indeed, says Benatar (2006: 100), human beings may be "engaged in a mass self-deception about how wonderful things are for us."

But it is unlikely that the vast majority of us are guilty of false consciousness. Benatar cannot possibly know this of every single human being who is happy to have been born. It is simply unfounded to deny the experience of literally millions of people who, for the most part, enjoy their lives and are happy to exist. Moreover, it is presumptuous for him to suppose that it is he (along with the few who may agree with him) who fully understands the human situation and has the appropriate response to it.

I conclude that, although the production of happy people does not make procreation so important that we have any obligation to engage in it (non-existent persons are not harmed or treated unjustly by not being created), at the same time, the fact that some people suffer, even greatly, does not make all procreation morally unjustified.

The Experiences of Pregnancy, Childbirth, and Breastfeeding

Other arguments for the value of procreation are founded upon the experiences of pregnancy, childbirth, and breastfeeding. Many women enjoy being pregnant and place a high importance upon having a physical connection with their future offspring from its conception. Some women appreciate the sense of power and accomplishment that can be achieved through labour and the delivery of an infant. Many women also enjoy breastfeeding, and there is ample evidence of its benefits both for the child and for the mother (Overall and Bernard, 2012). Pregnancy, child-bearing, and breastfeeding are unique events and like no others.

However, for several reasons I don't think the experiences of pregnancy, childbirth, and breastfeeding always make procreation better than adoption, and they certainly do not give procreation a special value in every case. First, it's worth mentioning that it is possible for some women who have not given birth to successfully breastfeed an infant, both in order to provide (some) nutrition for the infant, and for the sake of the shared experience for mother and child (e.g. Biervliet *et al.*, 2001; Szucs *et al.*, 2010). I'm certainly not claiming that all adoptive mothers of infants should attempt breastfeeding; I'm simply suggesting that the potential for breastfeeding is not necessarily confined only to birth mothers.

Second, not every woman who wants to be a parent also wants to be pregnant, give birth, or breastfeed. These experiences are not always positive and not every woman enjoys or values them. Pregnancy and birth create risks of illness, disfigurement, and even death that not all women are able or willing to undertake. Labour and delivery can be painful and sometimes the outcome is a caesarean section. So pregnancy, birth, and breastfeeding are not always positive for women who procreate.

But even when the experiences of pregnancy, childbirth, and breastfeeding are achievable and positive, claiming that procreation necessarily has great value because of them would be odd. What seems to be the widespread and growing romanticization of reproduction gives many women the idea that pregnancy, birth, and breastfeeding are ineluctably life-enhancing, indeed life-changing, and that they are a necessary prerequisite to one's role as a mother. But it's important to remember that pregnancy, childbirth, and breastfeeding take up a relatively small part of the woman's relationship to her child. Pregnancy is nine months at most; the mother's relationship with her child may last as much as another sixty years or even more of her life. Missing out on pregnancy, childbirth, and breastfeeding may indeed be a loss to those who seek those experiences, but the relationship itself endures for the rest of the parent's life, if all goes well. These experiences are also not necessary to a warm and supportive relationship with one's child. Non-gestators—such as the male or female partner of a pregnant woman—are able to have fine relationships with their children. Adoptive parents develop close connections to their children at whatever point in the child's life their relationship begins.

Procreation as a Way of Fulfilling Familial Duties and Conforming to the Bionormative Ideal

Another argument for the value of procreation points to the alleged importance of fulfilling duties to family: Having a child is seen as a way to honour one's family and one's upbringing. Procreators also thereby acquire the considerable social advantages of conformity to the bionormative ideal of the family (Witt, in this volume). This ideal defines womanliness in terms of child-bearing and manliness in terms of begetting. In some families and cultural groups, one is not even a "real" woman unless one has created a child. Having children is a means to achieve social normalcy and to demonstrate appropriately gendered behaviour. Procreators are not expected to justify their behaviour in the way that those who do not procreate often are.

But promoting procreation because it permits conformity to the bionormative ideal contributes to pronatalism, which is harmful both to those who have life goals other than (or in addition to) procreation and to those who are not capable of procreating. It unfairly disparages families created by adoption or step-parenting and treats as second-class all children who are not genetically related to their parents.

Moreover, to justify procreation out of a sense of duty to family or a desire to honour or please the future grandparents is to use the child as a kind of currency in the family exchange. Even in the absence of pronatalist pressures, having a child out of respect for one's parents is of doubtful value. It could lead to disappointment (since grandchildren don't always turn out the way one might hope). Choosing to have a child only to fulfil a duty to others is also unlikely to be good for the child himself. It doesn't seem to provide adequate motivation to sustain the great amount of commitment, work, and patience that go into childrearing—even if the grandparents will be around to help (many of them will be unavailable or will die before the children are grown). If the parents otherwise wanted to remain childless, they may even end up resenting their offspring rather than appreciating him for himself. Instead, if honouring one's parents is an important goal, there are other ways to do it: by encouraging one's parents in their own life endeavours; by supporting one's parents to perpetuate their own values and achievements; by caring for one's parents when they need it; and by making a good life for oneself.

Procreation as a Means of Perpetuating Family Property

It is unsurprising that many people want to pass on their property to their children. But to justify procreation for this purpose puts the cart before the horse. Handing down an inheritance benefits the children, but to have children *in order* to hand down an inheritance implies that one is having children in order to benefit the inheritance. The mere maintenance of property within a family line becomes the ultimate goal, instead of a way of supporting one's family. Instead of acquiring money and possessions in order to support one's offspring, one acquires offspring in order to support one's money and possessions. Doing so treats the child merely as a means to an end, and such treatment is morally unjustified. Furthermore, a person who has considerable wealth could usually create far more good by donating much of it to museums, libraries, schools, or charities that benefit people in need. Finally, if passing on property within the family is so important, there is no reason that it cannot be given to an adopted child (although adopting a child purely for purposes of inheritance once again treats the child as a means to an end).

Procreation as a Way of Meeting Religious Requirements

Some might argue that procreation is justified because God commands human beings, his creatures, to engage in it. For example, according to the book of Genesis, God blessed Adam and Eve, the supposed first human beings, and said to them, "Be fruitful, and multiply, and replenish the Earth, and subdue it: and have dominion over the fish of the sea, and over the fowl of the air, and over every living thing that moveth upon the Earth" (Genesis 1: 27–8, King James Bible). The value of having children is reinforced in an often-cited passage found in the book of Psalms: "Lo, children are an heritage of the Lord: and the fruit of the womb is his reward. As arrows are in the hand of a mighty man; so are children of the youth. Happy is the man that hath his quiver full of them: they shall not be ashamed, but they shall speak with the enemies in the gate" (Psalms 127: 3–5). Thus, procreation is valuable because God values it, and having children is a manifestation of religious obedience and devotion.

   There are two main objections to this argument for the value of procreation. There is first an epistemic problem. No one has direct access to the mind of God, and it is naïve to assume that the Bible or other scriptures provide unmediated evidence of God's wishes. Fundamentalists who cite scriptures as the source of their insights about God's wishes never offer independent reasons for thinking one scripture is more reliable than another or for interpreting scriptures in a particular way. They do not provide ways of reconciling the inconsistencies within particular scriptures or applying texts that are thousands of years old to the moral questions of the twenty-first century. Of course, many people would add that it is difficult, perhaps impossible, even to know whether God exists.

But set aside the epistemological doubts. Let us suppose that somehow we do know both that God exists and that God commands human beings to be fruitful and multiply—whatever that might mean for practical decisions about family-making in this century. But then a type of dilemma originally set forth in Plato's *Euthyphro* dialogue arises. Does God command us to procreate because procreation is valuable? Or is procreation valuable because God commands it? If the first disjunct is correct—that is, God commands us to procreate because procreation is valuable—then there is some other moral standard or purpose that is independent of God's command and justifies procreation. God's say-so is not the final word on procreation. We should therefore look for that other moral standard or purpose, whatever it might be, evaluate it, and see whether and to what extent it still applies to twenty-first-century life and how we make our choices about creating families.

If the second disjunct is correct—that is, procreation is valuable just because God commands it—then God has no reason for God's commands, and the supposed value of procreation is founded only upon God's fiat. But if God uses no moral touchstone for determining what to command human beings to do, then God's commands are morally arbitrary. God might command us to be fruitful one month, and the next month command every pregnant woman to have an abortion. Without any other moral standard there is no assurance about the direction that God's commands could take. But then human beings can have no more reason to obey God and to choose procreation than they would have to follow the dictates of a capricious human dictator. One might, of course, obey out of fear of God and dread of the consequences for disobedience, but one would not have any *moral* reason to accept God's values.

Because of this dilemma, and the epistemic difficulties of knowing God's supposed will, religious duties are an inadequate basis for the justification of procreation.

The Alleged Value of the Genetic Link

A frequently offered justification of procreation is the alleged importance of having a child of one's "own," a child who is genetically linked to the parents. Unlike the earlier argument that stressed the importance of pregnancy and childbirth, this approach puts a high value simply on begetting rather than gestating; that is, on the creation of children by means of one's own gametes, potentially independent of where the child's prenatal development occurs.[4]

For people who value the genetic link, having children is sometimes regarded as a way of achieving a kind of vicarious biological immortality:

> Humans are probably unique among species in their cognitive awareness of mortality, and particularly their conspicuous anxiety in anticipation of it. Humans are presumably also uniquely aware that "leaving something of oneself" for the future (despite mortality) can be accomplished by leaving genetic descendants. (Aarsson, 2007: 1769)

But the impression of immortality is, of course, an illusion. Aside from the fact that the human species will not last forever, it's possible that one's genetically related children will not reproduce, in which case one's direct and immediate genetic inheritance will come to a fairly swift end. At the same time, if the sheer perpetuation of a particular genetic line is of intrinsic value, it does not require the creation of children from one's own gametes: much of one's genetic material will be perpetuated for at least one generation as long as one's siblings have children. Some of it will be even be passed down if one's cousins procreate. Yet in all such scenarios, the size of one's supposed immortality will be limited: "In the normal case, our children share 50% of our genes. Thus, our grandchildren will each have 25% of our genes, and so on. In just a few generations, the genetic proportion for which we are causally responsible will be very low" (Levy and Lotz, 2005: 242).

But the assumption that one's genetic inheritance is so valuable that it must be preserved is dubious. After all, there are billions of people with intelligence, talents, good looks, athletic ability, nice personalities, and all the other characteristics that may form the basis for the idea that a particular genetic endowment is valuable. In that respect, no one is unique. Moreover, there is no guarantee that offspring will inherit their parents' positive attributes, or that, even having inherited them, they will decide to act upon them. We all know of famous writers, scientists, and athletes whose children lack either the talents of their parents or the motivation and perseverance to follow through on them. And although, undeniably, nature plays a role in how people turn out, so does nurture. So, it is possible to pass on some of one's gifts and abilities not only by having genetically related children, but also by adopting and raising children, by mentoring nieces and nephews or the children of friends, or by teaching and coaching.

---

[4] The not-so-subtle assumption underlying the focus on the genetic link is that it is better for well-off persons (who of course are also likely to be white and well educated) to increase their fertility rates, rather than persons whose offspring supposedly are less valuable to society and to humanity—persons who are likely to be poorer, less educated, impaired, and/or not white. Canada and the United States have a sorry history of engaging in the compulsory sterilization of members of certain groups, including some native people, poor people, people of colour, and people with impairments, having judged that such persons were unfit to perpetuate their own genes (e.g. Gould, 2002). The wrongness of those negative eugenicist policies is now recognized and in need of no further arguments, as also should be the wrongness of positive eugenicist policies that favour reproduction by some persons more than others.

Some people value a genetic link with their child because then, they think, the child will be clearly "their own." Yet it is not at all clear why a genetically related child should be or even feel more like one's "own." The sense of "own" cannot be the same as thinking of our clothes or our homes as our own. As human beings, children, however they are acquired, are not the kind of thing that can be owned; children are not property. So the child is not one's "own" in the sense of ownership. A child may be the parent's "own" in the sense that the parent has authority over the child, but the sense of ownness as authority also applies to adoptive parents, who have authority over their adopted child. It is not a function of the child's genetic connection to the parents. Perhaps calling a child one's own draws attention to the relationship: my son is my *own* in a way that my neighbour's or my brother's son is not. My son is my own because I have a parental relationship with him. But if so, then this sense of ownness is again not a function of the genetic connection, since adopted children are just as much their parents' own in a relational sense as genetic offspring are.

Perhaps for some people, thinking of a child as "their own" is a function of the view that "biological parenting is the expression and affirmation of a couple's love for one another" (Levy and Lotz, 2005: 244). And indeed, sometimes it is. Unfortunately, plenty of children are conceived not out of love but merely temporary attraction, or in some cases as a result of sexual assault. And plenty of men have thought of children as "their own" without knowing that the mother conceived the child with another man. Just as important, an adopted child can also be

the physical result and embodiment of a joint decision of the couple to commit, together, to the project of rearing and nurturing another human being. It is the decision to parent together that expresses the mutual love, trust, and commitment of partners in a relationship; and it does so irrespective of whether the resulting child is the biological product of the parents or not. (Levy and Lotz, 2005: 246)

Janet Farrell Smith (2005: 112) suggests that in some cases, "the special status of being 'one's own' derives from the fact that one biologically produced the child, and biological reproduction is prized as the primary, normal condition and foundation for parenting." As she indicates, the sense of "own" derives from the widespread *assumption* that procreation is the "normal" and even the best way of acquiring a child. Thus, the sense of "own" that is applied to genetically related offspring seems to assume precisely what it is trying to illuminate: that there is something special and unique about the relationship with a child that is the result of one's gametes.

People often suggest that noticing the similarities in offspring who are genetically related to other members of the family can be a source of pleasure. Perhaps the little boy has his father's eyes, or the girl has her mother's artistic abilities. Sometimes, in addition, the genetic connection at least appears to be a foundation for a stronger relationship or greater understanding: An offspring's quick temper may be more comprehensible if her grandmother has it too, and there may even be family lore about how to handle grandma's temper that applies to the daughter's tantrums.

But the genetic connection with offspring is no guarantee of similarities to parents or other members of the family. Just as often a child's proclivities and goals are a surprise to her parents, and the personality and individual needs of, for example, an introverted child may turn out to be a complete mystery to his extroverted parents. On the other hand, adoptive children may gradually reveal similarities to one or the other of the adoptive parents in ways that no one anticipated. Sometimes, indeed, familial similarities are not so much discovered as created, through the intimate associations of family life and the social environments provided by the family and the culture.[5] Thus, an athletic parent who provides many opportunities for her children to play lots of sports should not be surprised if her adoptive child loves soccer and is good at it. But equally, she should not be surprised—or disappointed—if her genetically related child does not.

Charlotte Witt suggests that some people believe both that "personal identity [in the sense of core defining characteristics] is determined in a substantive way by one's genes" and that

one's self-understanding requires a relationship with the source of one's genetic endowment, the birth family: If you thought that personal identity is substantially determined by genetic endowment, then you might also think that a relationship with the birth family would be central to an adequate self-understanding on the part of [a] child. (Witt, 2005: 137)

This is the view held by Velleman, who argues that "knowing one's relatives and especially one's parents provides a kind of self-knowledge that is of *irreplaceable* value in the life-task of identity formation" (2005: 357; my emphasis). Some might also argue that parents can learn more about themselves by seeing themselves "mirrored" in their genetically related offspring. From this point of view, then, procreation has value not only for the progenitors but for the progeny. That is, the supposed value of procreation would rest on the significance of the genetic link both for the parents and their offspring, both of whom come to understand themselves in terms of their genetic connections to each other.

I agree that there is something valuable, whatever the nature of one's family, in knowing about one's genetic origins and the familial similarities (and differences) that may (or may not) accompany them. Some, but not all, adopted children wish to be able to identify and even meet their genetic parents. This value contributes to the importance of open adoption, in which the identity of and information about his progenitors are not kept a secret from the child who is adopted into a new family. But it does not show that being raised by one's progenitors is essential to full personal development. Indeed, far from genetic origins being essential to who one is, "The contingent fact of adoption might be much more significant to a person's identity than a genetic predisposition to

---

[5] The kinds of similarities that parents, whether adoptive or genetic, are able to notice and are even allowed to notice may depend in part on the personal features that one's culture makes salient and definitive. Thus, because of the salience of racialization, it may be more difficult for a white mother to recognize her similarities to her black child.

develop flat feet" (Witt, 2005: 141), because, as Witt points out, who we are is consti-tuted not through bare genetic facts about ourselves but rather through our under-standing of ourselves. As Kimberly Leighton puts it, "social practices are *intimately* involved in one's own identity" (2005: 147; her emphasis). She adds, "identity (itself) and individual identities do not come from some essence of self but are instead the products of social, cultural, and historical forces" (Leighton, 2005: 148).

In addition, not knowing all the details of one's past does not inevitably make one less than a whole person (Leighton, 2005: 160). It's significant that most people who, through war or natural catastrophes, are separated completely from their genetic fami-lies as infants nonetheless manage to develop full and satisfying identities. People who, in the past, were adopted in closed systems where there was no knowledge of, let alone contact with, the genetic parents do not seem to have an inadequate sense of them-selves. (Indeed, adoptees may well enjoy a feeling of freedom from any need to con-form to the family patterns and tendencies.) Many people who have a fully developed sense of self were raised by fathers who, unbeknownst to both the children and in many cases the fathers themselves, were not genetically related to them. On the other hand, growing up with one's genetically related family is no guarantee of a secure identity and self-understanding. Some people have little in common with their genetic relatives and indeed may even be alienated from them. Therefore, being raised with and by one's progenitors is not only not always necessary to one's identity and self-understanding; it may even at times be irrelevant to it. So there is no argument for the value of procrea-tion to be found in an alleged universal necessity to establish a sense of self via living with those who created us.

There are also good reasons to reject the assumption that there is some sort of pri-mordial need to know one's genetic origins.[6] That need is at least partly created and manipulated. For example, people are *told* more and more frequently that parental genetic connections matter. Persuading people to cherish the genetic connection is, for example, a driving commercial force on behalf of DNA testing, as well as some forms of costly reproductive technology (such as intracytoplasmic sperm injection, in which a single sperm is injected into an oocyte) and reproductive services (such as contract pregnancy) that enable people to have a genetic connection with their off-spring. Perhaps the media attention to the Human Genome Project has played a role in creating a felt need to know about one's genetic origins. As a result, along with the welcome and justified movement from shamed and shameful secretiveness towards more disclosure about and to adoptive children, there may also be an expectation that persons who are adopted *must* search for their origins and *will* want to know the iden-tity of their progenitors.

---

[6] I am not arguing in favour of deliberately withholding knowledge about people's genetic origins. My point is simply that the supposed need for information about one's progenitors is unnecessarily exacerbated by social pressures.

Thus, people are urged to understand their identity primarily or even solely as the product of their genetic inheritance. Seeing oneself in that way is clearly a social product, not an instinctive response to facts about oneself. (Imagine a society in which the norm is to give offspring to genetically unrelated adoptive parents. Such a society might postulate that being adopted is a fundamental and constitutive characteristic of identity and that it is natural and inevitable for the non-adopted to want to know who their adoptive parents might have been. In such a society, an individual who is not adopted, and knows she is not, might very well have a recurrent desire and drive to explore and know more about who her adoptive parents might have been and who she would have become if in fact her family had, contrary to fact, conformed to the dominant social practice.)

The implication of my arguments here is that having a genetic link to children is not a good basis for choosing procreation. Moreover, emphasizing genetic connectedness may also have undesirable consequences. It may lead to a tendency by parents to have unrealistic expectations of their child, to see the child as their special creation, who must and will reflect well on them. If I have a child because I'm a marvellous clarinet-player and I want a daughter who will also be a marvellous clarinet-player, then I'm setting myself up for disappointment, and I may put a lot of pressure on my daughter that she will neither appreciate nor benefit from—especially if it turns out she'd rather be a potter or a plumber.

Choice and Control in Procreation

Some might argue that procreation has a special value because it is an easier and more reliable way of obtaining a child than going through the rigors of the adoption process. In adoption there is a formal evaluation of one's parental abilities, whereas in procreation—despite calls by some philosophers for the licensing of all potential parents (Tittle, 2004)—there is usually no formal evaluation of one's capacity to be a mother or father. Moreover, the availability of children for adoption in the West is low, and there are political and cultural difficulties in adopting a child from outside the West. In Canada, for example, international adoption reached a high of 2,180 in 2003, but as a result of growing restrictions, by 2010 the number had declined to 1,968. In the United States, the peak was 22,991 in 2004; by 2011 the number had declined precipitously to 9,320 (Pearce, 2012: A8). In addition, an adoptive child, especially an older one who has already spent months or years in an orphanage or with another family, can be a relatively unknown entity who may bring special medical or social challenges to her adoptive parents.

Hence, it may appear that procreation provides prospective parents with greater choice in and control over the process of acquiring a child. The parents can control the timing of parenthood; they influence the child not only from the time of birth but from the time of conception; they determine the child's genetic inheritance; they have some idea what the child will be like; and they may even have the opportunity (through preimplantation screening with *in vitro* fertilization, or prenatal testing) to diagnose impairments and genetic diseases in their potential offspring.

However, these ideas of control and choice are often a delusion, and hence procreation does not always carry an advantage in this respect over adoption. Prospective parents cannot always control when they have a child, or even, given the possibility of infertility, *whether* they have a child. The challenge is particularly acute for women, whose opportunity to procreate may conflict with the demands of employment; women can end up near or at menopause before they find a suitable partner or the requirements of their job permit them the time for pregnancy and breastfeeding. Nor can prospective parents be confident about avoiding, in their progeny, physical illnesses and impairments, cognitive deficits, or psychological problems, all of which may generate as many parenting challenges as those encountered by adoptive parents. A child is not predictable; he or she carries a genetic heritage from thousands of prior generations. And problems can develop during pregnancy or childbirth. Sometimes procreation just seems like the "luck of the draw," with some parents ending up with "easy," healthy children and others ending up, inexplicably, with kids who are ill, "difficult," or troubled. The idea that procreation provides control and choice is not always justified.

Conclusion

The point of my discussion has not been to claim that procreation has no value. We need not adopt the anti-natalism of Benatar. Procreation is the prerequisite for the continuation of the human species with all its activities, projects, and accomplishments (although it is doubtful that species continuity requires quite as much procreation as now occurs). For women, procreation provides the unique experiences of pregnancy and childbirth, along with an immediate relationship with the future child that begins at conception.

My point, rather, has been to indicate the ways in which most of the usual and familiar arguments for the supposed superiority of procreation over adoption are inadequate. The alleged intrinsic value of child-bearing and of human life and human beings; the creation of happy people; the fulfilment of familial duties and conformity to the bionormative ideal; the perpetuation of family property; the satisfaction of religious requirements; the alleged value of a genetic link between parent and child; and the supposed control and choice afforded by procreation: none of these provides a universally plausible reason for choosing procreation. Even more, they fail to demonstrate that procreation is always better than adoption. If anything, procreation is more costly to the planet in terms of resource depletion and waste generation. Hence, although it is a prerequisite for the practice of adoption, procreation is not more valuable than adoption as a means of family-making.

References

Aarssen, L. W. (2007). Some Bold Evolutionary Predictions for the Future of Mating in Humans. *OIKOS: Synthesizing Ecology*, 116, 1768–78.

Aksoy, S. (2004). Response to: A Rational Cure for Pre-Reproductive Stress Syndrome. *Journal of Medical Ethics*, 30, 382–3.

Benatar, D. (2006). *Better Never to Have Been: The Harm of Coming into Existence*. Oxford: Clarendon Press.

Biervliet, F. P., Maguiness, S. D., Hay, D. M., Killick, S. R., and Atkin, S. L. (2001). Induction of Lactation in the Intended Mother of a Surrogate Pregnancy: Case Report. *Human Reproduction*, 16, 581–583.

Gould, S. J. (2002). Carrie Buck's Daughter. *Natural History*, 111(6), 12–16.

Hursthouse, R. (1987). *Beginning Lives*. Oxford: Blackwell.

Leighton, K. (2005). Being Adopted and Being a Philosopher: Exploring Identity and the "Desire to Know" Differently. In S. Haslanger and C. Witt (eds), *Adoption Matters: Philosophical and Feminist Essays* (pp. 146–70). Ithaca, NY: Cornell University Press.

Lenman, J. (2004). On Becoming Extinct. In D. Benatar (ed.), *Life, Death, and Meaning: Key Philosophical Readings on the Big Questions* (pp. 135–53). Lanham, MD: Rowman and Littlefield.

Leslie, J. (1996). *The End of the World: The Science and Ethics of Human Extinction*. London: Routledge.

Levy, N., and Lotz, M. (2005). Reproductive Cloning and (a Kind of) Genetic Fallacy. *Bioethics*, 19, 232–50.

Overall, C. (2012). *Why Have Children? The Ethical Debate*. Cambridge, MA: MIT Press.

Overall, C., and Bernard, T. (2012). Into the Mouths of Babes: The Moral Responsibility to Breastfeed. In S. Lintott and M. Sander-Staudt (eds), *Philosophical Inquiry into Pregnancy, Childbirth and Mothering: Maternal Subjects* (pp. 49–63). New York: Routledge.

Parfit, D. (2004). Overpopulation and the Quality of Life. In J. Ryberg and T. Tännsjö (eds), *The Repugnant Conclusion: Essays on Population Ethics* (pp. 7–15). Dordrecht: Kluwer Academic Publishers.

Pearce, T. (2012). The Painful New Realities of International Adoption. *The Globe and Mail*, 18 Feb., A8–A9.

Plato. (1941). Euthyphro. In *The Collected Dialogues of Plato*, ed. E. Hamilton and H. Cairns (pp. 169–85). New York: Pantheon Books.

Smith, J. F. (2005). A Child of One's Own: A Moral Assessment of Property Concepts in Adoption. In S. Haslanger and C. Witt (eds), *Adoption Matters: Philosophical and Feminist Essays* (pp. 112–31). Ithaca, NY: Cornell University Press.

Szucs, K., Axline, S. E., and Rosenman, M. B. (2010). Induced Lactation and Exclusive Breast Milk Feeding of Adopted Premature Twins. *Journal of Human Lactation*, 26, 309–13.

Tännsjö, T. (2004). Why we Ought to Accept the Repugnant Conclusion. In J. Ryberg and T. Tännsjö (eds), *The Repugnant Conclusion: Essays on Population Ethics* (pp. 219–37). Dordrecht: Kluwer Academic Publishers.

Tittle, P. (ed.) (2004). *Should Parents be Licensed? Debating the Issues*. Amherst, NY: Prometheus Books.

Velleman, J. D. (2005). Family History. *Philosophical Papers*, 34, 357–78.

Witt, C. (2005). Family Resemblances: Adoption, Personal Identity, and Genetic Essentialism. In S. Haslanger and C. Witt (eds), *Adoption Matters: Philosophical and Feminist Essays* (pp. 135–45). Ithaca, NY: Cornell University Press.

Young, T. (2001). Overconsumption and Procreation: Are they Morally Equivalent? *Journal of Applied Philosophy*, 18, 184–92.

6

The Unique Value of Adoption

Tina Rulli

Adoption can provide a child with the critical resource of a stable, loving family, which institutional and foster care fail to provide. Absent a stable family and the benefits of constant care and attention, children are at risk for severe physical, cognitive, and emotional deficits. Adoption can not only prevent these deficits of institutional care, but for those children who experience neglect and abuse prior to adoption, it is the best cure (IJzendoorn and Juffer, 2006). In general, adoption is a good thing for children in need of a family.

But adoption offers unique value *for parents*, too. Though adoption is often considered a second best or even last resort for parents in making their families, this view fails to recognize the special value of adoption in its own right. This topic is almost entirely ignored in the philosophical literature. Thus, I will explore here the unique value of adoption. I begin by noting that the selective focus on the value of adoption for *only* those people pursuing assisted reproductive technologies employs the hidden assumption that adoption is second best to procreation. I will focus on the value of adoption for *all* prospective parents.

My discussion is driven primarily by reflection upon non-relative adoptions; that is, adoption of children not previously a part of one's extended family. Non-relative adoptions contrast with intrafamilial adoptions, where a grandmother adopts a grandchild, for instance, or a brother adopts the child of his sister. More generally, adoption is an alternative to procreation—with a notable exception. For some same-sex couples who use artificial reproductive technologies to create a child, the partner who does not contribute a gamete to the process must adopt the child.[1] My focus will be on non-relative adoptions that are not also procreative in this way. That is, the arguments offered here are guided by my reflections on adoptions that involve already existing children who are not related to their adoptive parents.

---

[1] The adoption requirement varies by jurisdiction. For discussion of this important challenge to the adoption–procreation binary, see Julie Crawford (in this volume).

In adopting a child, one typically has the opportunity meet a specific need that all children have—the need for a family. Clearly, meeting this need is valuable for adopted children, but adopted parents may also place value in this fact and be motivated, in part, to adopt for this reason. In contrast, procreation does not share this important moral value, for a child not-created is not a child in need of anything at all. After exploring the philosophical issue, I assess the empirical complexities of this claim, which have proven controversial in adoption and children's advocacy circles. While adoption practices, generally speaking, can play a role in addressing the needs of children worldwide, individual prospective parents face the complex task of determining where they can best contribute their efforts in order to help children in genuine need of families, while not contributing to harms or exacerbating existing injustices.

Next, I argue that since most of the reasons in favour of procreation are self-referential—i.e. they locate the value of having a biological child in the child's connection to one's own body or genes—adoption is valuable for the very opposite reason. Adoption provides a morally noble opportunity to extend to a stranger benefits usually withheld for one's genetic kin. In adoption, one's relationship to one's child is defined solely through a history of love and care rather than through bodily connection. As such, adoption offers a unique possibility in which impartial concern for an other can be the starting point for a lifetime of love and care. I discuss this possibility against the objection that adoptions involving a "rescue" motivation are problematic. Along the way, I demonstrate how adoption challenges a strictly dichotomous understanding of impartial and partial reasons for action. In the final section, I reflect on the transformative power that adoption can have for parents' own conception of self and family.

My goal is to highlight the unique value of adoption, challenging the widespread assumption that it has second-best status to procreation. Indeed, we'll see that adoption is oftentimes superior to procreation, providing a pure and exemplary model of what is most valuable about parenthood. However, making a superiority argument is not my primary aim here. I hope to show that adoption is a valuable option for all parents to consider and that it offers unique value of its own.

Focusing on Adoption's Value for All Prospective Parents

The majority of the sparse philosophical literature on adoption focuses on adoption as an option for infertile, subfertile, single, or homosexual people—a diverse group that I'll loosely refer to as *those who cannot easily procreate*. In this context, adoption is typically considered an alternative to using assisted reproductive technologies (ART). The narrow focus on adoption as valuable for those who cannot easily procreate expresses the widely held and largely undefended belief that adoption is a second-best alternative to biological procreation for having children. The underlying assumption is that only those who cannot biologically procreate without assistance would (or should) seriously consider adoption.[2] But, as I will show, many of the reasons for choosing adoption over ART apply *generally* to favour adoption over procreation. Adoption is a valuable option for all people who desire to parent children regardless of their fertility status.

One might think the narrow focus on the value of adoption for only those pursuing ART is justified because both ART and adoption require significant financial resources. This commonality makes adoption an obvious alternative to using ART. Consider, for instance, the argument that those who would use ART, in particular, have a *moral duty* to adopt children rather than spend resources on pursuing procreation (Petersen, 2002).[3] The argument relies upon the claim that resources spent on ART could be spent on adoption. Since only in adopting do these resources go to an existing child who needs them, some argue that people should adopt rather than pursue ART.[4]

This argument arises because the resources spent on ART are conspicuous. As such, the range of options—spend the money on ART or spend the money on adoption—is salient. Yet, anyone who chooses to become a parent has the *parental resource*—the money, time, emotional commitment, and care—to give to a child who needs it. This is the critical resource some existing children lack. One could put this resource towards a child of one's own creation or give it to an adopted child. That is, adoption is no less an alternative to easy procreation, though the option may be less salient. Granted, adoption may cost more than easy procreation, and these costs may trump a *duty* to adopt for many people; but I'm not defending a duty to adopt here. I'm arguing that adoption should be considered a valuable alternative to procreation more generally. It should not automatically be assumed to be second-best. For that reason, my discussion of the value of adoption applies to prospective parents generally, not simply to those facing the choice between adoption or ART.[5]

[2] This assumption is expressed in Smilansky (1995: 44), where he asserts that an argument for adoption instead of procreation (due to concerns about overpopulation) is not worth considering since there is no likelihood that it would be widely accepted. The view is further expressed in Hursthouse (1987: 309), where she states: "But it is, and would be, odd to want *to have a child* (i.e. be a parent) as an end in itself (i.e. not to secure the inheritance nor as a publicity stunt) without at all wanting *to have one's own child* (in the biological sense)."

[3] For evidence that this view is held by the public at large, see the comments sections of the following article/blogs, where the commenters frequently express variations on the opinion that the infertile should adopt rather than create children using ART. See Belkin (2009a and b) and Landau and Gumbrecht (2010).

[4] Cf. Rivera-López (2006), responding to Petersen (2002), rejects a targeted duty of *sub-fertile* parents to adopt children. He concludes that they are excused from this putative duty, given that the solitary focus on them is unfair. I do not share this conclusion; also consistent with fairness is a general expansion of the scope of a putative duty to adopt to include all prospective parents.

[5] I want to explicitly recognize that some same-sex couples may choose ART because they are, for all practical and legal purposes, prevented from pursuing adoption.

Helping a Child in Need

One of the greatest values of adoption is that one can help a child in need of a family. This is obviously valuable to the child; but this fact can be valuable to and valued by adoptive parents. One may be motivated to adopt out of recognition of this fact, and one may deeply value this feature of adoption. There is both a philosophical and an empirical aspect to the claim that adoption helps a child in need of a family. I consider them in turn.

But first, let me say more about the concept of *need* employed here. I have in mind Joel Feinberg's (1973: 111) definition of need, where "in a general sense to say that S needs X is to say simply that if he doesn't have X he will be harmed." This definition captures the important distinction between need and mere wants or desires. With unmet needs, a person comes to harm. Further, the need children have for a stable family is what I'll call a *critical need*. By this, I mean the fulfilment of that need is vital to the child's proper emotional and physical development.

All children *need* stable, loving families. Some however, have extant need, i.e. this need is currently unfulfilled or is in imminent danger of going unfulfilled. Many of the children with extant need have parents who are unable or unwilling to provide for them. In this way, they are *in need* of *new* families. Yet some children are orphans and have no existing families at all. Recognizing the difficulty of choosing an umbrella term for the children in question, I will speak of *children in need of families*. This is not to diminish the importance of the existing families of origin who may continue to play an important role in the children's lives. What these children need, even so, is a family that is able to raise them. Further, I indicate the specific need *for a family*, for these children may or may not otherwise be *needy*.

The opportunity to help a child in need of a family is a value unique to adoption; for in procreation one does not help a child in need of a family (or anything else). What one does is create a child, who is by her very nature vulnerable and needy in her dependence upon another to survive, *and then* she benefits the child with all the goods of parenthood. That is, one creates the need and then (hopefully) satisfies it. Only adoption helps an existing child with an unmet need.

Some will argue that in procreating you still *benefit* a child by bringing her into existence (Hare, 1975). Proponents of this unintuitive claim appeal to the more intuitively plausible possibility to *harm a person by bringing her into existence*. Many people think that creating a person who will endure terrible and incurable suffering harms that person. If this is possible, then for reasons of symmetry, we should at least grant the possibility that creating a child who will have a happy life benefits her. Further, the possibility to benefit a person by bringing her into existence can explain people's gratitude for their existence (Hare, 1975: 219). Many say the gift of life is the greatest benefit of all. Advocates of this view may argue that the benefit bestowed through procreation is similar to the benefit of adoption, undermining my claim that the adoption benefit is unique. They might say: *both* procreation and adoption provide *very important, large* benefits to a child.

Even if we grant that people can benefit others by bringing them into existence, this is no challenge to the claim that the adoption benefit is unique. For only in adopting do you meet an unmet need. That is, only in adopting do you alleviate an extant harm or prevent one from occurring. Should you choose not to procreate, there is no child who is harmed for not coming to exist; there is no child at all. And this difference matters: for not existing at all is not bad for "the person" who does not exist. But an existing person lacking or losing what he critically needs—in this case, a stable, loving family—is bad for that person. Thus, only in adopting can you respond to or prevent a very bad situation for a person. The opportunity to critically improve an existing person's life is the unique benefit offered by adoption as opposed to whatever other kind of putative benefit one can confer by procreation.

Attempting to diminish my claim, one might argue that the putative benefit of existence is necessary for and prior to the possibility of helping a person in need. One needs the benefit of coming to exist in order to enjoy any other benefits at all. Thus, my opponent may argue, I cannot so easily talk about the benefit of adoption without giving equal attention to the benefit of procreation, for existence itself precedes all other kinds of benefit.

Coming into existence is necessary for us to receive the other goods of life. But this does not mean that it is more important than those benefits that are possible only after a person exists, i.e. those that make her life worth living. For consider, without these other benefits—e.g. the benefits of food, shelter, love, and family—the putative benefit of coming into existence is no benefit at all. We are born vulnerable, dependent, and needy; we need more than existence alone to have happy lives. Coming-to-exist could only be counted a benefit (if one at all) if one receives the other benefits that make one's life worth living. *Bare existence* of a person is not by itself a benefit. In fact, absent any other benefits, bare existence is sufficient to ensure that the child is *harmed* by coming to exist. Thus, the benefits subsequent to coming to exist are what ultimately matter when we claim that existence is beneficial for someone.

It is clear that the critical benefit provided in procreation is not solely the putative benefit bestowed in creating a child—it is that bestowed in *parenting a child* and ensuring that she has all the goods that make her life worth living. The unique value in adoption arises out of recognition that we can give this benefit to *an existing* person in need of that exact good. In contrast, procreation doesn't meet needs; it creates them.

The unique value of adoption is further supported by comparison with the value of child-bearing. Rosalind Hursthouse (1987: 309) argues that *bearing* children is intrinsically worthwhile. She claims that the value of having children is "inextricably bound up" with the belief that death is evil, life is a benefit, murder is wrong, and each life is uniquely valuable. Our reverence for child-bearing is a reflection of the larger thematic belief in the "sanctity of life"—or in more secular terms, the idea that human life is intrinsically valuable (Hursthouse, 1987: 309–10).

Creating and then bringing a child to term in one's body is an activity that requires substantial sacrifice on the part of the pregnant woman. To "do it well," as Hursthouse

(1987: 315) says, requires "courage, fortitude and endurance." Bearing children takes considerable virtue in order to achieve the important end of new human life. Hursthouse (1987: 315) states: "It is in this connection that one can see why it is tempting to regard bearing a child as analogous to sacrificing a fair amount of time and effort to saving someone's life." In both pregnancy and saving a life, a person takes on considerable burden as a virtuous response to the intrinsic value of human life. In this case, the relation between the labours of pregnancy and those of life-saving sacrifice is metaphorical. Though no life is saved in creating and bearing a child, a life is preserved by and entirely dependent upon the pregnant woman who undertakes her pregnancy with virtue. Bearing a child *is like* saving a life.

Hursthouse (1987: 315) continues:

> What is done, is, I claim, not just worthwhile and significant but *morally* worthwhile and significant, because of its connection with, on the one hand, the value or sanctity of life and, on the other, with what I have roughly categorized as "family life"—the field of our closest relationships with other people. For these two areas are the concern of morality if anything is.

The value in child-bearing is not found merely in its relation to the sanctity of life, but also in its aim of love and family. Hursthouse (1987: 315) explains: "In bearing the child, the woman makes it particularly and peculiarly *hers, part of her* life-cycle, *her* family. In so doing, she enriches her own life and that of those who form part of it."

I will not evaluate or reject Hursthouse's account of the value of child-bearing. Rather, I want to leverage it as an argument for the value of adoption. If there is value in an activity that both expresses regard for the sanctity of human life and the value of love and family—making a person one's *own*—then adoption is a paradigm of such activity.

In many ways, adoption and maintaining pregnancy are morally similar. Both demonstrate deep regard for the sanctity of life. Adopting provides a benefit critical to a life going well. Maintaining a pregnancy ensures that the nascent life inside a woman's body will continue and flourish. In fact, parenting itself is one among this kind of activity, for in feeding, loving, and providing for our dependent children, we preserve their lives.

Yet, in many cases (not all, of course) the preserving of a life inside one's body during pregnancy is part of a greater decision to bring about that life in the first place. In such cases, it is more accurate to consider pregnancy as part of *creating a life*—making a life where previously there was none—rather than *saving a life—recognizing existing critical need and providing what is necessary to make that life go well.* There may be value in creating life, and this may be value that is tied to the greater theme of honouring the sanctity of human life. I'm not denying any of this. But the metaphor from pregnancy to saving lives is weakened.

In contrast, in adopting a child one is saving the life of an existing child. Though adoption is not always a *life or death matter;* it is a *critical* matter. It is about providing to a child a benefit that may make the difference between a life that goes well and one with exceptional hardship, struggle, and suffering. People consider their lives *saved* when someone provides them with critically needed support, helps them find the right path in life, or ensures their life is lived to its potential. In adopting, one undertakes considerable sacrifice to critically improve a life, to save a life in this way.

Moreover, we know that adopted children are considered their parents' *own children*; they are fully integrated into their parents' lives and families (Smith, 2005). If for Hursthouse, the deep value of pregnancy is through its connection to the saving of lives and the value of family, then adoption satisfies these criteria directly. Pregnancy, as a part of creation, satisfies the criteria only by the stretch of metaphor. This brings about an interesting inversion: adoption is the paradigm example of honouring the sanctity of life and the value of family. Adoption is not second-best. Morally speaking, it is the exemplar.

Let me now turn to the empirical criticisms of my claim that adoption helps a child in need of a family. In a popular exposé, E. J. Graff (2008: 59) proclaims that: "Westerners have been sold on the myth of a world orphan crisis." We are frequently told of the "millions upon millions" of orphaned children in the world by adoption agencies, who imply that by adopting a child we can do something to address this crisis. Graff counters this claim, noting that in fact there are waiting lists for adoption of healthy infants both in the United States and abroad. Prospective adoptive parents may be vying for the same limited pool of healthy infants. Thus, Graff's first criticism is that it is misleading to characterize adoption as helping children in need of families. The *adoptable* children—healthy infants, by Graff's definition—will be adopted one way or another, if not by you, then by one of the other many prospective adoptive parents.[6] Graff's second charge is more troublesome. Instead of a problem finding homes for children in need of families, there is a money-driven industry for finding children for adoptive homes (Graff, 2008).

We might attempt to address Graff's concerns by first agreeing on the number of *legally adoptable* children worldwide. But determination of an adoptability statistic is fraught with empirical complications. As a practical matter, an estimated 45 million births go undocumented each year in the developing world (Oreskovic and Maskew, 2008: 78). These children have no clear legal status, let alone any clear adoptability status. Also, many children institutionalized in orphanages have living biological parents, which can complicate or obscure their legal adoptability status. (Regardless, many of them have no *parents* in any practical or normative sense of the term: Bartholet, 2007: 95.)

Additionally, settling on an adoptability statistic is an inextricably value-laden determination. On one side, critics of international adoption worry that any adoptability statistic is inflated, since, they contend, it will count many children who would not be relinquished by their parents but for the "baby market" (Graff, 2008; Oreskovic

---

[6] See also Oreskovic and Maskew (2008, pp. 80-81).

and Maskew, 2008). On the other side, there are those who argue that the backlash against international adoption has rendered many children in need of adoption unavailable (Bartholet, 2007). Recently, the number of international adoptions, which consistently increased over the past several decades, has sharply dropped off (Carlson, 2010/11: 734). This is in part due to the closing of international adoption by some "sending" countries as a result of national pride and shaming;[7] it is partially due to national "subsidiarity," the view that local placement of children should take priority over international placement (Carlson, 2010/11: 735);[8] and it is partially due to the active campaigning of certain children's welfare groups that eye international adoption with suspicion due to the risks of child trafficking and exploitation of birth parents.[9]

Restrictions on adoptions for these reasons involve prior value judgements about adoption. There is the judgement, for instance, that preventing trafficking abuses should take precedence over placing children in adoptive families, i.e. that preventing active harms is morally more important than remedying harms through rescue. There is the view that children "belong" in their countries of origin, even if this means they will stay in subpar institutional or foster care. There is also the assumption by Graff and others that only healthy infants should count as *adoptable*, given the assumption that only they are desirable to prospective adopters.[10] The number of children available for adoption is directly impacted by prior value-laden opinions about adoption and adoptability.

If we are to assess the value of adoption by looking to the numbers of children who could be helped by widespread adoption practices, this number cannot already presuppose a judgement about the value of adoption. It may be true that there are waiting lists, but this is the artifact of adoption opposition from many sides. We cannot then cite this artifact of adoption opposition as an argument against the importance of adoption.

Though many will dispute the number of adoptable children, "the certainly true and important answer is that the number of children who would almost certainly benefit from adoption far exceeds the number of prospective adoptive parents" (Carlson, 2010/11: 735). An estimated 8 million children live in institutions (Secretary-General, 2006). Millions more children lack any form of stable parental care. If only a small

[7] E.g. South Korea restricted international adoptions after the 1988 Olympics in Seoul due to embarrassment about the perception that it was the world's leading "exporter" of children (Fisher, 2003: 344). Romania's complete ban on international adoption was connected to their bid for entry into the European Union (Carlson, 2010/11: 741). Most recently the Russian ban on American adoptions of Russian children was widely seen as a political response to the US passage of the Magnitsky Act (2012).

[8] Subsidiarity is endorsed in the UN Convention on the Rights of the Child, 1989. Subsidiarity, though open to some interpretation, sees intercountry adoption as a last resort for orphans. The Hague Convention moves intercountry adoption up one rung in priority for those countries that have signed the convention (only half of "sending" countries).

[9] UNICEF is one such prominent organization.

[10] It is notable that special needs adoptions constitute more than one quarter of all unrelated adoptions in the United States. See Fisher (2003: 339). There are waiting lists for adoption of children with Down's syndrome and for other children who were once deemed unadoptable. See Bartholet (2000: 180).

fraction of these children were available for adoption, now or with foreseeable policy changes, this would plausibly exceed the number of annual adoptions—only 30,000 international adoptions occur each year (Bartholet, 2007: 167). In the US alone, there were 107,000 children who were legally adoptable in 2010 (US Dept. HHS, 2011). Roughly only half were adopted. Many more are in the foster care system, unavailable for adoption given current legal and institutional barriers that view adoption as a low priority, second to keeping biological families intact. The numbers clearly reveal that a very large number of children could *benefit from adoption*. To focus only on those who are clearly, legally adoptable is to ignore the millions who have fallen through the cracks in the system.

Graff's first criticism does not support abandoning adoption, rather it supports changing adoption institutions and practices so that more children are helped by adoption. Graff cites the prevalence of older-aged children or children with special needs in the adoptability pool as a sober reminder to naïve, prospective adopters that it is not healthy infants who need rescuing. This fact is meant to quell the pro-adoption rhetoric. But this criticism takes such facts as inalterable features of our world. First, it overlooks the possibility of encouraging prospective parents to adopt older children or children with special needs and the possibility to provide institutional support to people who do so. Second, it ignores the potential for widespread adoption reform that would allow at least some of these children—whose older age and special needs can be aggravated by a sluggish and inadequate child welfare system—to find families at earlier ages, before some preventable cognitive deficits form (Carlson, 2010/11: 771). Instead of abandoning adoption as part of the solution, we need institutional reforms that will ensure more children in need of adoption are available for adoption.

But, in light of Graff's criticisms, what do we say to those people who are thinking of adopting now? First, the fact that a child has a good chance of being adopted by someone does not undermine the fact that, if you adopt that child, you will have met a critical need of that child. If you pull a drowning child out of a swimming pool, it is no less the case that you helped a child in need if there are also others willing to help out. Graff's concern about adoption demand does not undermine the main claim that adoption is valuable because it can help a child in need of a family. Yet we can characterize some children's need for a family as greater than others, if we take into account the alternatives readily available to them and their overall chance of being adopted. Cases involving children who are older or have special needs are the most urgent. Perhaps the reasons to help a child and the corresponding value of adoption will be greater for those children who have the most urgent need. Prospective parents can count the degree of a child's need for a family as one factor among many in guiding their decision to become parents. But any case of meeting critical need has important value.

Second, the willingness of parents to adopt children is a power that could be leveraged in changing institutions and laws so as to make more children available for adoption. We cannot simply wait for these changes. Adoption will not be seen as a part of the solution to children's needs if parents are not willing to adopt children at all. Convincing people of the value of adoption is in the service of this good.

This brings us to Graff's second concern about the role prospective adopters play in contributing to illicit "baby markets." There is a lively debate about the actual prevalence of illicit adoption practices and the appropriate response to them.[11] I cannot get into this debate here. But a genuine concern about illicit adoption practices is consistent with and endorsed by my position here: if a value of adoption is that it helps children in need of a family, clearly this value is not realized if any particular adoption is "helping" a child where no help was needed or if it is actively harming a child or her birth family. But I emphasize: concerns about unscrupulous adoption practices warrant closer scrutiny of those practices and vigilance by prospective adopters, not abandonment of the practices altogether. Prospective adopters have the responsibility to choose and support adoption practices that are ethical. Parents should not be naïve about the risks of exploitative adoption practices or baby-trafficking. Prospective adopters must take care in selecting the adoption agencies they will work with and scrutinizing the adoption practices in the country from which they will adopt.

Adoption critics raise important worries for prospective adopters to consider. This does not, however, undermine the possibility for adoption's unique value. Adoption can help a child in need of a family. As we've seen, this value is not a given and its value might be variable. We have a responsibility to promote adoption practices that reach the children in greatest need and that do not exacerbate existing injustices or create harms through illicit adoption practices.

Loving a Stranger as One's Own

Some of the reasons offered in favour of procreation as the best way to build a family can be leveraged in turn as reasons in favour of the unique value in adoption. People commonly appeal to the value in having children who share a biological relationship with the parents through the bodily connection a woman has with her child *in utero* and the genetic connection both parents share with their offspring.[12] A biological child is in some sense, they say, a part of each of them. The child shares with them a similar basis of genetic identity, and many think genes are predominantly what make us *who we are*. This is taken as a strong reason to favour procreation over adoption; for we should want our children to be connected to us in this specific way. Why this is so is typically left unexplained; perhaps it is self-evident for most people. Perhaps genetic similarity is intrinsically valuable. Some cite the putatively higher probability a genetically related child has of being physically and psychologically similar to her parents. The underlying assumption is that it is better that parents and children resemble each

[11] For a sample of this debate, see Oreskovic and Maskew (2008); Bartholet (2007); Carlson (2010/11).
[12] Works that raise some of the following themes or claims include: Velleman (2008); Kolodny (2010); Tooley (1999); Hursthouse (1987).

other in these ways. Others claim that we have a greater inclination to love and attach to children who are biologically and genetically connected to us.

I discuss these arguments in another paper, where I challenge the empirical and moral assumptions underlying the preference for biological children.[13] I won't repeat that discussion here.[14] Instead, I want to flip the argument on its end. If there is value in having genetically related children for the reasons offered above, then *for these same reasons*, adoption presents us with a unique and morally valuable prospect.[15] In adopting children, given that, putatively, none of these mentioned values are present, we have an opportunity to share one of the most intimate and loving human relationships with a stranger.[16] The adopted child is not attached to us by body or genetic identity; her existence is not the product of our actions or choices. We may not share the same personality traits, look, ethnicity, culture, or place of origin with this child. We may lack entirely a connection with this child other than that of common humanity. Yet, for all that, we may invite these children into our families.

For these reasons adoption is a practice of important and unique moral value. The parent–child relationship, typically and ideally conceived of (by some) as a relationship grounded in the similar genetic identities of each, is one of the most intimate personal connections humans can have. To willingly share this deeply intimate connection with a stranger is morally exemplary. It demonstrates the far range of possibility for human connection between strangers and the potential for intense, loving regard for an *other* in a context in which, typically, this very otherness is defined out of the relationship. Moreover, since adoption involves children—all of whom by their nature are needy and dependent upon adults for their care—adoption exemplifies the uniquely human capacity for responding to vulnerability, wherever it may occur.

In adopting a child, one is not limited in one's expectations about the child's future possibilities due to a narrow focus on the genetic determinants of a child's talents and personality.[17] A parent can stand witness to his child's development into her own person, a person bound to him in love, not in body. Indeed, an adopted child becomes one's own by relation and history only; not because she is linked to one's biological identity or is the product of one's own creation, or a natural possession of sorts. She becomes one's own through a relationship that is fostered over time, through care and love.

Adoption reminds us that it is this relation of intimacy that should ground our use of possessive speech when speaking of personal relations, i.e. when we say that a person is one of "ours" or is "mine" (Smith, 2005; de Gaynesford, 2010: 87). People who are *mine*

---

[13] In "Preferring a Genetically-Related Child," unpublished manuscript.

[14] For some other works addressing and rebutting these concerns see: Haslanger (2009); Lotz (2008); Witt (2005).

[15] This general idea was suggested to me in conversation with Sally Haslanger. What follows is my own analysis.

[16] Again, not all adopted children are unrelated or strangers. My focus here is on unrelated adoptions.

[17] This is not to suggest that parents genetically related to their children are necessarily so bound.

are not my property; rather that this person is *mine* means she stands in a special relationship with me that not all others share. Possessive speech is *relational* speech in two senses—as relating two people and as expressing personal closeness between them.[18] Understanding our use of relational speech in this way allows for a more expansive and inclusive application of "possessive" terminology and concepts. For instance, in the parent–child relationship, intimacy can be fostered in many ways: it may grow naturally from the bond a mother has to her child in pregnancy; but it may grow solely from a history of love, affection, and care (Kolodny, 2010). There is more than one way that a child can be one's own.

In sum, a unique value of adoption is in the transformation of a stranger to become a child of one's own, i.e. in choosing to love a child not previously connected to oneself through body or identity, but who will be one's child through a history of love and care. This value is independent of whether the child is in need of the relationship or whether the adoptive parents were motivated to adopt in part out of recognition of that need. One need not engage in moral reflection or deliberation to enact this possibility; it can be a natural and uncalculated reflex to extend compassion to a child in need of exactly that. The possibility for such generous and intimate love of an other is remarkable in itself.[19]

Yet, bringing the previous section to bear on this possibility, one *can* choose to enter into the parental relationship with a child out of recognition of that child's need for a family. One can let impartial, other-focused concern be the starting point for a lifetime of love and care for another person. In this case, adoption can have an other-focused starting point not shared by procreative parenthood.

I am not claiming that adoption is always or should always be a wholly other-focused act. For many people adopting a child fulfils a desire or need of theirs.[20] But I do want to draw the following distinction: arguments in favour or defence of procreation tend to emphasize the importance of the biological child's connection to oneself through genes or body. The value of procreation is located in the value of oneself. That is not to say it is primarily selfish or that procreative parenthood isn't also other-concerned, but the locus of value of this relationship is typically placed in self-referring terms. In contrast, those who adopt *can* locate the unique value of adoption in impartial concern for another. I can be motivated in part to make this child my child because she needs a family. I may also deeply desire to be a parent, but I may desire this for myself while being responsive to the moral reasons there are to share this relationship with a child who needs me. Adoptions that have some aspect of this other-concern I will call *altruistic adoptions*.

---

[18] The level of intimacy indicated by the use of possessive speech might vary with the type of relationship in question. My relation to my acquaintance differs in intensity and kind from my relation to my sister.

[19] This is one value that all adoptions of children not genetically or biologically related to the parent share.

[20] I thank Carolyn McLeod for prompting me to clarify this point. For more on the special value of the parent–child relationship *to parents* (procreative and adoptive), see Brighouse and Swift (in this volume).

Some worry that adopting a child out of an impartial "rescue" motivation is an inappropriate starting point for the parent–child relationship, which is not fundamentally impartial. Elizabeth Bartholet (1993: 66), adoption scholar and adoptive parent, notes with regret that adoption agencies often frown upon prospective parents whose primary motivation for choosing adoption is to rescue a child.

Against this concern, it's worth noting that people have biological children for far more trivial reasons, and these are rarely subjected to scrutiny.[21] Suspicion of adoptive parents' motives may be yet one more symptom of the deeply entrenched assumption that adoption has second-place status. The motivations for adoption are held to greater scrutiny, since adoption is considered by some people to be deviant from the norm. For them, people who would choose adoption must explain themselves. Despite my obvious scepticism, I will take some time to make sense of this objection. In the process, I can better illuminate the transformation that occurs when an adopted child becomes one's own child.

First, people may worry that rescue is the wrong reason to become a parent. Parenting is far too demanding, and well-intentioned adopters wanting to "rescue" children should not be so naïve about the demands of this particular kind of rescue. But of course one who wants to adopt children should also want to be a parent. Someone will not have helped a child in need at all if she gives her a family in name only, i.e. if she fails to give the child love and care that only a person dedicated to being a parent in the fullest normative sense can provide. This is no objection to my claim: I'm talking about the value of adoption *for prospective parents*—for those who want to dedicate a significant portion of their resources and time to raising a child. We would criticize the prospective procreative parent who is naïve about the extensive demands of parenthood. But that some people are problematically naïve about the demands of parenthood in this case is no objection to procreative parenthood generally. Likewise with adoptive parents, though the rescue motivation could be inappropriate if it is the sole motivation for becoming a parent, it is not obviously problematic for those with a realistic understanding of, preparation for, and desire for the demands of parenthood.

Perhaps the worry is that adopters-as-rescuers may pose heavy burdens of gratitude on their children.[22] Parents who rescue children may see their relationship with their children in a fundamentally different way than procreative parents, in a way that makes their children feel unduly indebted to them. This could negatively impact adopted children.

But seeing rescue as a reason for adoption rather than procreation does not mean that the rescue relationship must come to characterize our parental relationship with

---

[21] I suspect that criticism of adopters who are motivated to rescue children in need is an instance of *do-gooder derogation*—a phenomenon where some in the majority (with regard to a choice), due to anticipation of moral reproach, take a derogatory attitude towards those in a minority who claim to base their choice on moral grounds (Minson and Monin, 2012).

[22] This concern was presented to me by Marianne Novy (2010).

our children. Adoptive families form parent–child relationships that fit the familiar mould of parent–child relationship, characterized by the same filial duties, no more, no less than biological families. Moreover, biological families are not immune from an analogous worry: we often hear of biological parents burdening their children with the claim that they should be grateful to them for their very existence. Whatever filial duties may be grounded in this sentiment, we find it appropriate to criticize parents if they take this demand too far. The fact that some biological parents act inappropriately in this regard does not count as a reason against having biological children. It counts as a reason against expecting from one's children servile gratitude. Thus, the same response in the case of adoption applies: we ought to parent with compassion and an appropriate sense of what sorts of burdens ought not to be placed on children.

Ultimately, I believe the rescue objection arises due to a misconception of the relationship between the impartial and partial perspectives and the reasons generated by each. One might think that a person motivated by impartial reasons to adopt has moralized the parent–child relationship in a way that will interfere with her forming an appropriate partial, special relationship with the child. A deeper explanation of this concern will both assuage the worry and better illustrate the idea that an adopted child becomes fully his parents' own child.

Philosophers are engaged in an ongoing debate about the tension in morality between the impartial and partial perspectives.[23] On one hand, morality is in its very nature about the impartial concern an agent should have for other people. Morality requires that I have regard for other people as equal subjects of moral concern. In considering what I morally ought to do, I deliberate from the impartial perspective, taking all people into account. Moral reasons speak against favouring myself and my inner circle of people.

On the other hand, some paradigm moral behaviour is partial in nature. Parents should love *their* children, giving them extra care and attention. The fact that a person is *my* friend is a reason for giving her special attention I do not give to others. That somebody is *mine* sometimes gives me reasons to be partial towards her (de Gaynesford, 2010: 88). This is true even though all people are equally valuable. Beyond this, many believe that morality leaves room for or even requires some partial attention to ourselves. We may be permitted or required to live a good life that includes cultivating our talents and interests and pursuing our goals. Favouring one's own perspective, on this view, plays a prominent, if not essential, role in moral reasoning.

Our conception of morality is fraught with tension between the impartial and partial perspectives, for they often come into conflict. The starkest picture is one without a possibility for balancing the two perspectives: impartial morality forbids partial perspective-taking; or conversely, the privileged partial perspective cannot be overridden in any case by impartial concern for others.

---

[23] For an excellent collection of essays on the topic, see Feltham and Cottingham (2010).

But impartial concern as the starting point for altruistic adoption need not stand in conflict with concerns arising out of one's partial, personal perspective. The following discussion not only defends adoption against this charge, it shows adoption to be a counter-example to such a simplistic picture. Adoption provides an example of the possibility for reasons of partiality to proceed from impartial grounds.

The reverse case—where genuine impartiality proceeds from partiality—is instructive. Maximilian de Gaynesford (2010: 93) argues that the same grounds for partiality can also justify impartiality. The fact that you are a parent to your children provides grounds for partial treatment of them. *Because you are their parent* you ought to and are permitted to favour them in a range of circumstances over, for instance, the neighbour's children. Yet the very same reason grounding this relationship of partiality grounds reasons for *impartial* treatment *between* your children. Our normative conception of parenthood includes that you be fair and equal in your treatment of *your* children. You should do so *because you are their parent*. Thus, as de Gaynesford puts it, impartiality can proceed from genuine partial grounds. This possibility is testament to the complexity of the moral landscape, which is rigidly simplified by a strict impartialist/partialist dichotomy. We should not assume that impartial treatment always has an impartial grounding.

Altruistic adoption reveals the opposite possibility: reasons of partiality can be generated from impartial grounds. Prospective parents may make their decision on how to become parents by starting from an impartial standpoint. Whether they procreate or adopt, the reasons they have to do so *for* the sake of their child-to-be *must* be impartial reasons. No special relationship between the parents and their potential child exists to ground reasons of partiality to the child. Choosing to become a parent is in fact a choice *to create a special relationship* with a child where the special relationship itself is the partial benefit in question. Thus, the decision to become a parent is not made from a standpoint of partiality *to* a particular child. The decision may still be (and usually is) partial *to oneself,* privileging one's own preferences and values in making the decision. But the point is that as it pertains to one's reasons *vis-à-vis* the child one will parent, this can be an impartial decision.

Adoption shows us that what may have started as impartially driven concern for a stranger can seamlessly become a concern for another that is integral to and driven by one's own partial perspective. A child becomes one's own child through fostering a relationship of love and concern across time. When a person adopts a child, the care she will give to her child as a parent becomes central to her own identity and conception of self. In parenting her child, she is not rescuing that child at every moment, she is caring for *her* child. She is partaking in the parent–child relationship of special concern. The transition between the impartial and partial perspectives within the life of the agent cannot be starkly drawn.

This insight is critical. One objection to impartialist morality is that it fails to give sufficient weight to an individual's own concerns and interests, putting them on a par with the interests, needs, and demands of all other people. As such, critics claim that impartialist morality generates too many demands and sacrifices of an individual with regard to her own important projects. Some may think that altruistic adoption poses this problem—the impartial concern for another is incompatible with sufficient room for a person to privilege her own personal, partial sphere of action. I imagine the worry here is that a parent may come to resent the child that she "rescued."

But the objection to impartialist morality cannot be just about the *extent* or sheer burden of the demands of impartiality. For the demands generated from the partial perspective, such as those required in raising a child or being a good friend or family member, are extensive. Indeed, there is little else more demanding than parenthood itself. The crucial distinction must be that the demands of strangers are in *conflict with* the space from which one pursues one's important goals and projects. They are imposed upon one from outside, alien to one's important life projects. In contrast, answering the demands generated by one's special relationships in part *constitutes* a person's goals and projects. But we've seen that the picture is more complex: a stranger can be integrated into one's own personal sphere of concern, becoming one's own child. Thus, this criticism of altruistic adoption grounded in an objection to impartialist morality simply does not apply. It only arises if one thinks impartiality can never give rise to partial relationships. Adoption proves this view to be false.

In short, impartially driven concern for another in altruistic adoption need not entail a parent–child relationship characterized by rescue of another person. A stranger child in need of a family can quickly become a child of one's own, generating reasons of partiality rather than impartiality.

Personal Transformation

I've focused on the way someone can integrate an *other* into her own personal, partial perspective; but her own perspective and self-conception can also be importantly altered. Further, this can be a valuable and unique transformation. Transracial adoptions are a compelling example of this possibility. John Raible (2008: 95) reports that non-adopted, white siblings of adopted, non-white children experience "more nuanced and sophisticated understandings of the dynamics of race in our society, and a deeper appreciation for struggles against racism, both in history and in the lives of their adopted siblings, and ultimately, in their own lives." In effect, they are *transracialized*—gaining intimate and extended, vicarious experience of navigating the challenges of a racial hierarchical society.

Raible's exploration of transracialized identity generalizes to the experiences of transracial adoptive parents. Sally Haslanger, in her own elaboration of the idea, suggests that for parents, this transformation is first engendered through the bodily closeness that they share with their child. She explains: "This empathetic extension of body awareness, this attentiveness to the minute signals of another's body, [the] taking on the needs and desires of another body as if your own, perhaps especially if the other's body is marked as different, alters your own body sense" (Haslanger, 2005: 279). As a vivid illustration of this internalization of one's child's body, Haslanger (2005: 279 n. 14) shares the experience of a white mother of two Korean-born adopted children who, on a trip to Korea, expresses joy in being somewhere where "everybody looks like us".

Haslanger labels this phenomenon as one of having a "mixed" racial identity. If we think of racial identity as a map "that functions in a multitude of ways to guide and direct exchanges with one's social and material realities," those with mixed racial identities navigate their world by reference to more than one map—namely, that of their own race and the race of their adopted child (Haslanger, 2005: 283). This mixed identity may manifest itself in many ways. It may foster not just sympathetic but *empathic* understanding of racial injustice. Mixed or transracialized identity can facilitate a person's ease with and preference for social and personal relationships with different-raced individuals. A person's sense of community may fundamentally change—she may be less at home in non-diverse settings. She may find that same-race friends cannot relate to her specific concerns, for the experiences of her different-raced child have altered her perspective. None of this amounts to the claim that a transracialized or mixed-race identity entails that one comes to have the *same* racial identity of the adopted child; but as Haslanger (2005: 285) explains: "my day-to-day life is filled with their physical being and social reality, and by extension, the reality of their extended families and their racial community.... their realities have in an important sense become mine."

Though transracial adoptions are the starkest example of this possibility for integrating another's identity into one's own, arguably, more general *transpersonal* transformations occur in adoptions of all kinds. Foremost, the adoption experience challenges a deeply entrenched cultural conception of the family as bio-genetically based. Adoptive families come to have revised notions of kinship relations as those fostered by shared histories of concern and care. This can happen without denying the significance of bio-genetic ties; in open adoptions, families may come to include an adopted child's birth family. The possibility for transpersonal transformation is yet another benefit of adoption and testament to the unique moral value of adoption; it allows us to transcend the constraints of our own accepted identities and integrate into them what was once outside or foreign to ourselves. In a way, adoption makes us bigger than our original selves; it expands us beyond our original kin and community.

Conclusion

There is unique value for prospective parents to be found in adoption. This value is not limited to only those who cannot easily procreate. Adoption is a valuable alternative to procreation for all prospective parents.

When we highlight what is uniquely valuable about adoption, taking it out from under the shadow of procreation, we can see it in a new light. Adoption offers special value to parents: it can help an existing child in need of a family, whereas procreation creates a child with needs. It is a paradigm expression of human regard for the sanctity of life and the value of family. Adoption's unique value is in sharing an intimate special relationship with a stranger, in the process making her one's own. The impartial moral concern for another can be integrated into one's own personal perspective and reasons for action. Adoption has transformative powers over our relation to others and our own conception of self. For these reasons, adoption could hardly be considered second-best to biological procreation. In many ways, adoption is an exemplar for both the parent–child relationship and the human capacity for moral compassion.

Acknowledgements

I would like to thank the co-authors of this anthology for their spirited and immensely valuable engagement with my chapter. I am also grateful to Joe Millum for his helpful comments on an early draft.

References

Bartholet, E. (1993). *Family Bonds: Adoption, Infertility, and the New World of Child Production.* Boston, MA: Beacon Press.

Bartholet, E. (2000). *Nobody's Children: Abuse and Neglect, Foster Drift and the Adoption Alternative.* Boston, MA: Beacon Press.

Bartholet, E. (2007). International Adoption: Thoughts on Human Rights Issues. *Buffalo Human Rights Law Review,* 13, 151–203.

Belkin, L. (2009a) Too Many Ways to Have a Baby. *New York Times,* 9 Apr. Retrieved June 2013 from http://parenting.blogs.nytimes.com/2009/04/29/too-many-ways-to-have-a-baby.

Belkin, L. (2009b). The Guilt of Secondary Infertility: Motherlode. Adventures in Parenting. *New York Times,* 30 Apr. Retrieved June 2013 from http://parenting.blogs.nytimes.com/2009/04/30/the-guilt-of-secondary-infertility.

Carlson, R. (2010/11). Seeking the Better Interests of Children with a New International Law of Adoption. *New York Law Review,* 55, 733–79.

de Gaynesford, M. (2010). The Bishop, the Valet, the Wife, and the Ass: What Difference Does it Make if Something is Mine? In B. Feltham and J. Cottingham (eds), *Partiality and Impartiality: Morality, Special Relationships and the Wider World* (pp. 84–97). New York: Oxford University Press.

Feinberg, J. (1973). *Social Philosophy.* Englewood Cliffs, NJ: Prentice Hall.

Feltham, B., and Cottingham, J. (eds) (2010). *Partiality and Impartiality: Morality, Special Relationships and the Wider World.* New York: Oxford University Press.

Fisher, A. P. (2003). Still "Not Quite as Good as Having Your Own?" Toward a Sociology of Adoption. *Annual Review of Sociology,* 29, 335–61.

Graff, E. J. (2008). The Lie We Love. *Foreign Policy.* Retrieved June 2013 from http://www.foreignpolicy.com/articles/2008/10/15/the_lie_we_love.

Hare, R. M. (1975). Abortion and the Golden Rule. *Philosophy and Public Affairs,* 4(3), 201–22.

Haslanger, S. (2005). You Mixed? Racial Identity without Racial Biology. In S. Haslanger and C. Witt. (eds), *Adoption Matters: Philosophical and Feminist Essays* (pp. 265–90). Ithaca, NY: Cornell University Press.

Haslanger, S. (2009). Family, Ancestry and Self: What is the Moral Significance of Biological Ties? *Adoption and Culture*, 2, 91–122.

Haslanger, S., and Witt, C. (eds) (2005). *Adoption Matters: Philosophical and Feminist Essays.* Ithaca, NY: Cornell University Press.

Hursthouse, R. (1987). *Beginning Lives.* New York: Basil Blackwell.

IJzendoorn, M. H. van, and Juffer, F. (2006). The Emanuel Miller Memorial Lecture 2006: Adoption as intervention. Meta-analytic evidence for Massive Catch-Up and Plasticity in Physical, Socio-Emotional, and Cognitive Development. *Journal of Child Psychology and Psychiatry*, 47(12), 1228–45.

Kolodny, N. (2010). Which Relationships Justify Partiality? The Case of Parents and Children. *Philosophy and Public Affairs*, 38(1), 37–76.

Landau, E., and Gumbrecht, J. (2010). IVF Doctors, Families Celebrate Creator's Nobel Prize. *CNN Health*, 4 Oct. Retrieved June 2013 from http://www.cnn.com/2010/HEALTH/10/04/ivf.fertility.babies.nobel/index.html?hpt=C1.

Lotz, M. (2008). Overstating the Biological: Geneticism and Essentialism in Social Cloning and Social Sex Selection. In J. Thompson and L. Skene (eds), *The Sorting Society: The Ethics of Genetic Testing and Therapy* (pp. 133–48). Cambridge: Cambridge University Press.

Magnitsky Act [Russia and Moldova Jackson-Vanik Repeal and Sergei Magnitsky Rule of Law Accountability Act of 2012], 2012 HR 6156, 112th (2012). Retrieved July 2013 from http://www.gpo.gov/fdsys/pkg/BILLS-112hr6156enr/pdf/BILLS-112hr6156enr.pdf.

Minson, J. A., and Monin, B. (2012). Do-Gooder Derogation: Disparaging Morally-Motivated Minorities to Diffuse Anticipated Reproach. *Social Psychological and Personality Science*, 3(2), 200–7.

Novy, M. (2010). Newsletter for the Alliance for the Study of Adoption and Culture <http://www.adoptionandculture.org> (Fall).

Oreskovic, J., and Maskew, T. (2008). Red Thread or Slender Reed: Deconstructing Prof. Barthlet's Mythology of International Adoption. *Buffalo Human Rights Law Review*, 14, 71–128.

Petersen, T. S. (2002). The Claim from Adoption. *Bioethics*, 16, 353–75.

Raible, J. W. (2008). Real Brothers, Real Sisters: Learning from the White Siblings of Transracial Adoptees. *Journal of Social Distress and the Homeless*, 17(1–2), 87–105.

Rivera-López, E. (2006). The Claim from Adoption Revisited. *Bioethics*, 20(6), 319–25.

Secretary-General (2006). Report of the Independent Expert for the United Nations Study on Violence Against Children. Delivered to the General Assembly, 29 Aug. UN Doc A/61/299. Retrieved June 2013 from http://www.unicef.org/violencestudy/reports/SG_violencestudy_en.pdf.

Smilansky, S. (1995). Is there a Moral Obligation to Have Children? *Journal of Applied Philosophy*, 12(1), 41–53.

Smith, J. F. (2005). A Child of One's Own: A Moral Assessment of Property Concepts in Adoption. In S. Haslanger and C. Witt (eds), *Adoption Matters: Philosophical and Feminist Essays* (pp. 112–31). Ithaca, NY: Cornell University Press.

Tooley, M. (1999). The Moral Status of the Cloning of Humans. *Monash Bioethics Review*, 18, 27–49.

US Department of Health and Human Services (2011). Adoption and Foster Care Analysis Reporting System (AFCARS report #18), 20 June. FY 2010 data. Retrieved June 2013 from http://www.acf.hhs.gov/programs/cb/research-data-technology/statistics-research/afcars.

Velleman, J. D. (2008). Persons in Prospect II: The Gift of Life. *Philosophy and Public Affairs,* 36, 245–66.

Witt, C. (2005). Family Resemblances: Adoption, Personal Identity, and Genetic Essentialism. In S. Haslanger and C. Witt (eds), *Adoption Matters: Philosophical and Feminist Essays* (pp. 135–44). Ithaca, NY: Cornell University Press.

PART IV

Becoming a Parent:
State Interests

7

State Regulation and Assisted Reproduction

Balancing the Interests of Parents and Children

Jurgen De Wispelaere and Daniel Weinstock

Introduction

There was a time, not so long ago, when adoption of one of the many children abandoned in the care of civil or religious institutions was the only way that infertile persons wanting to become parents could realize their wish. Rapid progress in medical science and technology from the 1970s onwards has introduced a third option: prospective parents can now make use of a host of assisted reproductive technologies (ARTs) to assist them in giving birth to a child to which they are biologically connected.[1]

Social demand for ART has grown rapidly as the associated procedures have become safer and more socially acceptable. ART has now displaced adoption as the first port of call for couples unable to conceive naturally. While some critics decry the inherent biologism of this development (Bartholet, 1995, 1999), many hold that ART serves an important purpose, namely to grant prospective parents one of their deepest wishes—to bring "a child of their own" into the world.[2] In addition, many believe the fact that some couples find themselves unable to conceive naturally constitutes the sort of brute

---

[1] We use the term "biological" to refer both to shared genetic endowments and to the distinct biological process of gestation/giving birth. ARTs vary extensively, both in terms of intrusiveness of the procedure and the degree of genetic link between partners and child (Warnock, 2002), and each raises its own peculiar moral issues, heavily debated in the bioethics and policy literature (e.g. Harris and Holm, 1998). In what follows we abstract from these particularities and focus on ART more generally.

[2] The phrase "a child of one's own" is exceedingly ambivalent since with the more invasive ARTs, the child has no genetic or gestational connection with at least one, possibly even both, partner(s). Bartholet (1995, 1999) refers to these as "technological adoptions," although one characteristic (and complaint) is that these forms of reproduction bypass the legal process typical of regular adoption. For a fascinating account of current contradictions in the legal process of family-making, see Crawford (in this volume).

bad luck for which they should not be held responsible (Dworkin, 1981). On this view, ART is a means of redressing what otherwise would amount to an unfair inequality between fertile and infertile couples.

The question we want to address in this chapter is whether a policy of permitting, or even of actively promoting, ART through subsidization is justified, all things considered. Throughout this chapter we assume (but do not defend) the view that ART is *pro tanto* morally permissible. That is to say that we do not think there is principled reason to prohibit ART, nor that those seeking or giving reproductive assistance are doing anything morally wrong. But this does not mean that the state has no legitimate reasons to limit access to ART when taking into account the broader context of policies through which parenting is regulated. The purpose of this chapter is to examine the argument that the existence of large numbers of adoptive children in need of parents warrants the restriction of access to ART (Bartholet, 1999; Levy and Lotz, 2005; Allberg and Brighouse, 2010).[3]

In this chapter, we explicitly adopt the perspective of the regulatory state. That is, we are only tangentially interested in the ethics of ART, of adoption, or indeed of "natural" procreation. In recent years, philosophers such as David Benatar (2006), Christine Overall (2012 and in this volume), and Seana Shiffrin (1999) have probed deeply into the ethical implications of that most apparently natural and unproblematic of human activities, that of bearing children through sexual reproduction. Our interest in the context of this chapter is not to pronounce on the overall moral acceptability of different "child acquisition" practices. Rather, we are interested in defining how the state ought to regulate these practices, given certain minimal assumptions about human behaviour. We believe that the moral permissibility or impermissibility of a practice does not in and of itself settle the matter of whether the state should ban it, regulate it, or simply allow it without restriction. Even where a practice is deemed permissible when considered in isolation, there may be valid reasons for the state to nevertheless restrict its use once looked at in combination with other policies. In short, judgements about the morality of a practice tell us little about the specific ways in which it should be regulated in order to maximize attainment of the normative considerations militating in favour of regulation.

Taking a regulatory stance implies that the state should not regard sexual reproduction as morally equivalent to other forms of child acquisition. Though, all things being equal, sexual reproduction as a means of acquiring children to parent may stand in need of as much moral justification as do other forms of child acquisition (e.g. Cutas, 2009), it does not follow that it should be regulated in the same way as either ART or adoption. The state must assume that sexual reproduction will continue to be the method used by the overwhelming majority of citizens to acquire children to parent. Attempts to disrupt this pattern in the name of some abstract moral consideration

---

[3] See Post (1996) and Rulli (in this volume) on the ethics of adoption more generally.

raise massive practical issues, to say nothing of the risks posed to the liberty interests of citizens by a state that would purport, say, to place limits through coercive measures on the number of children that people can have through sexual reproduction. Thus, for example, while it makes sense to ask ourselves whether, and to what degree, the state should fund ART to attain public policy goals, it would not make analogous sense to ask ourselves whether the state (in the name of an ideal of moral neutrality as between various forms of child acquisition) should fund pre natal classes and maternal care. Given predictable patterns of human procreative behaviour, the public health consequences that would follow from failure to fund the latter would be catastrophic.[4]

Rights and Assisted Reproduction

Liberal societies typically insist that we should respect people's (notably women's) reproductive freedom. These claims are often cast in the form of two related rights. First and foremost is the *negative right* not to be interfered with in one's reproductive choices. This right implies that individuals themselves get to decide whether, when, and with whom to have a child through natural means (Robertson, 1994). This right is important, in part because of its close association to the right to bodily integrity, but it is nevertheless not limitless, as we shall discuss further. Two important sets of restrictions arise from protecting the interests of the prospective child and of the reproductive partner, respectively. Though liberal jurisdictions for the most part prescind from ascribing constitutional status to the unborn, they nonetheless engage in a great deal of non-coercive measures to protect the foetuses of mothers who have decided to carry them to term. They encourage prenatal exercise, healthy diets, frown on the use of alcohol and tobacco by pregnant mothers, and so on (see Kukla, 2005, for a critical perspective). Reproductive freedom is also restricted in that partners must respect each other's bodily integrity and interests over bodily resources such as gametes. This is why we restrict men's influence over whether a woman should abort a foetus or carry it to term, but equally prevent women from using a man's sperm without his consent (Strong *et al.*, 2000).

In the context of assisted reproduction, a negative right to reproduction implies that the state should not interfere with ART providers offering their services to those unable to conceive naturally and, equally, that it should not prevent individuals or couples from accessing their services. Of course such services come at a considerable cost, and some may not be able to afford them. Merely establishing a negative right appears insufficient from the point of view of infertile parents seeking to realize their interest in bringing a child into the world. The claim that there exists a *positive right* to reproduction would take matters a step further by demanding that the state offer full or partial subsidy of the associated costs, granting every citizen effective access to the means for

---

[4] We thank Serena Olsaretti for pressing us on this point.

ART. In what follows we will refer to those two types of rights to ART as the state "permitting" and "subsidizing" ART, respectively.

Is there a robust moral basis for thinking that a right to ART in its positive form exists, and if so, what are its limits? Note first that such a right does not necessarily follow from endorsing the right to reproductive freedom more generally. To the extent that the freedom to reproduce (naturally) is typically grounded in principles of bodily integrity, a state that refuses to establish ART facilities may not be interfering with a person's reproductive freedom in a strict sense. Such a person is free to do as she pleases with her body. She may, in particular, attempt to bring a child into the world, but she has in virtue of this freedom no obvious claim to assistance that might help her overcome the natural limits of her bodily abilities. To think that reproductive freedom demands a particular set of state services seems to require an additional argument.[5]

One such argument would be to think of ART as an aspect of reproductive health (Cook et al., 2003; Hughes and Giacomini, 2001; Mladovsky and Sorensen, 2010; Neumann, 1997). On this view, the state has an obligation to provide for assisted reproduction as part of a more general obligation to remedy a poor state of health. Conceiving of assisted reproduction within the reproductive health context implies that we regard a failure to reproduce naturally as a dysfunction of an important bodily function, and ART as a kind of medical intervention that is supported by the state in the context of its commitment to securing the basic health needs of its population.

There are, however, important limits to arguing for a positive right to ART on reproductive health grounds.

First, while some cases of infertility perfectly fit the idea of ART correcting for a medical problem (e.g. removing ovarian or testicular obstructions), in other cases, ART bypasses the functional limitation rather than correcting it. In these cases the defective health state remains unaltered and even when having conceived a child, the individual may still remain infertile in a strict sense.[6] This suggests that in those cases infertility may share more features with a "disability" than an "illness." The use of provider insemination or gestational surrogacy by same-sex couples, for instance, resembles a case of enhancing rather than correcting for species-normal functioning. A right to ART grounded in reproductive health might therefore conceivably be too restrictive in scope.

Second, establishing a right to ART on reproductive health grounds runs the risk of ART becoming unnecessarily entangled in debates about health resource rationing

---

[5] Would the logic of our argument saddle us with the unpalatable view that a right to abortion does not imply an obligation on the part of the state to fund safe abortions? We think not, for two reasons. First, the deleterious public health implications that would follow from reduced availability of safe abortions implies a strong moral obligation on the state to secure access to such services. Second, following Judith Jarvis Thomson's (1971) classic analysis, reduced availability of safe abortions directly affects a woman's negative right to bodily integrity. For these reasons, we firmly believe funding abortion is clearly mandated, even if funding ART were not. We thank the editors of this volume for having prompted us to deflect this unwanted—but thankfully unwarranted—implication explicitly.

[6] Extreme interventions such as human uterus transplantation may be an exception to this point (Catsanos et al., 2011).

and prioritization, with ART being displaced by more urgent health needs or interventions that have a larger aggregate positive health impact (Uniacke, 1987; more generally, Brock, 2002, 2004). This restriction would mostly limit the state's capacity to subsidize ART, but could also over time negatively affect ART service provision due to a decrease in aggregate demand.

Third, a reproductive health-based right to ART would likely give considerable weight to medical criteria in decisions of who should be granted access. Medical practitioners might unduly restrict access to ART by applying general criteria for the effective use of health resources (e.g. age rationing). Decisions as to who is granted access to ART might become based predominantly on medical criteria, leaving hardly any space for social considerations as to who should become a parent.[7]

In short, establishing a right to ART within a reproductive health perspective is likely to produce a right that is too restricted or too weak to be of any real value. We prefer an alternative approach that eschews a close relationship to health or medical needs and directly grounds the right to ART in people's interests in becoming parents.

Assisted Reproduction and the Right to Parent

The case for granting individuals a fundamental right to parent has recently been developed by Harry Brighouse and Adam Swift. Brighouse and Swift (2006 and in this volume) argue that one's interest in becoming a parent amounts to a fundamental albeit conditional and limited right. The right is limited in that it grants parents certain rights, but not necessarily others (e.g. the right to parent does not imply the right to bequest at will). The right is also conditional upon parents safeguarding the core interests of their children. This means, for instance, that the state may be justified in intervening when parents fail in protecting their children's interests in education, health, or adequate nutrition. But this right is nevertheless fundamental in that "it is owed to a person in virtue of their simply being a person, and its justification is grounded in the benefits it will bring to that person and not to others" (Brighouse and Swift, 2006: 87). The fundamental right to parent is grounded in a person's interest in being a parent, not in any instrumental role that parenting plays for others.

For Brighouse and Swift, it is the very special intimate relationship that parents have with their children that grounds a fundamental right.[8] This intimacy follows from the fact that the relationship itself is of a special kind (Brighouse and Swift, 2006: 91–101).

---

[7] Criteria that govern access to ART already differ considerably from those regulating access to adoption or fostering (Widdows and MacCallum, 2002). Relatedly, sociological evidence suggests the continued "medicalization" of reproduction explains individuals' or couples' reluctance to give up on medical solutions and instead consider adoption (Fisher, 2003; Appleton, 2004).

[8] Of course a similar interest might be said to ground children's rights, including perhaps a child's fundamental right to adoption (Woodhouse, 2005). In this chapter we refrain from making such a strong move, and instead opt for allowing potential adoptees' interests to counter the neutral stance a regulatory state would otherwise take in relation to competing means for parents acquiring children.

There are in their view four important aspects to parent–child relationships that warrant our viewing it as grounding a right. First, children are utterly dependent on parents for securing their basic needs and well-being, and therefore vulnerable to the choices and actions of parents. Second, children typically have no right or power to exit the relationship. Third, the love of children for their parents is (typically) spontaneous and unconditional, and this is the case even where parents are not fully "deserving" of their children's love. Finally, the parent is explicitly charged with the responsibility of taking care of the child.

Because of the form and quality of the intimacy associated with raising a child, many individuals develop a very strong interest in experiencing such a relationship:

Parents have an interest in being in a relationship of this sort. They have a nonfiduciary interest in playing this fiduciary role. The role enables them to exercise and develop capacities the development and exercise of which are, for many (though not, certainly, for all), crucial to their living fully flourishing lives. Through exercising these capacities in the specific context of the intimately loving parent–child relationship, a parent comes to learn more about herself, she comes to develop as a person, and she derives satisfactions that otherwise would be unavailable. The successful exercise of this role contributes to, and its unsuccessful exercise detracts from, the success of her own life as a whole. (Brighouse and Swift, 2006: 95)

The special nature of this intimate relationship makes parenting genuinely unique: the right to parent is not just valuable, but it is also hard to imagine any social activity that could genuinely substitute for the lack of a parenting role. Many interests, even fairly fundamental ones, allow for multiple realizability. The development of one's intellectual abilities, for example, can be realized through maths or through literature. Professional achievement too can be attained in a variety of ways (De Wispelaere and Weinstock, 2012). But for those whose conception of the good life includes parenting, it is difficult to see how it could be substituted for without loss for those who are unable to procreate without assistance.

The combination of "value"—parenting is an activity of great value for those who choose to partake in it—and "nonsubstitutability"—it cannot be replaced without significant loss—seems to ground a particularly strong obligation on the state to promote this fundamental right. In the case where individuals face barriers to reproduce naturally, it would seem that (at least) permitting or perhaps even subsidizing ART constitutes one important way in which the state can discharge its putative correlative obligations.

Before we move on to discuss the limits to the resulting *pro tanto* right to ART, we need to briefly discuss one complication. For note that the intimate relationship account provides us with a fundamental right to parent a child, but remains silent on which child this right applies to. Anca Gheaus (2011) has recently argued that the arguments put forward by Brighouse and Swift are insufficient to prevent us from redistributing children after birth. More to the point for our concern, the intimate relationship account is silent concerning the means by which parents acquire the child they subsequently develop an intimate relationship with. On this view, all that is required is that prospective parents be given an opportunity to develop a stable parenting relationship with a particular child, independent of whether this child is one they conceived naturally. The interest in parenting can be realized through relationships with children obtained through any one of the many forms of reproductive assistance, or indeed through social opportunities for parenting—notably adoption.[9]

Some scholars no doubt will want to go further and insist that the general right to parenting we adopt here is too weak: instead, we should conceive of parental rights as rights to parent one's biological baby. One reason to think parental rights should take the biological link more seriously is by focusing on the importance of a genetic connection between parent and child. However, as Levy and Lotz (2005) have decisively argued, many reasons that have been advanced as to why genetics should matter seem utterly implausible.[10] A more promising route to arguing for a stronger biological connection is offered by Gheaus's (2011) argument that the intimate relationship already starts during pregnancy, such that parents may have a right to parent their baby in virtue of morally relevant aspects of the gestational process.

But while refocusing the point at which the intimate relationship between parent and baby begins tells us about important differences between the various means of becoming a parent, it does not imply that this difference translates into a distinct set of rights. It may well be, as Gheaus argues, that the burdens associated with giving birth facilitate the development of the intimate relationship between parent and child, but it would be hard to see how this feature on its own justifies a right to have one's parental interest satisfied through a fundamental right to conceive a child to whom one is biologically related. To be clear, we take issue here not with Gheaus's claim that once a person or couple has gone through the process of giving birth, this may give rise to a right to keep that particular baby, but merely with the possible implication (which, we hasten to add, Gheaus does not make) that such a process would grant a more general right to access the very opportunity of becoming a parent in this manner. In addition, we should of course keep in mind that not all forms of ART are associated with the types of gestational burdens that underpin such claims.

The upshot of these considerations is that, in our view, individuals have a general fundamental right to become a parent. The opportunity to become a parent can be satisfied through natural reproduction, through access to ART, but also, importantly, through adoption. The right to parent does not require that there be a biological connection between parent and child. None of this is to say that there are no important differences between parenting a "biological" or an "adoptive" child. For one, many

---

[9] We leave fostering out of the equation because fostering relationships in many cases are too short-term to be sufficiently similar to the aforementioned. This is not to say we should not take account of the opportunities of satisfying important interests of both children and some parents through fostering.

[10] More generally, Daniel Friedrich (2013: 7) argues that prospective parents' "desire for biological children is often based on false beliefs and the failure to put salient upsides into perspective".

parents would value the experience of pregnancy independent of the intimate relation-ship with their child, although this point should not be taken for granted in all cases (Friedrich, 2013). In addition, there may be important differences between the child one ends up parenting if one elects adoption: adoptive children in some cases may be older, they may have special needs (not in the least if they have spent part of their lives in a care facility), and in the case of international adoptions specific cultural concerns may arise. Our view, however, is that these are not the sort of factors that determine the existence of a right to parent. Rather, they have an impact on the quality of the experi-ence associated with exercising one's right to parent. While we may give some value to such factors, in part because they are valued by individuals, they are not sufficiently important to outweigh important interests of children that we discuss in more detail in the next section (see also Ahlberg and Brighouse, 2010).

With these qualifications in place, we can now turn to examining whether there are legitimate reasons for the state to limit individuals' unrestricted access to ART. Our main focus will be on outlining the extent to which the alternative of adopting a child constitutes a sufficiently good reason for the state to regulate access to ART, either by not offering (partial) subsidy or by refusing permission outright.

Adoption and Limits to Assisted Reproduction

If individuals have a fundamental right to parent, grounded in a strong interest in the sort of intimate relationship that parenting entails, it would appear the state has a cor-relative obligation to facilitate individuals becoming parents. Specifically, we might think that in the case of individuals or couples who have difficulty conceiving nat-urally, this implies an obligation on the part of the state to facilitate access to ART. But as mentioned earlier in this chapter, ART is not the only way in which the right to parent can be satisfied. Adoption is another important parent-generating device. Several scholars have argued that adoption in fact should be preferred over access to ART (Anderson, 1990; Bartholet, 1995, 1999; and in the specific case of cloning Levy and Lotz, 2005; Ahlberg and Brighouse, 2010).

The argument runs as follows. Following Brighouse and Swift's (2006) intimate relationship account, individuals have a fundamental right to become parents. However, that right does not guarantee one a child to whom one is biologically related. This means that, as far as the parental perspective is concerned, the state ought to be *pro tanto* neutral between any means for satisfying parental interests, whether through ART or adoption. The state's preference for one method for acquir-ing children to parent over others will thus depend on how its policy affects the interests of third parties.[11]

---

[11] A second reason why the state may legitimately disregard its neutral stance is where one type of repro-duction requires significantly more public resources to be spent than its next-best alternative (resources that could be better spent elsewhere, perhaps to invest in better institutional child care arrangements).

While there are no morally relevant reasons for the state to accede to the preferences that some people might have for parenting children to whom they are biologically related—for example, by heavily subsidizing ART for infertile individuals—by contrast there exist relevant third parties who would be hugely affected by the state's decision to depart from its neutral stance. Those relevant third parties are potential adoptees:

Precisely, the people whose interests would be damaged by a regime in which reproductive [assistance] is allowed are potential adoptees. Any society contains newborns and young children who have no parents. Those children have a very powerful interest in acquiring parents, because, in modern industrial democracies, we have been unable to develop alternatives to families that serve children's needs adequately. Children need both immediate provision of shelter, care, nutrition and affection, and the security of long-term relationships with responsible adults who will supervise their moral, cognitive, physical and emotional development. In other words, they need parents. So it is important, for the sake of orphaned or abandoned, or very severely abused children, that there be a pool of potential adoptive parents. (Ahlberg and Brighouse, 2010: 550)

This argument assumes that children are typically better off living in a family setting rather than in an institution, or moving from one foster family to another. While we accept there may be specific cases where this assumption does not hold (e.g. where the parents are abusive or neglectful), we agree with Ahlberg and Brighouse that, on the whole, families are vastly superior to institutional alternatives, even in developed modern societies.

This argument points to an important asymmetry between ART and adoption:

In the case of adoption, need fulfilment is symmetrical: an existing child's need for a parent is fulfilled by an adoptive parent whose own need (or preference/desire) for a child is, in turn, satisfied. By contrast, [ART] involves unilateral and asymmetrical need-satisfaction: of the parents for a (biological) child. (Levy and Lotz, 2005: 247–8)

Prospective adoptive children have an interest in restricting the availability of competing reproductive technologies as this affects their chances of finding an adoptive family. Children who owe their existence to ART, on the other hand, have no such interest, for if ART were not made available to prospective parents these children would simply not exist. In other words, where "adoptive children" can be harmed by promoting ART, "ART children" cannot be similarly harmed by promoting adoption.[12] This asymmetrical dimension of ART versus adoption gives the state a legitimate reason to promote adoption over ART, a reason that is furthermore proportional to the extent that ART directly affects adoptive children's interests. At the same time, prospective parents have no valid reason for complaint, for their fundamental right to parent is satisfied through adoption even when they are denied access to parent their own biological child through ART.

---

[12] But see Søbirk Petersen (2002) for a careful philosophical analysis of the many complications surrounding this view.

A complication arises, however. It follows from the previous point that the legiti-macy of restricting access to ART is dependent on the availability of sufficient adoptive children to satisfy the parenting rights of those unable to reproduce naturally. There are many reasons affecting the number of adoptive children: natural disasters, epidemics, and war produce vast numbers of orphans, while changes in cultural norms pertaining to birth control and the availability of abortion in turn may decrease the number of children given up for adoption. Some variability notwithstanding, there is a good case to be made for the claim that, once we expand our perspective to include the interna-tional dimension, "the number of children who would almost certainly benefit from adoption far exceeds the number of prospective adoptive parents" (Carlson, 2011: 735). This appears to favour the view that access to ART should be heavily restricted, and that those who are unable to reproduce naturally should routinely employ adoption as an alternative means for acquiring children to parent. We should resist this conclusion, however, for at least three reasons.

First, some children in need of adoptive parents require such a specialized envi-ronment to accommodate their special needs that allowing them to be adopted by anyone who is suitable to parent children would be beneficial neither for the child nor the parents. These children are of course entitled to the best familial environ-ment available, but the reality is that fewer families interested in adopting a child would be regarded as sufficiently suitable. To impose on such a child an adoptive family that is ill-prepared or unmotivated to take care of her is unacceptable; the child might not be better off than if she had grown up in an institution, and at least in some cases presumably the child will end up back in institutional care. Equally, in some circumstances, we would not think it fair to prospective parents that they should face a choice between parenting a special needs child or remaining childless. Should parents find themselves in such a situation, a case could be made that satis-fying their fundamental right to parent might require granting them access to ART. The argument here turns on the view that giving a special needs child to parents who are ill-prepared (or unwilling) to accommodate his needs makes it exceedingly unlikely that the parents will achieve the intimate relationship that underwrites their fundamental right to become a parent; whereas the mere fact that a child is not biologically related to his parents does not face this challenge. Of course, we hasten to add that nothing we write above implies that special needs children do not have an important interest in finding a suitable family; our point is merely to suggest that imposition of any type (including where parents decide to go down this route because the ART alternative is unavailable) would be fraught with difficulties for both parent and child.

Second, leaving aside intra-family adoptions, from the 1990s onwards international or intercountry adoptions have grown considerably. Framed in international law by the United Nations Convention on the Rights of the Child (CRC) and The Hague Convention on Protection of Children and Co-operation in Respect of Intercountry Adoption, for a number of years a considerable number of children from orphanages throughout the world found family homes in the US and beyond. However, for a variety of reasons, intercountry adoption is now frowned upon by some groups (Bartholet, 2010, 2011; Selman, 2012). Some child rights organizations, including UNICEF, have taken a very negative view of what they view as a violation of a child's heritage right, while some countries have asserted cultural and subsidiary principles to counter what they regard as the taking of their most precious resources (Bartholet, 2007; Carslon, 2011).[13] International adoption policy is a complex matter (Jones, 2010), but what matters for the purposes of our argument is that in recent years access to intercountry adoptions has become increasingly difficult, with waiting lists growing exponentially. In addition, intercountry adoptions increasingly involve older children with special needs (Selman, 2012). Given this reality, parents may insist that they be given access to ART in order to satisfy their fundamental right to parent.

Third, the availability of adoption as a genuine alternative to ART is highly constrained in many jurisdictions by the existence of stringent screening requirements on adoptive parents (De Wispelaere and Weinstock, 2012; McLeod and Botterell, in this volume). By contrast to natural parents, and even those who engage in ART, adoptive parents are subjected to intense scrutiny in order to determine their overall suitability as parents, and even to assess whether they are the most suitable parents for a given child. The screening process varies considerably across jurisdictions but will typically include a review of formal criteria (e.g. age and marital status) and other detailed information, interviews with partners and other children, house visits, and so on. Criteria for determining suitability again may differ considerably but are likely to favour those who best approximate the stereotypical nuclear family. The justification for such screening is typically put in terms of safeguarding children's well-being, with the state taking responsibility for ensuring that the child's new living environment meets strict standards of good parenting.[14]

However, while strict screening may protect some children from being adopted into an abusive or neglectful family, this process unfortunately also decreases the overall opportunities for children to find a suitable adoptive family in two ways. To begin with, extensive screening is likely to include significant numbers of false positives (adoptive parents who fail the test but would nevertheless be perfectly acceptable parents). Thus, at least some suitable prospective parents who are inclined to go down the (often tortuous) adoption route are screened out. What's more, a screening process that is expensive, time-consuming, burdensome, and intrusive may cause many prospective parents to forgo the adoption route altogether (Søbirk Petersen, 2002). This outcome affects the interests of adoptive children: some may end up spending much more time in subpar institutional care, while others may lose the opportunity to become adopted

---

[13] For instance, Romania, formerly a leading "exporter" of adoptive children, was required to discontinue intercountry adoption as a condition upon joining the EU.

[14] The child well-being justification of course begs the question why such screening should be restricted to adoptive parents only (LaFollette, 1980, 2010). We leave this question open here but have discussed some of the problems with licensing natural parents in De Wispelaere and Weinstock (2012).

altogether. This would seem to negate the very reason for which a state might think it appropriate to restrict access to ART in favour of promoting adoption. Finally, it might end up being the case that, from a parental perspective, excessive screening means adoption can no longer be considered a genuine equivalent to ART. This might lead to a situation in which, in the absence of positive state support for ART, the state would no longer be securing the fundamental right to parent at all.

Each of these three complications weakens the argument that adoption legitimately restricts access to ART. These constraints are of a contingent, empirical nature, and the state might respond to these challenges in different ways. With respect to the first difficulty, while the state cannot directly reduce the number of special needs children awaiting adoption (or at least not in a morally acceptable manner), it can accommodate families by offering sufficient public assistance, which may induce some prospective parents to opt for adopting such a child. Nevertheless, presumably there will remain prospective parents who are reluctant to adopt a special needs child no matter how much assistance is involved; in fact, too much state support might raise concerns about the appropriate motivations for parenting a special needs child. In the second case, the state may engage in legal and political actions to overcome the objections of sender countries against intercountry adoption, thus again opening the prospects of increasing access to adoption for some prospective parents. But here too we must acknowledge the limitations of state action in the short or even medium run, given apparently pervasive opposition to intercultural adoption from the likes of UNICEF.

Finally, in the case of screening, the state could equalize the current disparity between very strict adoptive screening and comparatively lax screening for ART (Bartholet, 1995, 1999). This may require relaxing adoptive screening to the point where adoptions are facilitated, while simultaneously introducing greater screening into the ART process. In our view this would constitute a major improvement for both parental and children's rights. Note further that, while the state faces external constraints in terms of how it can affect the first two cases, it is by contrast entirely free to alter (or even dispose of) the process that screens adoptive parents. However, pragmatics suggest this option too may run into practical and political difficulties. All it takes is one case of severe child neglect or abuse by adoptive parents to derail any attempt at relaxing screening protocols. And we should also not underestimate the resistance of the cottage industry of adoption placement agencies or various government departments who have a direct interest in maintaining the status quo. The reality is that the state often finds it difficult to change existing practices. In light of these considerations, we argue the question of how to bring the current adoption reality to bear on the provision of ART remains pertinent.

To conclude, access to adoption by and large remains too complex and burdensome for the time being to be able to justify significant restrictions on the availability of ART for those prospective parents who cannot reproduce naturally. At a minimum, current circumstances seem to imply that the state should permit access to ART for those individuals or couples who can afford the cost. But this still leaves the important question whether ART should be more actively promoted, through the state (partially) subsidizing access to assisted reproduction.

Who Gets ART? Equal Treatment of Prospective Parents

The previous section set out a general case for determining under what conditions access to ART can legitimately be restricted by the state, against the normative background of each person having an intimate relationship interest in parenting that is sufficiently strong to ground a fundamental right. We endorsed in principle the argument that the existence of large numbers of children awaiting adoption operates as a legitimate constraint on providing access to ART. But we also qualified this argument by suggesting that the current reality of adoption policy is too burdensome for prospective adoptive parents, which implies that adoption should not be regarded as equivalent to ART when it comes to satisfying the fundamental right to parent. In particular, the extensive screening requirements the state currently imposes on prospective adoptive parents appears not only to drive a wedge between ART and adoption (as currently practised), but equally to fail to take all children's well-being seriously.[15]

The combination of these considerations appears to endorse a mixed policy in which the state can introduce a number of measures aimed at promoting adoption, but is not justified in prohibiting access to ART for those who prefer this option. Imagine for a moment that the state can somehow regulate access to ART to optimize for the level of adoptive children finding a home. This leaves us with a serious distributive problem: who is to receive access to ART as opposed to adoption? We can imagine a reasonable number of prospective parents will prefer adoption, in particular if the screening process is reformed to better accommodate the concerns mentioned in the previous section. But it is fair to say, given the sociology of adoption, that the number of individuals opting for ART, *ceteris paribus*, will outstrip what the adoption restriction would deem acceptable (Fisher, 2003). In short, some of those who want ART will get it, while others must either remain childless or consider adoption. In what follows we briefly examine two alternative ways of addressing the distribution problem: a licensing procedure (not unlike the screening of adoptive parents) and a pricing mechanism (ability to pay). In the first scenario, we screen out some prospective parents through a set of formal criteria (e.g. age or family composition); in the second scenario, we allow everyone to access the service but the state only partially subsidizes ART, thus effectively preventing those without sufficient means from using it.

---

[15] Elsewhere we describe this problem in terms of child–child trade-offs (De Wispelaere and Weinstock, 2012).

Allocation through Licensing?

Several considerations come into play when evaluating the use of a licensing scheme to allocate access to ART. Private providers and state regulators may want to deny access to ART to individuals or groups—e.g. to persons within a particular upper age limit—for medical reasons. For instance, the intervention may impose an unacceptable health risk or the treatment itself may be less effective at an advanced age, so that the efficient use of scarce resources suggests certain rationing criteria.[16] In addition, in a reversal of current policy, we can imagine access to ART being regulated through the imposition of certain social criteria, not unlike access to adoption today. Here we might think of advanced age not as a medical limit of access to ART, but as a social limit in a society that frowns upon older parents giving birth to a newborn (Bowman and Saunders, 1994).

It is hard to imagine a reason why access to ART should not at least be governed by the same criteria as adoption. To the extent that screening is motivated by ensuring parents are minimally competent (LaFollette, 1980, 2010), surely the same reasoning holds for prospective parents opting for ART. The question is whether the state can legitimately require prospective ART parents to satisfy stricter criteria of suitability as part of a policy that aims to optimize family placement for adoptive children. One argument refers to the fact that ART implies the creation of a new person, which entails a higher degree of responsibility on the part of those involved because of the ability to cause extensive harm (Benatar, 2006; Shiffrin, 1999). Intuitively this seems quite plausible. We have little reason to deny an adoptive child, who grew up in a succession of care homes, an adoption into a family of moderately competent and caring parents, for the simple reason that this will likely make her significantly better off than she would otherwise be. But in the case of the newborn ART child, we could (and perhaps should) be tempted to impose a higher standard, for the denial of ART does not imply the child reverts back to a worse-off situation but rather that she will never be born (which does not constitute a harm). It follows that a situation that would significantly improve the well-being of an adoptive child might nevertheless be judged insufficient to warrant bringing a new child into the world.

Despite its prima facie plausibility, we do not think that licensing schemes are an appropriate way in which to respond to the problem of how best to regulate access to ART. First, the licensing scheme under plausible empirical assumptions would violate an important ethical stricture, which holds that the state's policies should not embody (or express) the view that one method that infertile persons might employ in order to realize their fundamental right to parent is intrinsically superior to the other. Consider what would happen to a prospective parent or set of parents that were screened out by a licensing procedure restricting access to ART. If the rejected prospective parents

---

[16] The latter argument is more common where the treatment is publicly funded or part of an insurance package.

now opt for the adoption route because they still want a child to parent, this conveys the view that the adoption process is a mere *faute de mieux*, a second best to which one resorts when one has been unable to acquire the more "desirable" kind of child. The expressive harm for prospective and currently adopted children is pretty self-evident.

Second, a licensing scheme just isn't a solution to the problem that we have identified above. A licensing scheme, whether for adoption or ART, should be designed to select individuals that are deemed "good enough" to parent a child. Our concerns have been about the ways in which access to ART can be limited to ensure that all children who are available for adoption are actually adopted. Given the priority that the interests of such children have in our scheme (relative to the preferences that parents might have for children to whom they are biologically related), it is important in our view that access to ART not be too easy. But setting the screening criteria in a manner designed to generate the optimal mix of adoptive parents and ART parents involves something of a category mistake. It involves using a technique that is appropriate to screen out inadequate parents for the inappropriate purpose of achieving the desired numerical balance. The technical challenges that would be faced by the designers of such screen-ing practices would be daunting. But the main reason to reject such a method is ethi-cal: parents would be rejected on the basis of qualitative criteria (and thus given to believe that they were somehow not "good enough"), whereas the real reason for rejec-tion has to do with the availability of children for adoption. This time the expressive harm affects prospective parents.

Allocation through Pricing?

In the light of these difficulties, it seems like the solution might be to manage demand for rather than access to ART. This could be achieved through the use of an appropri-ately designed pricing mechanism. Let us start by mentioning at the outset one impor-tant advantage of this kind of mechanism. To the extent that access to ART is supposed to vary as a function of the supply of adoptive children (in order to accommodate both the right to parent and prospective adoptive children's interest in living in a family setting), a price mechanism seems a suitably fine-grained instrument to optimize the allocation of prospective parents to ART and adoption. The state could offer partial subsidies to those parents who have insufficient means to pay for ART, with the rate of subsidies increasing or decreasing in line with the effective demand for adoptive fami-lies. This option would appear to satisfy the right to parent (all prospective parents will have a genuine opportunity to become a parent), while also safeguarding the interest of adoptive children as much as possible, taking into account the practical constraints mentioned in the previous section.

Would this solution nevertheless be regarded as unacceptable because of the perva-sive social division it creates or maintains? This objection can be cashed out in differ-ent ways. We might object to the inequality of status expressed by reproduction now being directly associated with the distribution of wealth. But then similar associations are pervasive in societies permeated with vast social and economic divisions, which begs the question why we would specifically object to the use of prices for allocating this particular type of "good." Perhaps more seriously, we might think this solution constitutes a form of unfairness because the fate of adoptive children is a collective responsibility that is now being discharged by only part of society, a blatant case of freeriding—or so it would seem. But this objection only carries force to the extent that adoption is regarded as a burden rather than a genuine opportunity to experience the intimate relationship goods associated with parenting a child.[17] Third, when we regard adoptive parenting somehow as being of "inferior quality" because of the sort of child that is involved, the fact that only some can afford to opt for the "superior" reproductive option may constitute yet another form of unfairness. Leaving aside the particular circumstances of special needs children or international adoptions, the most common way in which adoption is regarded as a second-best reproductive solution is because the child one parents is not "one's own" in any biological sense.[18]

The response to this last objection is to treat the preference for ART over adoption as a private expensive taste (Dworkin, 1981) for a biological link between parent and child, the cost of which should be borne by those who pursue this option. The state incurs an obligation to safeguard the strong parental interest in raising a child by providing equal opportunity for all to become parents, but it is the opportunity for developing an intimate relation with a child that is the core interest at stake, not the biological connection. This means that unequal access to ART does not affect a core interest in cases where access to adoption is a reasonable alternative. Where this is not the case, as we discussed before, the state may be required to further facilitate access to ART by (partially) funding it, but only to the degree that all prospective parents' interests are satisfied, irrespective of whether this is achieved through natural reproduction, adoption, or ART.[19] Aside from its practical virtues, in contrast to the screening approach discussed earlier, such a pricing mechanism would have the additional advantage of conveying the "right" moral message. The state would in effect be saying that its responsibility is to ensure that all people are able to enjoy the goods of parenting. Those who desire to experience these goods through the parenting of a child to whom they are genetically related should be willing to pay for this expensive taste, rather than expecting society as a whole to subsidize it.

What about the fact that contemporary society promotes a norm that instils in us the expectation to value our biological children disproportionately (Bartholet, 1995,

---

[17] Parents might also value adoption for other (moral) reasons independent of the intimate relationship with the future child (see Rulli, in this volume).

[18] It should be noted here that those perceptions are very rarely shared by people who actually have first-hand experience of adopting a child. In addition, the views of reproductive clinicians and the wider community might vary considerably (Miall, 1996; Post, 1996).

[19] A refusal to adopt the view that ART constitutes a form of expensive taste implies either a denial of the fact that intimate parental relations represent the core interest motivating a right to ART, for instance by insisting that there is something special about the biological connection between parent and child (a possibility we ruled out in a previous section), or an affirmation that adoptive children themselves are somehow inferior.

1999)? Would such a norm not entail that people are rightly justified in understanding their parental interests more narrowly as being fundamentally concerned with bringing "a child of their own" into the world, be it through natural or assisted reproduction? The existence of such a norm seems to suggest the "expensive taste" for a biological child is not entirely private, but instead socially constituted. This in turn suggests "biological parenting" may be a legitimate object of state support after all, without having to rely on some form of naïve biologism or on casting adoptive children as inferior *per se*. An answer to this final objection begins by trading on the different perspectives of individuals and the state that is central to our chapter: while the social argument means we should not penalize (e.g. ostracize or tax) individuals who do nothing wrong by seeking to reproduce a biological child, it does not entail an obligation on the part of the state to actively support such a wish, given the strong countervailing interests of adoptive children.

Practically this means that, while the state is wrong to interfere with individuals seeking ART (or professionals offering ART), no strict requirement to subsidize such services exists. This position implies a form of inequality, but not objectionable injustice or unfairness, provided (1) each person has a real opportunity to secure the core interest of becoming a parent; (2) those who prefer to parent their own biological child are not actively prevented from doing so; (3) and those unable to reproduce naturally or via ART because of its expense are given access to a perfectly reasonable alternative, adoption. The added advantage of this approach is that adoptive children are given the best possible chance at securing their interest in being raised in a family. And last but not least, to our mind this solution constitutes a workable scheme that commits the state to minimal licensing or screening of parents while periodically adjusting the cost of ART in accordance with the number of children available for adoption (without any major overhaul of the rules and regulations governing ART or adoption).

Conclusion

In this chapter we examine whether the state has legitimate reasons to restrict prospective parents' access to ART. We outline what we believe to be the most promising starting point for justifying a right to reproductive assistance, through affirming individuals' fundamental right to becoming a parent in view of their interests in the non-substitutable relationship good that comes with parenting a child. For those individuals who are infertile, the state can promote their right to parent through two means: ART or adoption. Although the state in principle should adopt a neutral stance between either of these options, we argue this neutrality must give way once the central interests of adoptive children in finding a family are taken into account. The asymmetry between assisted reproduction, where no child's interests play a role before actual conception (or even birth) has taken place, and adoption, where children's interests are affected by any policy diminishing their chances of being adopted, clearly favours restricting access to ART.

However, we believe this conclusion is incomplete, for the presumption in favour of adoption over ART must be qualified by virtue of the fact that current adoption policies prevent prospective parents from easily securing their parenting interests through adoption. Several features of the contemporary context complicate the use of adoption to satisfy individuals' right to parent. Many parents will legitimately decide adoption is not equivalent to ART, or even feel adoption may not be a genuine alternative to natural reproduction. In response to these real constraints, the state may have to relax its opposition to ART, endorsing at least the permissibility of individuals seeking ART instead of adoption.

The further question whether the state should also actively promote ART by (partially) subsidizing access to expensive assisted reproduction for those who cannot afford it is a question complicated by distributive justice and fairness considerations between different prospective parents. In this chapter we briefly examine two mechanisms for allocating restricted access to ART. We find the first option—selective access through screening prospective parents for their suitability—inherently flawed. Instead we argue for the merits of a pricing mechanism which uses subsidy rates for ART as a tool that allocates prospective parents to ART and adoption through their ability and willingness to pay. We defend this option against the charge that this solution is objectionable for allowing rich infertile individuals to produce a "child of their own," while relegating poor infertile persons to taking on the responsibility of giving adoptees a family home. We argue that partial subsidy of ART indeed generates (or maintains) inequality but that such inequality is nevertheless not objectionable (as it does not violate people's fundamental right to parent) and offers a practical solution between the competing interests of infertile parents and adoptees.

Acknowledgements

An earlier version of this chapter was presented at the University of Wales, Newport, Dalhousie University, Halifax, and the GRIPP seminar at McGill University. We are grateful to the participants at these events for stimulating discussion, and to Françoise Baylis, Gideon Calder, Anca Gheaus, Mianna Lotz, Carolyn McLeod, Leticia Morales, Serena Olsaretti, Adam Swift, Tina Rulli, and Vardit Ravitsky for additional comments and suggestions.

References

Anderson, E. S. (1990) Is Women's Labor a Commodity? *Philosophy and Public Affairs*, 19(1), 71–92.

Appleton, S. F. (2004). Adoption in the Age of Reproductive Technology. *University of Chicago Legal Forum*, 393–451.

Ahlberg, J., and Brighouse, H. (2010). An Argument Against Cloning. *Canadian Journal of Philosophy*, 40(4), 539–66.

Bartholet, E. (1995). Beyond Biology: The Politics of Adoption and Reproduction. *Duke Journal of Gender Law and Policy*, 2(1), 5–13.

Bartholet, E. (1999). *Family Bonds: Adoption, Infertility, and the New World of Child Production.* Boston, MA: Beacon Press.

Bartholet, E. (2007). International Adoption: The Child's Story. *Georgia State University Law Review*, 24(2), 333–79.

Bartholet, E. (2010). International Adoption: The Human Rights Position. *Global Policy*, 1(1), 91–100.

Bartholet, E. (2011). International Adoption: A Way Forward. *New York Law School Review*, 55(2010/2011), 687–99.

Benatar, D. (2006). *Better Never to Have Been: The Harm of Coming into Existence.* Oxford: Oxford University Press.

Bowman, M. C., and Saunders, D. M. (1994). Community Attitudes to Maternal Age and Pregnancy After Assisted Reproductive Technology: Too Old at 50 Years? *Human Reproduction*, 9(1), 167–71.

Brighouse, H., and Swift, A. (2006). Parents' Rights and the Value of Family. *Ethics*, 117(1), 80–118.

Brock, D. (2002). Health Resource Allocation for Vulnerable Populations. In M. Danis, C. Clancy, and L. Churchill (eds), *Ethical Dimensions of Health Policy* (pp. 283–309). Oxford: Oxford University Press.

Brock, D. (2004). Ethical Issues in the Use of Cost Effectiveness Analysis for the Prioritization of Health Care Resources. In S. Anand, F. Peter, and A. Sen (eds), *Public Health, Ethics, and Equity* (pp. 201–24). Oxford: Oxford University Press.

Carlson, R. (2011). Seeking the Better Interests of Children with a New International Law of Adoption. *New York Law School Law Review*, 55(1), 733–79.

Catsanos, R., Rogers, W., and Lotz, M. (2011). The Ethics of Uterus Transplantation. *Bioethics*, 27(2), 65–73.

Cook, R. J., Dickens, B. M., and Fathalla, M. F. (2003). *Reproductive Health and Human Rights.* Oxford: Clarendon Press.

Cutas, D. (2009). Sex is Over-Rated: On the Right to Reproduce. *Human Fertility*, 12(1), 45–52.

De Wispelaere, J., and Weinstock, D. (2012). Licensing Parents to Protect our Children? *Ethics and Social Welfare*, 6(2), 195–205.

Dworkin, R. (1981). What is Equality? Equality of Resources. *Philosophy and Public Affairs*, 10(4), 283–345.

Fisher, A. P. (2003). Still "Not Quite as Good as Having your Own"? Toward a Sociology of Adoption. *Annual Review of Sociology*, 29(1), 335–61.

Friedrich, D. (2013) A Duty to Adopt? *Journal of Applied Philosophy*, 30(1), 25–39.

Gheaus, A. (2011). The Right to Parent One's Biological Baby. *Journal of Political Philosophy*, 20(4), 432–55.

Harris, J., and Holm, S. (1999). *The Future of Human Reproduction: Ethics, Choice, and Regulation.* Oxford: Oxford University Press.

Hughes, E. G., and Giacomini, M. (2001). Funding In Vitro Fertilization Treatment for Persistent Subfertility: The Pain and the Politics. *Fertility and Sterility*, 76(3), 431–42.

Jones, S. (2010). The Ethics of Intercountry Adoption: Why it Matters to Healthcare Providers and Bioethicists. *Bioethics*, 24(7), 358–64.

Kukla, R. (2005). *Mass Hysteria: Medicine, Culture, and Mothers' Bodies*. Lanham, MD: Rowman & Littlefield.

LaFollette, H. (1980). Licensing Parents. *Philosophy and Public Affairs*, 9(2), 182–97.

LaFollette, H. (2010). Licensing Parents Revisited. *Journal of Applied Philosophy*, 27(4), 327–43.

Levy, N., and Lotz, M. (2005). Reproductive Cloning and a (Kind of) Genetic Fallacy. *Bioethics*, 19(3), 232–50.

Miall, C. E. (1996). The Social Construction of Adoption: Clinical and Community Perspectives. *Family Relations*, 45(3), 309–17.

Mladovsky, P., and Sorensen, C. (2010). Public Financing of IVF: A Review of Policy Rationales. *Health Care Analysis*, 18(2), 113–128.

Neumann, P. J. (1997). Should Health Insurance Cover IVF? Issues and Options. *Journal of Health Politics, Policy, and Law*, 22(5), 1215–37.

Overall, C. (2012). *Why Have Children? The Ethical Debate*. Cambridge, MA: MIT Press.

Post, S. G. (1996). Reflections on Adoption Ethics. *Cambridge Quarterly of Healthcare Ethics*, 5(3), 430–9.

Robertson, J. A. (1994). *Children of Choice: Freedom and the New Reproductive Technologies*. Princeton, NJ: Princeton University Press.

Selman, P. (2012). The Global Decline of Intercountry Adoption: What Lies Ahead? *Social Policy and Society*, 11(3), 381–97.

Shiffrin, S. (1999). Wrongful Life, Procreative Responsibility, and the Significance of Harm. *Legal Theory*, 5(2), 117–48.

Søbirk Petersen, T. (2002). The Claim from Adoption. *Bioethics*, 16(4), 353–75.

Strong, C., Gingrich, J. R., and Kutteh, W. H. (2000). Ethics of Sperm Retrieval After Death or Persistent Vegetative State. *Human Reproduction*, 15(4), 739–45.

Thomson, J. J. (1971). A Defense of Abortion. *Philosophy and Public Affairs*, 1(1), 47–66.

Uniacke, S. (1987). In Vitro Fertilization and the Right to Reproduce. *Bioethics*, 1(3), 241–54.

Warnock, M. (2002). *Making Babies: Is There a Right to Have Children?* Oxford: Oxford University Press.

Widdows, H., and MacCallum, F. (2002). Disparities in Parenting Criteria: An Exploration of the Issues, Focussing on Adoption and Embryo Donation. *Journal of Medical Ethics*, 28(3), 139–42.

Woodhouse, B. B. (2005). Waiting for Loving: The Child's Fundamental Right to Adoption. *Capital University Law Review*, 34(2), 297–329.

8

"Not for the Faint of Heart"

Assessing the Status Quo on Adoption and Parental Licensing

Carolyn McLeod and Andrew Botterell

Introduction

The process of adopting a child is "not for the faint of heart." This is what we were told the first time that we, as a couple, began this process. Much of the challenge presented by making a family by means of adoption lies in fulfilling the licensing requirements for adoption, which, beyond the usual home study, can include mandatory participation in parenting classes. Overcoming these hurdles is not for the faint of heart, we discovered, because it is time-consuming and frustrating (not because it is intellectually challenging[1]). Prospective adoptive parents invest significant amounts of time and money without always having a clear sense of why the requirements are necessary or why they are imposed on adoptive parents alone.

In this chapter, we consider the question of whether licensing adoptive parents is morally justified. In brief, should the state have the right to screen prospective adoptive parents by subjecting them to potentially intrusive background checks and vetting? Should the state require that such individuals participate in educational programmes from which "natural parents" are exempt?[2] Should it have the right to deny prospective adoptive

---

[1] To the contrary, some parts of the process were intellectually numbing, in particular the parenting classes, which appeared to be directed at individuals who lack basic common sense concerning what children need.

[2] The term "natural parents" in the literature refers, it seems, to parents who are not currently licensed. The default assumption is that these people will care for their children properly—they are natural or naturally parents—and do not need to be licensed. Notice that the class of parents who are not now licensed includes not just biological parents, but also step-parents and the social (and predominantly heterosexual) parents of children created through gamete donation. Natural parenthood is therefore not equivalent to biological parenthood.

parents the opportunity to parent if it has concerns about their parental competency? In our view, while strong reasons exist in favour of licensing adoptive parents, these reasons are not unique to adoptive parents: they support the licensing not only of adoptive parents but of all or some subset of natural parents as well. We therefore conclude that the status quo with respect to parental licensing, according to which only adoptive parents need to be licensed, is morally unjustified.[3]

After discussing parental licensing in some detail (e.g. the nature of it, what actual systems of licensing are like), we turn to various arguments in favour of licensing adoptive parents. In our view, each of these arguments fails to justify licensing only these parents. Due to limitations of space, we do not survey the many arguments against licensing adoptive parents specifically or against licensing parents generally.[4] Thus, we cannot conclude that any parents, adoptive or otherwise, ought to be licensed. Rather, our goal is simply to challenge the common assumption that a licence to parent is morally justified in the case of adoption but not in the case of (assisted or unassisted) biological reproduction.

Licensing

Let us begin by doing three things: first, defining "parental licensing"; second, presenting a sample licensing scheme for adoptive parents (the one that we experienced); and third, describing what we take to be the status quo on licensing and adoption.

Parental licensing refers primarily to restrictions imposed on people's ability to be a parent *before* they become a parent in the social sense.[5] In other words, typically such licensing involves restrictions that are in the first instance prospective rather than retrospective. When the state interferes retrospectively with people's ability to parent—that is, only after they have become parents—such interference typically does not involve licensing people to parent.[6] The state normally interferes with natural parenthood only to this extent, for example, and only then if the parenting is disastrous

---

[3] We will go on to elaborate on this description of the status quo.

[4] These arguments are significant. In our view, the most significant among them rests on the claim that we cannot screen for bad parents, because we lack the tools needed to do so. Hence, we should not have parental licensing. The necessary tools include tests that can predict whether someone will satisfy a certain standard of parenting and good criteria for meeting this standard. There is disagreement in the literature about whether reliable tests that can predict whether someone will be a bad parent exist (see e.g. Mangel, 1988; Sandmire and Wald, 1990). There is also concern about whether we can devise appropriate criteria for being a good or decent parent (De Wispelaere and Weinstock, 2010).

[5] As Hugh LaFollette (1980: 188) notes, parental licensing, like all licensing, involves "prior restraint": "a parental licensing program would deny licenses to applicants judged to be incompetent even though they had never maltreated children." Similarly, we deny licences to drive cars to people who have never caused an accident. One has to show some competency in the area before being licensed.

[6] On the other hand, parental licensing requirements *could* be merely retrospective. As we go on to suggest, in a discussion about children with special needs, we can imagine situations in which retrospective restrictions alone function as a form of licensing.

(De Wispelaere and Weinstock, 2010). By contrast, most people who wish to adopt children must fulfil state-mandated requirements before taking a child into their care. The state licenses them to be parents.

Parental licensing schemes are therefore alike in that they primarily involve prospective restrictions on becoming a parent. Otherwise, these schemes can differ substantially. For example, they may or may not restrict parenthood retrospectively; that is, by requiring that families in which parents are licensed be monitored. (In our case, after adopting our son, we had four mandatory visits with our social worker, who then had to report her findings to our provincial government.) Systems of licensing can also vary in what sort of parenting they aim to exclude—disastrous, not good enough, or suboptimal parenting—which in turn should influence what kinds or degrees of restrictions they impose, *ex ante* or *ex post*. To illustrate these variations in kind and degree, respectively, consider that licensing need not involve mandatory parenting classes (these classes are a relatively recent requirement in our province, Ontario) and the frequency of monitoring families after they have been created can, of course, differ.

The general definition of parental licensing just offered identifies what is at the core of parental licensing; however, it does not explain why being licensed is often onerous for prospective parents. To gain a sense of why actual licensing schemes, those that target adoptive parents, may not be for the faint of heart, consider the scheme currently in place in Ontario.[7] The licensing system here is somewhat unusual only in that it requires prospective parents to take parenting classes. Anyone who is a resident of Ontario and who wishes to adopt a child domestically (either publicly or privately) or internationally must have a home study conducted and must complete parenting classes called PRIDE: Parent Resources for Information Development and Education.[8] The stated purpose of these requirements is to determine whether prospective adoptive parents are ready to become adoptive parents and to determine what sort of child they might be willing to adopt. The home study requires references and criminal background clearances. It includes details concerning one's personal and family background; significant people in one's life; one's family relationships; why one wants to adopt a child; one's expectations for the child; one's parenting skills; how one plans to integrate the child into one's family; one's current family environment; one's health, education, employment, and finances; and the social worker's recommendation. The home study must be updated if one's adoption is not completed within two years. An update is also required if one wants to adopt another child. Thus, one is licensed for one adoption at a time, and that adoption has to occur within two years.

---

[7] In our province, adoption is regulated by two pieces of legislation: the Child and Family Services Act, RSO 1990, c. C-11, and the Intercountry Adoption Act, 1998, SO 1998, c. 29.

[8] See http://www.children.gov.on.ca/htdocs/English/topics/adoption/how/index.aspx.

There are challenges in meeting the above requirements in terms of time, and often money and privacy. Collecting all of the data for the home study takes a significant amount of time. For example, one must schedule multiple meetings with a social worker, undergo a physical examination by a physician, and apply in person for criminal background checks. The parenting classes are also time-consuming: the total class time is twenty-eight to thirty-two hours.[9] The financial costs for PRIDE training are $700 per individual or $1,400 per couple, and for the home study, upwards of $2,500. These costs are borne by the prospective adoptive parents in the case of an international or private domestic adoption, while they are borne by the state in the case of a public domestic adoption. As we have noted, the investment of time or money in being licensed for adoption is frustrating when prospective parents lack a clear sense of why they alone have to meet these requirements. The process can also seem intrusive, given that one has to provide information about one's self and one's family that is normally deemed private (normally, that is, if one is a natural parent).

We have said that, according to the status quo, only adoptive parents have to satisfy licensing requirements such as the home study. However, this description of the status quo needs to be unpacked, in part because there are different forms of adoption and not all forms involve licensing. The key distinction for our purposes is between what we will call *family member adoption* and *non-family member adoption*.[10] As we take it, with few exceptions,[11] the status quo requires licensing only in the context of non-family member adoptions.[12] There are two types of family member adoption: step-parent adoptions, where an individual who has remarried wishes to adopt the children of his or her new spouse; and relative adoptions, which occur where parents can no longer care for a child and a relative wishes to adopt that child.[13] Focusing again on Ontario, where all parties—prospective adoptive parent, current parent, and child—reside in Ontario, the process of family member adoption entails simply submitting an application to an Ontario Family Court. No agency is involved and no licensing is required.

---

[9] See https://secure.adoptontario.ca/pride.main.aspx.

[10] This is the terminology used in Ontario. Other jurisdictions may employ different language to draw the same distinction.

[11] An exception in Ontario is a family member adoption where the child resides in another country; for explanation, see n. 14.

[12] The most common forms of non-family member adoptions are public or private domestic adoptions, where a private or a government agency, respectively, facilitates the adoption of a child residing in the same country as the prospective adoptive parents; and international or intercountry adoptions, where a private adoption agency licensed by the government of the prospective adoptive parents' country of residence assists in the adoption of a child residing in a foreign country.

[13] "Relative" is a legal concept. According to Ontario's legislation, a relative is a grandparent, uncle, or great-uncle, or an aunt, or great-aunt. Thus, for the purposes of relative adoption, a relative need not be related to the adoptive child by blood, since an individual could adopt, via relative adoption, the child of his or her spouse's sister.

In summary, when we refer to the status quo on licensing and adoption, we have in mind licensing only for non-family member adoption.[14] Our research tells us that this is the status quo in most Western jurisdictions, including Canada, Australia, New Zealand, the UK, and the USA.[15] Thus, we focus in what follows on whether there should be licensing for prospective *non-family member* adoptive parents, but not for other prospective parents. (For simplicity's sake, we call the parents who require licensing according to the status quo simply adoptive parents.) To be clear, our aim is to determine whether licensing of some form or other, for these parents alone, is morally justified; our goal is not to examine what form parental licensing should take. Thus, we consider arguments in favour of licensing adoptive parents that support different kinds of licensing schemes. Let us turn to those arguments now.

The Arguments in Favour of Licensing Adoptive Parents

We are interested in arguments in favour not of licensing parents generally, but of licensing adoptive parents specifically. An example of the former type of argument comes from Hugh LaFollette. He contends that the problem of disastrous parenting provides the state with a reason to license all parents (LaFollette, 1980). Others[16] would claim, by contrast, that because disastrous parenting is more likely to occur in adoptive families, the state is justified in licensing adoptive parents alone. This latter view is clearly specific to the adoption context. We consider four arguments that are intended to be specific in this regard. They are based either on (1) harm to prospective adoptive children; (2) the feasibility of licensing adoptive parents; (3) the transfer of parental responsibility that occurs during an adoption; or (4) the lack of a claim by prospective adoptive parents to a specific child. Our goal is to show that insofar as these arguments purport to be adoption-specific, they fail either because they are more general than they might initially appear to be and so do not support licensing only adoptive parents,

---

[14] Which is not to say that all family member adoptions are exempt from licensing requirements. For example, family member adoptions in which the child lives outside of Canada are treated in Ontario like international or intercounty adoptions. Similarly, if a prospective adoptive parent wishes to adopt a child to whom she is related in a way that is not captured by Ontario's legal definition of a relative, she must follow the same steps as for any other (non-family member) adoption. See http://www.children.gov.on.ca/htdocs/English/topics/adoption/how/index.aspx.

[15] One exception that we know of is the licensing in the UK for the intended parents of children created through embryo donation and contract pregnancy (see http://www.hfea.gov.uk/1424.html#5). Here, there is licensing with some assisted biological reproduction. (Also note that sometimes with assisted reproduction, parental screening is mandated not by the state, but by a fertility clinic or embryo "adoption" programme. See Krueger, 2011; Frith *et al.*, 2011.)

[16] By "others," we simply mean members of the general public, whom we think are likely to have the beliefs described that would support the view that disastrous parenting is more common among adoptive parents. An example is the belief that the lack of a biological tie signals a lack of natural affection.

or because they do not establish that licensing is justified for adoptive parents, and so do not justify the licensing requirement in the first place.

Harm to Children

The first of the above arguments revolves around concerns about harm to children. Some would argue that adopted children are at greater risk of harm from bad or incompetent parenting than are non-adopted children. Screening prospective adoptive parents is therefore important insofar as it reduces this risk. The reasons one might give in favour of such a view are twofold: (1) the lack of a biological tie between parent and child in an adoption increases the possibility of harm to the child; and (2) adopted children have special needs that not everyone is competent to satisfy; these children will be harmed unless the state ensures that their parents possess the relevant competence. Our analysis of each of these reasons shows that they fail to justify the status quo on parental licensing.

Why would the lack of a biological connection between parent and child increase the likelihood of abuse or neglect in the parent–child relationship? For many, the answer to this question is obvious: parents who are not biologically related to their children have no "natural affection" for them and so are more likely to harm them (LaFollette, 1980: 195). The thought here, presumably, is not that adoptive parents cannot feel affection for their children—that would be absurd—but that the affection is not instinctual in the way in which it supposedly is with biological parents. Such ideas explain why we have screening for adoptive parents in the first place, according to Elizabeth Bartholet (1993: 81).

However, the claim that adopted children are at heightened risk of abuse from their parents is unsubstantiated, as Bartholet and others, including LaFollette, explain (see Bartholet, 1993: ch. 8; LaFollette, 1980, 2010). In a recent paper in which he revisits the topic of parental licensing, LaFollette cites a study on the "importance of biological ties for parental investment" which concludes that adoptive parents are as invested in their children as biological parents are (LaFollette, 2010: 336; citing Hamilton et al., 2007). He also writes that adopted children "are less than half as likely to be maltreated compared to children reared by their biological parents" (LaFollette, 2010: 336). Additionally, in her work on the value of biological ties, Sally Haslanger (2009: 105) cites research showing that, compared to non-adoptees, adoptees "do not suffer in developing a core self" (i.e. a self imbued with worth, confidence, and the like), a result that would be unlikely if abuse from adoptive parents was more common than from non-adoptive parents.

However, perhaps there is insufficient evidence of adopted children being at heightened risk of abuse from their parents precisely because adoption licensing is effective at screening out people who would abuse their non-biological children. If people could simply walk into an adoption agency and walk out with a child, then abuse rates might be quite high within adoptive families. In responding to this concern, one must ask, first, whether the rates of abuse would be higher than they are with biological parents, and second, whether those higher rates would justify screening one group but not the other. In our view, one should not simply speculate about what the answers are. And regardless, one cannot assume that current screening of adoptive parents explains why levels of abuse by adoptive parents are low. The screening might not be effective. (It *is* not effective, according to Bartholet, and it *could not now* be effective, according to Michael Sandmire and Michael Wald, who argue that we lack reliable tools for predicting who will abuse children.) Thus, for all this argument says, it may be that people who want to adopt children are simply not likely to abuse them.

In short, arguments in favour of licensing based on harm to children and a lack of biological relatedness in adoptive families may not be empirically valid. But more important for our purposes is the observation that, even if such arguments were empirically valid, they would not be specific to the adoption context and so would not support the status quo. The arguments justify licensing for a larger group of parents than just adoptive ones.[17] The reason, of course, is that some non-biological (more specifically, non-genetic[18]) parents are not adoptive parents. Examples include parents whose children were conceived using donor gametes. If the lack of a biological tie signals a heightened risk of abuse and a need for parental screening, then these parents ought to be screened along with adoptive parents.

A different sort of argument in favour of licensing that is based on harm to children focuses not on the lack of a biological tie but rather on the special needs of adopted children and the harm that parents can cause if they are not competent to meet these needs. For example, David Archard maintains that, because adopted children "may present particular and possibly serious difficulties... arising from the fact that they have been rejected or abused by their natural parents," regulating adoption is justified (Archard, 1993: 146; cited in Engster, 2010: 253). (The suggestion that if the natural parents of children who are adopted have not abused these children then they have at least rejected them is seriously problematic, given that these parents may act in their child's best interests by giving them up for adoption. Nonetheless, some adopted children may *feel* rejected and have special needs as a result.) Adopted children could have special needs for reasons beyond having been abused or feeling abandoned by their natural parents. Some people—such as David Velleman (2005)—suggest that these children face special challenges insofar as they lack an ongoing connection with biological relatives (see Witt, in this volume). These same people oppose the donation of gametes for others' reproductive use because the practice deprives children born as a result of it of their biological identity. Others would worry that adopted children's lives are especially difficult by virtue of them having a non-biological (and possibly transracial) family in a culture that is biased against such families. As Bartholet

[17] And they support licensing for some adoptive parents who are not currently licensed, such as adoptive step-parents.

[18] Some maintain that the class of biological parents includes gestational but non-genetic parents who develop a biological connection with their children (i.e. the children they gestate) during the gestational process (e.g. Feldman, 1992). The argument we are considering now does not include these biological parents.

(1993: ch. 2) explains, our society has a "biologic bias;" that is, a bias in favour of biological families, which are presumed to be more "real" somehow than adoptive families. Licensing—especially if it includes an educational component—can serve to ensure that adoptive parents are competent to parent children with such needs.

Our response to the above argument is unsurprising: if children's special needs justify licensing adoptive parents, then they must also justify licensing some natural parents, namely those who have children with special needs. But while some children who are not adopted have special needs, we do not license their parents. Moreover, the needs of adopted children in general do not seem so dire that we need to ensure that *their* parents, but not the parents of other children with special needs, are competent to care for them. In our view, the point about special needs speaks in favour only of education for—not necessarily licensing of—adoptive parents. The education should also not be mandatory, at least not unless similar educational programmes are mandatory for other parents of children with special needs.[19] Finally, the educational programmes in question should not be driven by views about the needs of adopted children that support and entrench the biologic bias of our society. The claim that these children need ongoing contact with biological relatives in order to form a stable identity, *regardless* of whether they live in a society that privileges biological families over other families,[20] is arguably a case in point. We agree with others who suggest that such claims are purely speculative (Haslanger, 2009; Kolodny, 2010: 61 n. 33; Lotz, in this volume; Witt, in this volume).

In short, arguments based on harm to children that favour the licensing of adoptive parents cannot explain the status quo on parental licensing. For it is not at all clear that adopted children are at *special* risk of harm from parents who are either abusive or not competent to satisfy these children's needs.

Feasibility

Now consider a different argument in favour of licensing adoptive parents: while all parents ought in principle to be licensed, licensing adoptive parents is feasible or manageable in a way in which licensing natural parents is not. Because the latter would have too many negative effects, particularly on women and children, licensing adoptive parents is a "next best" solution. Daniel Engster (2010: 248) makes this sort of argument,[21] and in doing so emphasizes the "unequal impact" that licensing

---

[19] Some might worry that adoptive parents will not care enough about their children to seek out these educational opportunities. However, this objection is predicated on the view that adoptive parents lack affection for their children, which we have noted is unsubstantiated.

[20] The biologic bias might explain the need that some adopted children feel to know their biological relatives (see Haslanger, 2009: 113). Our society's response to this need (e.g. in the education we give to adoptive parents) should differ depending on whether we understand it to be a byproduct of a culture that stigmatizes adoptive families (Haslanger, 2009: 114).

[21] "Above all, the main reason for supporting a screening and licensing program for adoptive parents but not for biological parents is because of the greater viability of the former" (Engster, 2010: 253).

biological parents would have on women compared to men. For example, women without a licence to parent a child who experience an unplanned pregnancy would be forced to consider having an abortion or having a child with whom they were in a close physical relationship for nine months taken away from them (Engster, 2010: 247–8). To try to avoid this second outcome, they could try to hide their pregnancy, which would involve forgoing assistance from healthcare professionals during pregnancy and birth; but, of course, this option would pose serious risks to their health and to the health of their child (Engster, 2010: 247–8). On the other hand, men would not face choices as distressing as these as a result of parental licensing. Even after the birth of their child, their situation would not be the same as that of women who give birth. For example, given the unequal economic status between women and men, the choice between staying with a partner who is denied a parental licence and keeping one's child would not be as difficult, on average, for heterosexual men compared to heterosexual women (Engster, 2010: 248). In general, men leave women without experiencing as much financial hardship as women do when they leave men. For Engster (2010), the disproportionate burden that widespread parental licensing would have on women is too great. He also worries about harm that would befall children who are created but do not have a home to go to because their parents do not have a parenting licence and so lose custody of them at birth (Engster, 2010: 250).

We think the feasibility argument is very important; however, our complaint, again, is that the argument does not justify licensing adoptive parents alone. The reason is that parental licensing is as feasible for people who reproduce with assisted reproductive technologies (ARTs) as it is for people who create families by adopting children. To explain, consider the point made by Engster that if we were to license biological parents, then ideally applications for licenses would occur before women become pregnant. But there is an obvious problem with timing the applications prior to pregnancy: namely that "many pregnancies are unplanned" (Engster, 2010: 246). What Engster fails to notice, however, is that the problem of planning does not affect all biological parents equally. In particular, it does not arise with biological parents who reproduce using ART. (There will be no unplanned pregnancies in this group!) By Engster's own lights, then, licensing is an option for these biological parents, since the women involved would not have to make difficult decisions about abortion or about having to avoid pre natal care. In sum, the feasibility argument does not support the status quo, in which infertile people are free from scrutiny as prospective parents before they attempt technologically assisted conception. To the contrary, the feasibility argument encourages a system in which access to ART is available only to people with a parenting licence.[22]

---

[22] One could run a similar argument that focuses on procreative rights, rather than on feasibility. That is, one could insist that women's procreative rights prevent the state from interfering in their pregnancies, but not with their access to ART. This conclusion presumably follows if procreative rights are merely negative rights. We have a separate project in which we examine procreative rights and how they should influence the way the state treats biological parents compared to adoptive ones See our paper, "Can a Right to Reproduce Justify the Status Quo on Parental Licensing?" (forthcoming in S. Hannan, S. Brennan, and R. Vernon (eds), *Permissible Progeny*. New York: Oxford University Press).

Transfer of Responsibility

A third argument in favour of licensing adoptive parents hinges on certain important facts about the adoption context: before an adoption occurs there exists a child for whom someone—the state or an actual person—is responsible, and during an adoption, this responsibility is transferred to someone else, namely the child's adoptive parent(s).[23] The argument then proceeds as follows: the transfer of parental responsibility in an adoption ought to occur in a morally serious manner, but that can happen only if the party or parties relinquishing responsibility for the child can reasonably expect that the child's future will be good (or at least good enough). Licensing provides such assurance to the child's pre-adoptive guardians and is justified for this reason.[24]

Although we find the transfer of responsibility argument intriguing, we doubt that it supports the status quo on licensing. To see why, consider first that transfers of parental responsibility may not be unique to the adoption context; they might occur in certain assisted reproductive contexts as well. For example, David Benatar insists that such transfers happen whenever people donate their gametes for others' reproductive use.[25] He makes this claim in "The Unbearable Lightness of Bringing into Being" (Benatar, 1999), which provides the main impetus for the transfer of responsibility argument in favour of adoption licensing, as we have constructed it. Benatar argues as follows. In providing gametes to others for their reproductive use, one is transferring childrearing responsibilities that one has in any children "who result from" one's gametes (Benatar, 1999: 174). Moreover, transferring childrearing responsibilities is a morally serious business: no one should do it lightly. However, most gamete donors *do* do it lightly, according to Benatar. Typically, they do it anonymously and thus give little to no consideration to whether the people who will obtain their gametes are competent to parent a child. Hence, their actions are morally wrong. The "lightness" with which gamete donors allow beings created from their gametes to come into being is unbearable to Benatar. He does not insist that people who reproduce using donor gametes be licensed, which could allow gamete donors to transfer their responsibilities in a morally serious way. The transfer of responsibility argument that we have been considering would certainly support this conclusion, however. Thus, the argument extends to the licensing of gamete donors, assuming that Benatar is correct about their moral situation.

Another possibility (one that we think is more likely) is that transfers of parental responsibility occur in the context of contract pregnancy, which would mean that people

---

[23]   For ease of exposition, we will assume in the remainder of this section that responsibility is being transferred to two parents, rather than one.

[24]   The reasoning here is straightforward. If we think—as we surely do—that *delegating* responsibility for a child to someone else, such as a nanny, should occur in a morally serious manner—e.g. that we ought to do so only after having met the nanny and seen her references—then surely we ought to conclude that *transferring* responsibility for a child from a pre-adoption guardian to a post-adoption guardian (i.e. the child's adoptive parent) should occur in that way as well. A version of the transfer of responsibility argument can be found in Engster (2010: 253): he argues that the state has "a special responsibility to assure the biological parents of an adoptive child" that their child is well placed.

[25]   See Weinberg (2008) for a similar discussion about gamete donation and parental responsibility.

who become parents through this practice should also be licensed. For what it is worth, we find it unlikely that gamete donors transfer parental responsibility to the social parents of the children produced from their gametes, for we are not convinced by Benatar that they acquire any such responsibility.[26] However, we are more persuaded that women who gestate children for others assume parental responsibility that they must then transfer to the child's social parents at birth. The relationship formed between the pregnant woman and her foetus during gestation arguably grounds parenthood better than mere genetics.[27] Of course, so long as either a genetic connection or gestation gives rise to parental responsibilities, transfers of these responsibilities will occur outside of adoptions and within what has been called "third party reproduction." In addition, if the transfer of responsibility argument is sound, then some or all people who reproduce with the assistance of third parties will have to be licensed, just as adoptive parents are licensed. Again, the status quo finds little support within the argument we are considering.

One could further question whether the transfer of responsibility argument justifies the status quo on licensing by making the following observation. According to this argument, *transfers* of parental responsibility require the kind of scrutiny that licensing provides; however, a similar claim could also be made about *acquisitions* of parental responsibility.[28] Do we really think that more thought and information should go into transferring or delegating responsibility for children than goes into acquiring that responsibility in the first place?[29] If, as we think, the answer is "no," then the considerations that led us to accept parental licensing when there are transfers of parental responsibility should prompt us to accept it more generally.

Thus, it is not at all obvious that the transfer of responsibility argument can ground a system of licensing in which only adoptive parents are licensed. One could use the reasoning found in this argument to show that some (or all) people who have children as a result of third party reproduction need a parenting license. Moreover, one could use similar reasoning to prove that LaFollette is correct in thinking that all

[26] Benatar's (1999: 173) specific claim is that "[p]eople have a presumptive responsibility for rearing children who result from their gametes." He explains this responsibility in terms not of mere biology (i.e. of biological parenthood), but of the exercise of one's *reproductive autonomy* (Benatar, 1999: 174). Reproductive autonomy involves the freedom to make reproductive decisions, and with this freedom comes responsibility for—as Benatar (1999: 174) puts it—one's "reproductive conduct". He argues that if the conduct results in a child, then one is responsible for rearing this child, i.e. unless one transfers this responsibility to someone else. To apply this argument to gamete donors: their conduct involving their gametes results in a child, making them presumptively responsible for rearing this child; however, they defeat the presumption when they transfer the responsibility to the recipients of their gametes. The weakness in Benatar's argument for why gamete donors have this childrearing responsibility is his failure to spell out what counts as reproductive conduct. He seems to exclude donations of gonadal tissue from this category e.g. (even though gametes are part of the tissue; Benatar, 1999: 175), but gives no explanation for why these donations would not count, while donations of sperm and eggs do count, as "reproductive conduct."

[27] On the "grounds of parenthood," see Bayne and Kolers (2006).

[28] We owe this point to Reuven Brandt.

[29] Of course, we think that people ought to take the question of whether to bring a child into the world very seriously. We think it irresponsible of people to engage in unprotected sexual intercourse when there is a risk of pregnancy but no desire to create a child.

parents—whether they acquire their parental responsibility on their own or have others transfer it to them—should be licensed.

But it might be objected that, thus far, we have ignored an important fact about transfers of responsibility for children who are adopted: that these transfers inevitably involve the state. We have also ignored an important distinction: namely between the state's doing something and the state's allowing something to happen. In short, one might insist that when the state *does* something—namely transfer responsibility to adoptive parents—it acquires an obligation to ensure that those individuals are or will be minimally competent parents, an obligation that it discharges by the licensing requirement. On the other hand, when the state merely *allows* something to happen, as when it allows individuals to engage in procreative sex or to engage in third party reproduction, it acquires no such obligation, and so the licensing of those individuals is not necessary. It is this distinction—between the state's doing something and the state's merely allowing something to happen—that explains why licensing adoptive parents is permissible but licensing non-adoptive parents is not.

This is an interesting argument. All the same, in our view it can and should be resisted. For one thing, the doing/allowing distinction is notoriously slippery, and it is unclear how much normative weight it can bear.[30] But even if we are prepared to accept the distinction for present purposes, it by no means follows that appeal to it can justify the status quo. With this in mind, note that, because there are different ways in which transfers of responsibility occur in the adoption context, there are different ways in which the state can be implicated in those transfers. For example, in some cases responsibility for a prospective adopted child is transferred from the biological parents to the state, which then transfers responsibility for that child to the adoptive parents. In this sort of situation the state is, at some point, legally responsible for the child in its care and is therefore directly implicated in the transfer of responsibility. And it might be argued that it is this fact that justifies the state's licensing of adoptive parents: because the state is actively *doing something* by transferring *its* responsibility for a child to somebody else, it has an obligation to ensure that those prospective parents are minimally competent. Public domestic adoptions clearly conform to this model.[31]

The problem, however, is that not all cases of adoption involve the state assuming and transferring parental responsibility. For in many cases the state never assumes legal responsibility for the child, although it may play a role in overseeing the transfer of responsibility for the prospective adopted child from the biological parents or other legal guardians to the adoptive parents. In international adoptions, for example, two states are involved: the foreign state where the prospective adopted child resides and the domestic state where the prospective adoptive parents live. In an international adoption, however, responsibility for the prospective adopted child is never

---

[30] For discussion see Quinn (1989) and Thomson (1976).

[31] In such cases, responsibility for the child is initially transferred, sometimes voluntarily, sometimes involuntarily, from the biological parent to the state—which is to say that the child becomes a ward of the state—after which the state transfers that responsibility to the child's adoptive parents.

transferred to the domestic state. What the domestic state does is ensure that the foreign adoption process conforms to its own domestic laws and regulations and perhaps to universal conventions, such as the Hague Convention.[32] The domestic state signals its approval of this process by agreeing to issue various government documents, such as a visa and/or a citizenship certificate for the adopted child. Similarly, in the case of private domestic adoptions, the prospective adopted child never becomes a ward of the state, and so *the state* never assumes responsibility for the child, and so never transfers that responsibility to someone else.

What does the foregoing indicate about the transfer of responsibility argument? It shows that the argument fails to justify the status quo on parental licensing, if the normative principle underlying the argument is that, with adoption, the state inevitably *does* something normatively significant, namely transfer *its responsibility* for a child to somebody else. The problem is that only some adoptions, namely public domestic adoptions, are captured by this principle. Hence, the transfer of responsibility argument, so understood, can only justify licensing prospective parents in these adoptions.

Conversely, the claim that the state takes no responsibility or ought not to do so when it merely allows (assisted or unassisted) procreative activity to occur seems doubtful. The state *does* sometimes intervene in such contexts, as when it removes a child from a biological parent who has a history of child abuse or has previously had a child apprehended and removed from his or her care. Moreover, if the following is true—that transfers of responsibility occur in the context of contract pregnancy, that there is no relevant normative distinction between them and transfers that happen in private domestic or international adoptions, and that the state is prepared to require that prospective parents in these adoptions be licensed—then the state ought to require that people who become parents via contract pregnancy are licensed. But this point makes trouble for the transfer of responsibility argument framed using the doing/allowing distinction, since in cases of contract pregnancy the state merely allows a certain kind of procreative activity to occur. This way of framing the argument is therefore not persuasive.

Finally, consider the suggestion that the transfer of responsibility argument should be based instead on the following normative principle: that prospective adoptive parents should be licensed because the state, in *overseeing* the transfer of responsibility, has an obligation to ensure that children have parents who are minimally competent. Here again it is plain that the transfer of responsibility argument cannot justify the status quo, since this normative principle applies to adoptive parents who are currently exempt from the licensing requirements, such as step-parents, and also, arguably, to the social parents of children born through contract pregnancy. In addition, the principle must extend to the overseeing of *acquisitions* of parental responsibility (e.g. through procreative sex and pregnancy), if one accepts the moral equivalence of transfers and acquisitions of this responsibility (see above). In that case, the principle applies to all prospective parents, adoptive and non-adoptive alike.

[32] I.e. the Hague Convention of 29 May 1993 on Protection of Children and Co-operation in Respect of Intercountry Adoption. See http://www.hcch.net/index_en.php?act=conventions.text&cid=69.

In short, the problem with the transfer of responsibility argument is that its core normative principle results in an argument that is either underinclusive—since it does not apply to all adoptive parents—or overinclusive—since it applies to adoptive and non-adoptive parents equally. Either way, the argument does not justify the status quo.

No Claim to a Specific Child

The final argument we will consider in support of the status quo points to what is taken to be a unique feature of the situation of prospective adoptive parents: unlike people who reproduce with or without assistance, prospective adoptive parents do not have a legitimate claim to a particular child. That is to say, there is no child that they could possibly say is *theirs* before the adoption occurs. According to this argument, the same is not true of natural parents, not even those who reproduce with assistance from third parties. For example, people who initiate contract pregnancies can claim to be the parents, or at least to be included among the parents, of the resulting child. (They are what have been called the child's "intentional" parents. See Hill, 1991.) The argument then proceeds as follows. Natural parents have a legitimate complaint to make against a system of licensing that might deny them the opportunity to parent, given that the child it would prevent them from parenting is their own. The situation is different, however, for prospective adoptive parents, since licensing them does not ignore or interfere with any claim they have to a specific child (De Wispelaere and Weinstock, 2010). Licensing them is therefore permissible.

However, can the fact that prospective adoptive parents lack a claim to a specific child justify licensing them, and them alone? In our view, the answer is "no," for two reasons. First, the view that prospective adoptive parents lack a claim to a specific child is, for some adoptive parents, simply false. For example, in many jurisdictions, lesbian couples that have children via donor insemination or IVF are treated differently from heterosexual couples that have children in the same way. For while the lesbian partner who gives birth to the child is, as a result of that biological fact, considered to be the child's legal parent, the other lesbian partner is not, and must adopt the child to become its legal parent. (The law treats homosexual and heterosexual couples differently in this regard because of the parentage presumption: the legal principle that says where a heterosexual couple has a child, the husband is presumed to be the father of the child. See Crawford, in this volume.) Consequently, the situation of lesbian couples constitutes a case where a prospective adoptive parent *does* have a legitimate claim to a specific child, or where there is a specific child who *is* hers prior to the adoption. So our first objection is that the argument that prospective adoptive parents do not have a legitimate claim to a particular child is open to counter-examples.

But let us set that worry aside. For even supposing that prospective adoptive parents do not have a claim to a specific child, we cannot get from this fact an argument that licensing adoptive parents is justified. All that we can get from it is that a particular kind of complaint against licensing is not available to prospective adoptive parents: that the licensing might prevent them from parenting *their* child. Something other than the fact that there is no child that is in any sense theirs must explain why prospective adoptive parents must undergo screening. Consider that one could respond to prospective adoptive parents having no claim to a specific child by simply matching them with a child or by allowing them to pick one. That is, one could meet the challenge of the no-claim-to-a-specific-child argument without testing the parental competency of prospective adoptive parents at all. Thus, the conclusion that adoptive parents ought to be licensed cannot be based on the fact that there is no specific child to which they have an antecedent claim.

Yet perhaps one could strengthen the above argument in favour of licensing adoptive parents by adding the following premise to it: children who are not tethered to a particular adult should get good parents or even the best possible parents.[33] Indeed, it is arguable that the reason we do not ensure that all children have good parents or the best possible parents is that redistributing children at birth violates the rights of adults who have an interest in rearing their own biological children.[34] In other words, while we might think that all children are owed good parents, there are other interests that must be taken into account when children have biological parents who want to care for them.

The question that naturally arises, then, is whether the interest many people have in raising their biological children trumps the interest these children have in being raised by good parents. The answer must be "yes" for the strengthened argument set out above to work. Yet finding a good argument to support this conclusion is difficult. Like Harry Brighouse and Adam Swift (2006: 97–8), we are unaware of such an argument in the literature. But without one, it is impossible to say with confidence that the fact that natural parents have a claim to a specific child is what precludes the state from licensing them, while the fact that prospective adoptive parents have no such claim explains why they can legitimately be licensed.

Conclusion

In this chapter we have canvassed a number of arguments—the best we could think of—that are designed to establish that the status quo on parental licensing is justified. As we have tried to show, none of these arguments succeeds in this task, since all of

---

[33] One might add that children available for adoption are in short supply, so "why not select those who will do the best job," or at least a good job, at parenting them (Bartholet, 1993: 82)? Bartholet rightly responds that these children are not in short supply, given in particular "the untold millions" of children around the world who live in orphanages. (See also Rulli, in this volume.) If our experience of visiting orphanages in Africa is any indication of the current supply of prospective adopted children, then the supply is large indeed.

[34] In an influential paper on parents' rights, Harry Brighouse and Adam Swift (2006: 98) argue that redistributing children at birth can violate the rights of adults, but only those who "would be adequately good parents." Their view is not incompatible with licensing natural parents.

them support licensing for a larger group of parents than just (non-family member) adoptive parents. The relevant larger group is, for the arguments based on the risk of harm to children, either all non-biological parents or all parents who have children with special needs; for the feasibility argument, all people who reproduce using ART and all adoptive parents; and for the transfer of responsibility and no claim to a specific child arguments, most likely all parents.

We do not, and cannot, conclude that parental licensing ought to occur for any of the above groups. To do so, we would need to show that all arguments *against* licensing these parents are flawed, or are at least not as strong as arguments in favour of licensing them, and this is not something we have tried to do. We are also uncomfortable concluding that those who engage in unassisted reproduction should be licensed, because the feasibility argument speaks so strongly against licensing these people. On the other hand, from our perspective, it seems unfair to license only prospective adoptive parents and people who reproduce using ART. Thus, we are unsure what system of licensing, if any, we would endorse.

What we are fairly certain of, however, is that the differential treatment of adoptive and non-adoptive parents with respect to licensing is unjustified. Let us end by explaining why it is important to challenge this practice. A core issue, as we have noted, concerns fairness. Simply put, it strikes many as unfair that one class of prospective parents is subject to potentially intrusive licensing and screening requirements while other groups of prospective parents are not. Again, what is it about adoptive parents that make them, and them alone, subject to licensing requirements? Our answer, in a nutshell, is: nothing. A deeper concern, however, is the following. The current licensing regime serves to reinforce the belief that biological families are superior to (more natural, less likely to be dysfunctional, than) adoptive families; it promotes, in other words, the biologic bias. For this reason, requiring licensing for adoptive parents alone may in fact harm adopted children and their families, given that it expresses, either explicitly or implicitly, the view that these families are normatively suspect.

Acknowledgements

For their insightful comments on earlier versions of this chapter, we would like to thank the other authors in this volume as well as Reuven Brandt and Anthony Skelton. McLeod presented our argument at Trent University in their colloquium series, at the 4th International Conference on Adoption and Culture of the Alliance for the Study of Adoption and Culture, at the 2012 Congress of the International Network on Feminist Approaches to Bioethics, and in the Health Law and Policy Seminar Series at Dalhousie Law School. Thanks to the audiences at these venues for their helpful feedback. McLeod is also grateful for funding from the Graham and Gale Wright Fellowship program at Western University and the Canadian Institutes for Health Research (grant FRN 102516).

References

Archard, D. (1993). *Children: Rights and Childhood*. New York: Routledge.

Bartholet, E. (1993). *Family Bonds: Adoption, Infertility, and the New World of Child Protection*. Boston, MA: Beacon Press.

Bayne, T., and Kolers, A. (2003). Parenthood and Procreation. *Stanford Encyclopedia of Philosophy*. Retrieved June 2013 from http://plato.stanford.edu/archives/spr2003/entries/parenthood.

Benatar, D. (1999). The Unbearable Lightness of Bringing into Being. *Journal of Applied Philosophy*, 16(2), 173–80.

Brighouse, H., and Swift, A. (2006). Parents' Rights and the Value of the Family. *Ethics*, 117(1), 80–108.

Child and Family Services Act, RSO 1990, c. C-11 (1990). Retrieved July 2013 from http://www.e-laws.gov.on.ca/html/statutes/english/elaws_statutes_90c11_e.htm.

De Wispelaere, J., and Weinstock, D. (2010). Licensing Adoptive Parents. Presented in a panel on "Childish Subjects" at a meeting of the American Philosophical Association.

Engster, D. (2010). The Place of Parenting within a Liberal Theory of Justice. *Social Theory and Practice*, 36(2), 233–62.

Feldman, S. (1992). Multiple Biological Mothers: The Case for Gestation. *Journal of Social Philosophy*, 23(1), 98–104.

Frith, L, E., Blyth, M. P., and Berger, R. (2011). Conditional Embryo Relinquishment: Choosing to Relinquish Embryos for Family-Building through a Christian Embryo Adoption Programme. *Human Reproduction*, 26(12), 3327–38.

Hamilton, L., Cheng, S., and Powell, P. (2007). Adoptive Parents, Adaptive Parents: Evaluating the Importance of Biological Ties for Parental Investment. *American Sociological Review*, 72(1), 95–116.

Haslanger, S. (2009). Family, Ancestry and Self: What is the Moral Significance of Biological Ties? *Adoption and Culture* 2(1), 91–122.

Hill, J. L. (1991). What does it Mean to be a "Parent"? The Claims of Biology as the Basis for Parental Rights. *New York University Law Review*, 66(2), 353–420.

Intercountry Adoption Act, 1998, SO 1998, c. 29 (1998). Retrieved July 2013 from http://www.e-laws.gov.on.ca/html/statutes/english/elaws_statutes_98i29_e.htm.

Kolodny, N. (2010). Which Relationships Justify Partiality? The Case of Parents and Children. *Philosophy and Public Affairs*, 38(1), 37–75.

Krueger, A. (2011). An Overview of Embryo Donation in Canada. *Infertility Awareness Association of Canada*. Retrieved July 2013 from http://www.iaac.ca/en/584-269-an-overview-of-embryo-donation-in-canada-by-angela-krueger-spring-2011.

LaFollette, H. (1980). Licensing Parents. *Philosophy and Public Affairs*, 9(2), 183–97.

LaFollette, H. (2010). Licensing Parents Revisited. *Journal of Applied Philosophy*, 27(4), 327–43.

Mangel, C. (1988). Licensing Parents: How Feasible? *Family Law Quarterly*, 22(1), 17–39.

Quinn, W. (1989). Actions, Intentions, and Consequences: The Doctrine of Doing and Allowing. *Philosophical Review*, 98(3), 287–312.

Sandmire, M., and Wald, M. (1990). Licensing Parents: A Response to Claudia Mangel's Proposal. *Family Law Quarterly*, 24(1), 53–76.

Thomson, J. (1976). Killing, Letting Die, and the Trolley Problem. *The Monist*, 59(2), 204–17.

Velleman, D. (2005). Family History. *Philosophical Papers*, 34(3), 357–78.

Weinberg, R. (2008). The Moral Complexity of Sperm Donation. *Bioethics*, 22(3), 166–78.

9

On Non-Biological Maternity, or "My Daughter is Going to be a Father!"

Julie Crawford

This chapter takes its subtitle from my father's comment upon learning that my long-term same-sex partner, Liza, was pregnant. While many friends expressed outrage on my behalf ("But you're going to be a *mother!*"), I appreciated his attempt to acknowledge the intimate, originary, even quasi-biological, role I would play—indeed was already playing—in the life of our baby. While I was well aware of the extent to which motherhood is both mystified and considered the supreme role in the life of the child, and of the deep investments in both gender and biology that undergird such beliefs, I became curious about what happens when the gender function remains (to almost everyone, I would be a mother, not a father) but the biological is absent. Or rather, in my understanding of the process in which we were engaged, when the "biological" is dispersed or scattered across a range of registers and functions.

I was also deeply curious about what it meant that my originary, quasi-biological, father-analogous role was nonetheless irrelevant to the legal definition of parenthood in the country in which I live, and that despite our use of Assisted Reproductive Technology (ART)—technology that automatically affords non-genetic *fathers* the legal right of parenthood in a process that the legal theorist Elizabeth Bartholet calls "technologic adoption"—I, as a non-genetic *mother*, would be afforded no such automatic legal rights (1995: 5). Thus having been afforded the secondary status of "second parent," I was also, perforce, going to have to adopt the child for whom I was co-creator under precisely that aegis. (In the United States the process is called "Second Parent Adoption.") This chapter is thus my attempt to examine the epistemological status and ethical implications of parents who, like me, are at once procreative (more on this below) *and* adoptive parents. The case of non-biological lesbian mothers thus troubles the binary that scholars often posit between ART and adoption—work that understands these means primarily as alternative, and even opposed, methods of family-making. In addition, this chapter also has implications for thinking both about the parenting model on which ART is based and about the ethical implications of what Bartholet (1995: 9) calls the "restrictive regulatory regime" of adoption. Do those of us who become parents through *both* ART and adoption have a particular role to play in debates about alternative family-making? Does our petitioning for certain parental rights based on our procreative, rather than parental, intentions further cement the biologic bias of our society and further affirm the second-class status of adoption as a mode of family-making? Has the increasing normalizing of gay and lesbian life in the last twenty years affected how we, and society as a whole, understand the concept of "chosen family"? While one chapter cannot answer all of these questions, it can nonetheless help us think through the myriad issues and biases involved in lesbian family-making.

Biological Dispersal

Among other things, the category of the "other mother," "co-mother," "second mother," or "social or psychological mother" illustrates, with some poignancy, how much of the weight of biology in parenthood is borne by mothers. As the agents of biological reproduction, fathers and mothers are distinguished by gender as well as biological function, and while the father's role is essential to any story of baby-making, it is the mother's that is central. While my father's understanding of my parental role was informed by his sense of the structural significance and even necessity of fatherhood, it was also informed by his sense of the overwhelming biological responsibility borne by mothers—in this case, in his view, by Liza, and not by me. The "second" or "other" mother, as the chaser adjectives suggest, always follows along behind the main biological agent.

Historically, maternity has been far less easily dispersed, or spread out among a range of functions and actors, than paternity. While donor insemination (DI) has been used for centuries, it was not until recent advances in ART that more than one woman could play a role in the biological creation of a child. With egg retrieval, for example, lesbian mothers can create a child using one woman's egg and the other's womb (or, indeed, those of other women altogether). One could argue, however, that there has always been a more general form of maternal biological dispersal under the sign of gender. Indeed, in many ways, the naturalness—the biological weight—of maternity is based not only on the biological function of mothers *per se,* but also on the purportedly natural propensity all women have to mother. One of the many strange initiation ceremonies involved in becoming a parent is that you realize how many otherwise sensible people harbour strange and deeply naturalized views about gender and maternity: men, for example, who rhapsodize about the mystical and often impenetrable (to and by them) bond their children have with their mothers, wilfully ignoring the difference that their own comparative absence or disinvestment makes in the quality of parent–child attachment. At a recent conference I ran into a colleague at the end of the first day, pulling her suitcase behind her. "Are you leaving already?" I asked. She stopped briefly to wipe the sweat from her forehead. "I have to get home," she said. "You have no idea what it's like. You have *another mother* at home." When older women, or even some peers, are at our house they often marvel when one of us is, for example, feeding one child, and the other is getting the other child ready for bed: "Wow," they say, "this is what it would be like with two mothers." It is only (misguided?) politeness that keeps me from saying, "No, this is what it's like with two *parents*." To many people, all women are, at least to some extent, natural mothers, and motherhood is something for which we are always already suited. Biological childbirth is simply the most direct instantiation of a biological and cultural given.

There has been a lot of good press about lesbian parents recently—how well-adjusted, open-minded, successful, etc. our children are. (See e.g. the US National Longitudinal Lesbian Family Study (NLLFS) and the work of Bos *et al.*, 2008.) But despite the good PR this research provides, and the temptation it offers to claim lesbian *superiority*, the research itself suggests that the success of lesbian parents has less to do with their same-sexness or gender *per se* than with the quality and, crucially, the *equality* of their parenting. Any argument, however implicit, about the "obvious" or "natural" superiority of lesbian mothers is troubling on feminist grounds. A celebration of lesbian mothers that is predicated on their femaleness (as opposed to the quality and equality indicated above) relies on assumptions about biology and nature that are far more troubling to me than any comfort I might take in people valorizing my parenting—or, indeed, my sexual orientation. In a similarly double-edged sword, many lesbian mothers, particularly butch or otherwise masculine-identified ones, experience normative gender valorization for the first time in their lives through their maternity (see Lewin, 1993).[1] This valorization is often based on the heteronormative assumption that if there's a baby, heterosexuality must be close by, but it is also based in the assumption that maternity is the place where gender necessarily lines up with biology, even to the extent that biology can trump, or make invisible, the patently *non*-normative gender presentation of the woman in question. (Liza likes to point out that while her normally womanly figure never prevented other women from directing her short-haired, butch self in no uncertain terms *away* from the women's bathroom—"the men's room is *that* way!"—her pregnant belly and babe-in-arms solicited smiles and the welcoming holdings of doors to those very same bathrooms. Her maternity, that is, made her gender make sense.) If my father's response to our news was partly motivated by a recognition of the weight that women bear in biological reproduction, part of my appreciation for

---

[1] Corinne P. Hayden succinctly summarizes Ellen Lewin's (1993) book *Lesbian Mothers*: "By her own account exceeding the goal of her earlier work on maternal custody strategies—showing that lesbian mothers are 'just as good' as heterosexual mothers—Lewin finds that 'motherhood' in American culture constitutes a defining feature of womanhood that indeed supersedes the 'difference' of lesbian identity" (Hayden, 1995: 3).

his response was that its automatic default was not based in a mystified valorization of my biologically determined place in the dispersed, always-immanent maternity of womanhood, nor in a wilful reassertion of my partner's (or my own) rightful gender role. "My daughter is going to be a father" suggested that my desire to be a parent was a plan, not a given, and that its most meaningful volitional motivations came from some other place than sex or gender.

The ability many people have to see lesbians qua women as mothers, however, contrasts with their inability to see them as *reproductive couples*. Unlike heterosexual couples, lesbians receive no normative presumption of nor normative pressure towards reproduction. (While the question was constantly asked of my heterosexual brothers and their wives, no one ever asked me and Liza "when" we were going to have a baby.) Despite the dispersed maternity of women, then, the absence of a father prevents both the presumption of reproduction and the presumption of parenthood for lesbian mothers. This, once again, is one of the reasons I appreciated my father's response: it presumed the constitutive centrality of my role.

While I was neither a genetic nor a gestational mother, I was an integral part of my child's conception, in all the other senses of the term, from the very beginning. I was, in a very real sense—indeed, a definitional sense—a *procreator*: one who "begets, produces offspring"; in extended use, "one who brings into existence, who gives rise to" (*OED*, 2012). My partner and I chose to have a child together, and we chose the sperm provider—with all the quasi-eugenicist, biomedical, and legal implications that the process entails—together. Indeed our choice to use a clinic and an unknown gamete provider, rather than inseminate at home or in a clinic with a known donor, was a choice with both medical and legal implications. (In the American context, as I mentioned above, clinical gamete providers are not considered the legal father of the child, but known sperm donors are, or can be (see Nordqvist, 2011).[2] The clinical setting itself is thus a place of what Charis Thompson rather memorably calls "ontological choreography": a situation in which many aspects of conception of different ontological orders are separated out and brought into a carefully managed dance. In Petra Nordqvist's summary of Thompson's argument, "ontological choreography" ensures "that the potential transgression of traditional family, parenthood and kin discourse embedded in new reproductive technologies and assisted conception is carefully managed through a range of techniques which coordinate medical and scientific issues with the legal status of parenthood" (Nordqvist, 2011: 116). Above all, it helps to manage issues of gender and kinship. Such "ontological choreography," including the anonymization of the sperm donor, enables the social and legal construction of the couple pursuing clinically assisted conception as the proper parents of the conceived child. It also

[2] Clinical donors are not known to the couple as they undergo treatment, and such donors are not considered the legal father of the child (but children conceived after 1 Apr. 2005 can access information about their donor at the age of 18). However, for couples who self-arrange donor conception, the donor is considered in law to be the 'natural' father of the child, and as such, he can claim parental responsibility.

enables their recognition as the conceived child's *only* parents, "protecting the normative discourse that a child has (exclusively) two parents," and, crucially, that those parents are one man and one woman (Nordqvist, 2011: 118). Thus in this very crucial sense, my father's daughter was not, in fact, going to be a father—or even, at least in the eyes of the law, a parent at all.

Yet despite its ultimately restrictive outcome, the clinical setting is nonetheless a place in which biology is dispersed in very material ways. Indeed the whole process of ART, in which bodies, wombs, eggs, and gametes can come from a range of people with differing degrees of investment in the resulting child, can be understood as a practice of biological dispersal. The social scientist Katrina Hargreaves has examined families with children conceived using donor gametes in order to see what happens in a context that Helena Ragoné calls "collaborative reproduction": when there is "a group of procreators whose relationship to one another and to the child *is contained in the act of conception rather than in the family itself*" (Hargreaves, 2006: 262; emphasis added). This mode of reproduction, Hargreaves argues, illustrates the extent to which the natural facts that purportedly undergird social arrangements like family are themselves contingent and mutable, even constructed. Kinship of any kind is constituted by the division and combination of social and biological facts, but it really happens, Hargreaves argues, "at their intersection" (2006: 272). And the intersection of social and biological facts for lesbian mothers is a particularly busy one. It is an intersection, moreover, always arrived at by choice—by *willing* biology to meet sociality. There is no such thing, in other words, as an accidental lesbian pregnancy. The intersection where lesbian and gay kinship and biological reproduction meet is a place, as the critic Laura Mamo (2007: 225) puts it, where "sex without reproduction meets reproduction without sex"; where what Kath Weston (1991: 35) calls "nonprocreative sexual identities" procreate via (often) non-identity-based biology. It is a place where "chosen families" choose biology.

In some broad sense the intersection between gay and lesbian kinship and biology has always existed. As Weston pointed out in her ground-breaking book on gay and lesbian kinship, *Families we Choose*, the very idea of "chosen family" is only meaningful in the context of a powerful cultural belief in the significance of biological or blood ties. In her review of Weston's book, the anthropologist Marilyn Strathern points out that perhaps the fundamental critique enacted through chosen families is that they make "explicit the fact that there was always a choice as to whether or not biology is made the foundation of relationships" (1992b: 3). There is no singular notion of biology through which kinship is made, in other words, nor any fixed degree to which blood ties become relationships in a given family. Biology is at once foundational to kinship and its greatest variable. Yet David Schneider's (1984: 37–8) argument that kinship is a "folk theory of biological reproduction" organized around the potent symbol of heterosexual intercourse is nonetheless crucial to any understanding of the issues. (As Strathern puts it, "the relationship of the sexual act to conception is not… simply technical": 1992a: 4.) One could argue, for example, that my father's understanding of my parental role is haunted by the potent symbol of heterosexual intercourse; that my status as "father" marks the *conceptual and structural* presence of heterosexual intercourse even in its material absence—that "the father" must exist, because reproduction, and thus kinship, are legible in no other way.

The category of the "other mother," "co-mother," "second mother," or the "social" or "psychological" mother illustrates, again with some poignancy, precisely this epistemological crisis. In her research on lesbians who create families through donor insemination (DI), Corinne P. Hayden illustrates the extent to which lesbian mothers employ biology as an important symbol in their family-making via both the figurative and literal sharing of blood. Lesbians engage in a wide range of practices of what she, too, calls "dispersal of the biological tie" (Hayden, 1995: 45). They include choosing a donor from the non-biological mother's family (i.e. her brother); choosing a donor "whose physical characteristics in some way resemble" the non-biological mother; using the same donor gametes for multiple children; and, most metaphorically, hyphenating the children's names (Hayden, 1995: 50–53). Whether literal or implied, genetic continuity is an integral unifying resource for the lesbian families in her study, one that seeks, moreover, "to locate the metaphor of biological, generative power in the co-parent" (Hayden, 1995: 51). The "other mother" is thus invested with a central role in the child's creation—with authorship or "generative power"—in a way usually reserved for fathers. This authorizing role, however, is not simply analogous to that of the father; indeed Hayden's theory of authorizing generation corresponds in many ways to the "procreative" role I discussed earlier. (A *procreator*, in extended use, is "one who brings into existence, who gives rise to.") Lesbian generation, Hayden argues, is less a genetic concept *"than a kinetic one*; it is less an issue of the ownership of biogenetic substance than one of *placing this substance in motion, or being responsible for starting a process"* (Hayden, 1995: 52; emphasis added). In such a context, biology is abstracted and dispersed in a way that challenges the cultural assumption of the primary role of the male genitor or seed, placing the authority in the generative or procreative co-mother.

For many lesbians, moreover, the biogenetic substance is kept separate from the identity of the provider. As Petra Nordqvist argues, lesbians often undertake a psychic washing process in their use of DI, both when securing and processing the sperm itself, and in clinically—and thus legally—managing its ontological significance. (Making sure, in other words, that it does not run the risk of becoming a metonym for "the father.") Notably *unlike* lesbians, many women in heterosexual relationships who use DI see it as a kind of adultery, or see the sperm donor or provider as a third party in their relationship with their male partner (Weston, 1991: 171). Biological dispersal is thus not primarily the result of the DI technology itself, but rather of the generative, procreative involvement of the other parent. Lesbian mothers' strategies to gain symbolic legitimation for their families effectively disperse the biological connection as it has traditionally been conceived. But far from "depleting its symbolic capital," the dispersal of the biological tie in lesbian reproduction highlights its elasticity "within the symbolic matrix of American kinship" (Hayden, 1995: 45). Even as it informs the importance of biology in the process of family-making, it illustrates the ways in which a biogenetic connection is a contingent feature of relatedness rather than its immutable ontological grounds.

The Presumption of Parenthood

Yet no matter what forms of biological dispersal lesbian mothers practise in order to legitimate their families, the "ontological choreography" of American ART is, as I indicated at the beginning of this chapter, nonetheless conceived of as a dance for two people—and those two people are a man and a woman. Even if the genetic father is ultimately erased in the clinical biological dispersal of American ART, that is, the social father seamlessly takes his place. In the US, clinical ART is organized around one mother and one father per baby. A man and a woman can walk out of a given scene of "ontological choreography" as parents because the impregnated mother's male partner *automatically and legally* becomes the baby's father. (This is what Bartholet (1995: 5) calls "technologic adoption.") It is this reality that explains how someone like myself, a non-genetic but generative, procreative, and, in Hayden's terms, kinetic mother, must also become an adoptive one.

Regardless who is doing it, all the ontological choreography and biological dispersal involved in ART is part and parcel of what critics have called the "biologic bias" and pronatalism of our society (see Bartholet, 1993, for "biologic bias"). This bias conceives of and valorizes families in biological terms; it presumes not only that biological family-making is the norm, but that it is superior to other forms. As a number of chapters in this volume point out, this biologic bias has deeply affected both the status of adoption as a means of family-making, and how adoption itself works. Not only do adoption procedures favour two (heterosexual) parents, but they are organized in tight relation to biology, the (presumed) trauma of a child's separation from its biological parents, and, increasingly of late, to the rights of adopted children to know their genetic origins. (Increasingly, as Kimberly Leighton, 2013, points out, the biologic bias is also affecting the ways in which gamete donation is understood and regulated in the US, endowing donors with father status, and insisting on the rights of children conceived with donated gametes to know their biological origins.) Thus lesbians who choose ART are, like their heterosexual peers, participating in the biologic bias of family-making, and, it could be argued, reaffirming the supremacy of biological reproduction over adoption as a means of family-making. Yet the case of lesbian parents also reveals that the either/or argument of "alternative" family-making—that one can either use ART or adopt—is a chimera. Lesbians participate in the biologic bias of technological reproduction for similar reasons as heterosexuals, but they also avail themselves of the technology because its free-market ethos knows, as yet, little homophobic discrimination—at least on the technological side. ART, that is, presents far fewer barriers to lesbian couples who want to be parents than adoption does. ART does not discriminate on the basis of sexual orientation whereas adoption practices, both domestic and international, do. (Procedural and legal impediments, of course, vary from country to country and state to state.) Same-sex parents, that is, enjoy the privileges of a biologically biased, bionormatively driven, pronatalist, free-market, (relatively) unhomophobic, technoreproductive consumer society—but only to a certain extent.

Like heterosexual parents, lesbian parents can walk out of ART with a baby, and thus they automatically have certain rights *vis-à-vis* the baby they have created. (Although I had no legal parental rights, no one stopped me from accompanying my son home from the hospital.) Yet unlike American heterosexual parents who use ART, American lesbians do not have the rights of "technologic adoption": "arrangements that result in the social equivalent of either step-parent adoptions or full adoptions, where the child is produced in order to be raised by one or more parents who will not be genetically or biologically related," presumptions and processes which "do not involve the legal process required for such adoption" (Bartholet, 1995: 5). (Legal scholar Susan E. Dalton calls these "summary adoptions": 2001: 207.) For American lesbians, the "free market" of ART is only free in terms of achieving pregnancy; it's restricted in terms of enabling the full rights of parenthood. A child conceived with DI is not the child of its non-genetic procreative mother in the same way that it is the child of its non-genetic procreative father. (This, then, is yet another way in which I was not my son's father.)

Moreover, if I am not a biological mother, but rather part of a dispersed biological field of production—a player in an ontological choreography who nonetheless has no place in the legal structure of parenting that is its result—I am also not an adoptive mother in the way that most legislators and scholars think of the term. I am not an adoptive mother, crucially, in the positive sense that many philosophers promote: someone who shares social goods with someone whose existence pre-existed this sharing, and who already had needs before my adoption was able to answer them. My child did not "already exist" and have "material social and emotional needs" before I became his parent (Overall, in this volume). Indeed much of the writing about adoption simply does not apply in cases like mine. The claim that "before an adoption occurs there exists a child for whom someone—the state or an actual person—is responsible, and during an adoption this responsibility is transferred to someone else" is not true in my case (McLeod and Botterell, in this volume). Nor is the definition of adoption as "the legal placement of children who have been abandoned, relinquished or orphaned with an adoptive family" (Juffer and van Ijzendoorn, 2009). Nor is the claim that adoption "involves the transfer of a child from its birth parents to new parents with whom there is no biological link" (Bartholet, 1995: 7). Similarly, the claim that "prospective adoptive parents do not have a legitimate claim to a particular child. That is to say, there is no child that they could possibly say is *theirs* before the adoption occurs" (McLeod and Botterell, in this volume) is also patently untrue in cases like mine. (Lesbian ART also makes it clear that some people who become parents through ART *are* in fact licensed by the state, even if this licensing occurs after, rather than before, the material practice of parenting is underway. Lesbian ART thus serves as a counter-indicator to the claim that only adoptive parents are licensed, while those who use ART escape this aspect of the "restrictive regulatory regime" of adoption.)

Like *all* adopted children (as Overall crucially points out in this volume), my child meets the definition of a (hitherto presumptively biological) child as the "physical result and embodiment of a joint decision of the couple to commit, together, to the project of rearing and nurturing another human being" (Levy and Lotz, 2005: 246). And, despite its creepy connotations, I fit at least the latter part of the following claim (as would parents adopting through surrogacy): "For women, procreation, unlike adoption, provides the unique experiences of pregnancy and childbirth, *along with an immediate relationship with the future child that begins at conception*" (Overall, in this volume; emphasis added). My experience of adoption is simply not analogous to that of parents who adopt a child who already had an existence outside of their "conception." In no sense, in other words, was my son adopted in a way that accords with the first definition of "adopt" in the *Oxford English Dictionary* (2012): "To take (any one) voluntarily into any relationship (as *heir, son, father, friend, citizen*, etc.) *which he did not previously occupy*" (emphasis added).

In the United States, non-genetic procreative mothers must also be adoptive mothers because they do not enjoy the legal benefits of what is known as the "parentage presumption." The parentage presumption is a doctrine that presumes that a child born during a marriage is the child of the mother's husband (even, crucially, if that child was not his biological child). Indeed the parentage presumption continues to be curiously indifferent to issues of biology. Courts, as the legal theorist Jennifer L. Rosato (2006: 75) points out, "have even ignored accurate positive results of a paternity test, [continuing to] apply the presumption in situations where a husband finds out, through DNA testing, that the child he has been raising with his wife is not his biological child." In both the US and Canadian contexts, the parentage presumption includes children conceived by heterosexual parents through DI. The presumptive father has all the rights and responsibilities of a biological parent, including child support, custody, and visitation, and parental obligations are imposed regardless of whether the parents stay together, divorce, or move to another state. The presumption is based "on the well-established principle that children need the stability of two parents, and that *an individualized determination of paternity by a court should not be necessary*" (Rosato, 2006: 75; emphasis added). No licensing required.

Rosato (2006: 75) has made the "modest proposal" that "the children of same-sex couples need the parentage presumption that married couples receive as a matter of right." (In 2009, Ontario's Expert Panel on Infertility and Adoption affirmed precisely this right: "The birth mother, and a person with whom she shares a conjugal relationship, whether of the same or opposite sex, may jointly register the child's birth with a Vital Statistics registry showing themselves to be the child's parents. They do not have to go to court to get declarations of parentage": p. D4, para. 28.) Rosato argues that, although a parentage presumption through marriage provides the most comprehensive protection, the state does not need to legalize same-sex marriage to give the presumption efficacy. (The Canadians' use of the term "conjugal relationship" deftly avoids this dilemma.) At a minimum, she argues, children should receive the benefit of the parentage presumption as an integral part of the state's custody law or domestic partnerships/civil unions law. The presumption can be based "on the existence of the partners' committed relationship, or the quality of a partner's relationship with the biological parent's child" (Rosato, 2006: 75). It is the former case, of course, that informs the concerns of parents like me.

Unlike in Canada, in the United States the parentage presumption does not apply to children of same-sex couples as a matter of course. There are a few notable exceptions: if the couple is able to marry, or if the couple enters a domestic partnership or civil union in a state that recognizes a parentage presumption on these grounds. (Even then, however, the recognition of rights and responsibilities may not extend beyond the state's borders.) Rosato points out that the need for the parenting presumption is exacerbated by the fact that the number of children affected will increase alongside the increasing demand for ART, and by the fact that there has as yet been little widespread attempt to limit the access of gay and lesbian couples to ART. As such, she points out, there is "a tension" between such couples' seemingly unlimited access to ART and "the denial of the children's needs once they are born" (Rosato, 2006: 79). If we "allow these children to be created," as she rather infelicitously puts it, society has a

collective responsibility to provide them with the optimal environment for their physical and psychological development. The best way to protect these children would be for states to include a statutory-based parentage presumption that would provide the children of same-sex couples benefits *as a matter of right*. (Rosato, 2006: 79)

While some states have a "parentage presumption" that includes same-sex parents,[3] most do not, and this is where "second parent" adoption comes in. Until very recently, state statutes limited the right to adopt to individuals and to married couples adopting jointly. Given that adoption officially terminates the existing parent's legal rights, many states enacted a step-parent exception to this law, allowing heterosexual, legally married spouses to adopt their step-child(ren) (assuming the other biological parent had died or had relinquished his or her rights) *without* terminating the existing parents' rights. "Understanding that children of gay and lesbian couples deserve the legal protections of a two-parent home," as the National Center for Lesbian Rights (NCLR, 2003b) website puts it, some states began to allow what are known as "second parent adoptions." Much like step-parent adoptions, second parent adoptions permit the homosexual partner of an adoptive or biological parent to adopt without terminating

---

[3] An effective example of this approach has been adopted in Vermont, which created civil unions in 2000. That statute specifically states that "[t]he rights of parties to a civil union, with respect to a child of whom either becomes the natural parent during the term of the civil union, shall be the same as those of a married couple, with respect to a child of whom either spouse becomes the natural parent during the marriage" (cited in Rosato, 2006: 80).

the existing partner's rights. Today, about half of the states allow second-parent adoptions, either by statute or through case law.[4] (In February 2002, the American Academy of Pediatrics issued a policy statement supporting gay and lesbian second parent adoption (see HRC, 2013); on 4 February 2002 the *New York Times* carried the story on the front page: Haller, 2002. In August 2003, the American Bar Association approved a resolution to support laws and court decisions permitting second parent adoptions: Haller, 2002; NCLR, 2003a.)

Yet, despite this outpouring of support for second parent adoption, it has its problems. Dalton points out that the fact that the adoption system has a long history of dividing adoption petitioners into two mutually exclusive categories, married couples and single individuals, created an impasse for lesbian and gay couples, "technically two single adults" who have "a pressing need for access to joint adoptions without the benefit of marriage" (2001: 205). Thus state oversight or regulation of adoption resulted in the creation of what Dalton calls a three-tiered adoption hierarchy. At the top of the adoption hierarchy are men who are married to women who give birth to children within the context of their marital relationships. At the second tier of the adoption hierarchy are men and women who marry partners who have children from previous relationships and make use of step-parent adoption procedures. At the third tier are individuals who adopt children from outside their immediate families: i.e. the independent adoption, which differs from both summary and step-parent adoptions. While states generally do not get involved in summary adoptions at all and are involved in step-parent adoption only minimally, they are very involved in independent adoptions (e.g. by conducting an in-depth investigation into the adopting adults' home and family lives).

The case of gay and lesbian parents has resulted in what Dalton calls a "fourth tier" in the adoption hierarchy: second-parent adoption, which combines the procedures of the step-parent adoption with the independent adoption. Like step-parent adoption, second-parent adoption allows a parent to extend their parental right to another adult, but adults seeking second-parent adoptions are subjected to the same in-depth investigation as adults seeking independent adoptions. Unlike other forms of adoption, moreover, which are clearly spelled out in state adoption statutes, second-parent adoptions have received no formal support from state legislatures. Same-sex couples seeking to build two-parent families through adoption bear an extra burden of proof

---

[4] Currently, according to the NCLR (2012), appellate courts in the following states have approved second-parent adoptions: California, the District of Columbia, Illinois, Indiana, Massachusetts, Pennsylvania, New York, New Jersey, and Vermont. Trial court judges in one or more counties of the following states have also granted second-parent adoptions: Alabama, Alaska, Delaware, Hawaii, Iowa, Louisiana, Maryland, Minnesota, Nevada, New Mexico, Oregon, Rhode Island, Texas, and Washington. California, Connecticut, and Vermont have statutes expressly permitting second-parent adoptions. Appellate courts in the following states have held that second-parent adoptions are not permissible under their respective adoption statutes: Nebraska, Ohio, Wisconsin, and Colorado. Mississippi and Utah have laws that effectively ban second-parent adoptions, while Florida prohibits gay men or lesbians from adopting children under any circumstances (NCLR, 2012). See also Dalton (2001).

and an extra load of labour: they must convince the judges hearing their cases that the proposed adoption is both in the best interest of their children and permissible under that state's law. In addition, the legal rights and protections that second-parent adoptions provide are "remarkably limited, especially when compared to those afforded heterosexual couples through marriage" (Dalton, 2001: 212). The parents remain "legal strangers" to one another; outside of Vermont, Massachusetts, and New York, no one really knows if any given second-parent adoptions will withstand a court challenge; it is a costly way to make a two-parent family (marriage certificates cost $50 while second-parent adoptions typically cost $4,000–$6,000—per child); it subjects families to substantial scrutiny by the adoption system; it has the potential to delimit the parental rights of those lesbians who do *not* obtain adoptive status; it is underpublicized so as not to attract the attention of those who might challenge it in the courts, thus limiting its ability to offer legal protection to those it seeks to protect (Dalton, 2001: 212–15). Finally, the non-genetic mothers seeking second-parent status are vulnerable during the ten months it commonly takes to complete a second-parent adoption. (Dalton quotes one mother's response to the waiting period: "if [my partner] had dropped dead or been hit by a truck, her parents could have broken up our family": 2001: 215.) What second-parent adoption provides, in other words, are all the restrictions of the "restrictive regulatory regime" of modern adoption, plus homophobia.

As it does for Rosato, it seems clear to me that the best way to address the problems with second-parent adoption is to press for the parentage presumption to be extended to same-sex parents. With such an extension, same-sex parents would not have to seek individualized determination of parentage by a court, and thus would not be subject to state scrutiny and unfair economic burdens. Relying on evidence of same-sex "partners' committed relationship" as the grounds for the parentage presumption certainly reaffirms a normative two-parent model of family-making. (As Michael Warner [2000] points out, the trouble with normal is that it insistently demotes all other kinds of family formations.) But the existence of the parentage presumption would nonetheless keep same-sex parents from helping to promote a normative (and punitive) model of family values that defines "good" lesbians as those who get married or obtain second-parent adoptions and "bad" lesbians as those who do not. The rights conferred through the parentage presumption are based in the child's needs, that is, rather than in marriage as an institution.

Yet at the same time, the current situation in which non-genetic lesbian mothers find themselves—welcome to make use of ART but subject to the restrictive regulatory regime of adoption—highlights precisely the problems addressed in this volume. That is, while cases like mine may well reveal that ART and adoption are not always *alternative* means of family-making, they nonetheless illustrate the extent to which the restrictions of the current system of adoption are based on the perceived distance a prospective parent is from the central myth of biological kinship. This leaves me in an ethical dilemma: to press for the "parentage presumption" on the basis of my own biologically dispersed, procreative, generative, intentional maternity thus seems potentially to work against the advocacy of adoption rights that are based in something *other than* biology. In what way are my "procreative" role and "intentions" substantively different from those of any other adoptive mother? Perhaps my indignation is best directed at the restrictions of adoption *per se* rather than that of "cases like mine." Second-parent adoption is a burden not only because of homophobia, but because of the intense regulation of adoption of any kind in America. At the most fundamental level, such regulation presumes that those who cannot reproduce "naturally" must prove their fitness through other means. It thus not only reaffirms the biologic bias, but continues to put the burden of proof on chosen families. In a historical moment in which gays and lesbians are increasingly appealing to, and receiving rights from, the state, I for one don't want to see those rights come at the expense of activism on behalf of more equitably distributed rights for the forms of alternative family-making that many Americans, queer and otherwise, have been engaged in for many years.

References

American Civil Liberties Union (ACLU) (1998). In the Child's Best Interests: Defending Fair and Sensible Adoption Policies, 30 Apr. Retrieved June 2013 from http://www.aclu.org/lgbt-rights_hiv-aids/childs-best-interests-defending-fair-and-sensible-adoption-policies.

Bartholet, E. (1993). *Family Bonds: Adoption and the Politics of Parenting*. Boston, MA and New York: Houghton Mifflin Co.

Bartholet, E. (1995). Beyond Biology: The Politics of Adoption and Reproduction. *Duke Journal of Gender Law and Policy*, 2(1), 5–14.

Bos, H. M. W., Gartrell, N. K., van Balen, F., and Peyser, H. (2008). Children in Planned Lesbian Families: A Cross-Cultural Comparison between the United States and the Netherlands. *American Journal of Orthopsychiatry*, 78(2), 211–19.

Dalton, S. E. (2001). Protecting our Parent–Child Relationships: Understanding the Strengths and Weaknesses of Second-Parent Adoption. In M. Bernstein and R. Reimann (eds), *Queer Families, Queer Politics: Challenging Culture and the State* (pp. 201–20). New York: Columbia University Press.

Expert Panel on Infertility and Adoption, Ontario, Canada (2009). Raising Expectations: Recommendations of the Expert Panel on Infertility and Adoption. Retrieved June 2013 from http://www.children.gov.on.ca/htdocs/English/infertility/report/index.aspx.

Haller, K. (2002). The American Academy of Pediatrics Coparent or Second-Parent Adoption by Same-Sex Parents Policy Statement: Its Science, its Implications. *Journal of the Gay and Lesbian Medical Association*, 6(1), 29–32.

Hargreaves, K. (2006). Constructing Families and Kinship through Donor Insemination. *Sociology of Health and Illness*, 28(3), 261–83.

Hargreaves, K., and Daniels, K. R. (2007). Parents' Dilemmas in Sharing Donor Insemination Conception Stories with their Children. *Children and Society*, 21(21), 420–31.

Hayden, C. (1995). Gender, Genetics, and Generation: Reformulating Biology in Lesbian Kinship. *Cultural Anthropology*, 10(1), 41–63.

HRC (Human Rights Campaign). Professional organizations on LGBT parenting. Retrieved June 2013 from http://www.hrc.org/resources/entry/professional-organizations-on-l gbt-parenting.

Juffer, F., and van Ijzendoorn, M. H. (2009). International Adoption Comes of Age: Development of International Adoptees from a Longitudinal and Meta-Analytic Perspective. In G. M. Wrobel and E. Neil (eds), *International Advances in Adoption Research for Practice* (pp. 169–92). New York: Wiley.

Leighton, K. (2013) To Criticize the Right to Know we Must Question the Value of Genetic Relatedness. *American Journal of Bioethics*, 13(5), 54–6.

Levy, N., and Lotz, M. (2005) Reproductive Cloning and a (Kind of) Genetic Fallacy. *Bioethics*, 19(3), 232–50.

Lewin, E. (1993). *Lesbian Mothers: Accounts of Gender in American Culture*. Ithaca, NY: Cornell University Press.

Mamo, L. (2007). *Queering Reproduction: Achieving Pregnancy in the Age of Technoscience*. Durham, NC: Duke University Press.

NCLR (National Center for Lesbian Rights) (2003a). American Bar Association Votes to Support Second Parent Adoptions. Retrieved June 2013 from http://www.nclrights.org/site/ PageServer?pagename=press_2parentadopt081303.

NCLR (National Center for Lesbian Rights) (2003b). Second Parent Adoptions: A Snapshot of Current Law. Retrieved June 2013 from http://cdm15025.contentdm.oclc.org/cdm/singleitem/ collection/p266301coll9/id/46/rec/12.

NCLR (National Center for Lesbian Rights) (2012). Adoption by LGBT Parents. Retrieved June 2013 from http://www.nclrights.org/site/DocServer/2PA_state_list.pdf.

Nordqvist, P. (2011). "Dealing with Sperm": Comparing Lesbians' Clinical and Non-Clinical Donor Conception Processes. *Sociology of Health and Illness*, 33(1), 114–29.

Oxford English Dictionary (OED) (2012). Retrieved June 2013 from http://www.oed.com.

Ragoné H. (2000). Of Likeness and Difference: How Race is Being Transfigured by Gestational Surrogacy. In H. Ragoné and F. Winddance Twine (eds), *Ideologies and Technologies of Motherhood* (pp. 56–76). New York and London: Routledge.

Rosato, J. L. (2006). Children of Same-Sex Parents Deserve the Security Blanket of the Parentage Presumption. *Family Court Review*, 44(1), 74–86.

Schneider, D. (1984). *A Critique of the Study of Kinship*. Ann Arbor, MI: University of Michigan Press.

Strathern, M. (1992a). *After Nature: English Kinship in the Late Twentieth Century*. Cambridge: Cambridge University Press.

Strathern, M. (1992b). *Reproducing the Future*. New York: Routledge.

Thompson, C. (2005). *Making Parents: The Ontological Choreography of Reproductive Technologies*. Cambridge, MA: MIT Press.

Vanfraussen, K., Ponjaert-Kristoffersen, I., and Brewayes, A. (2003). Family Functioning in Lesbian Families Created by Donor Insemination. *American Journal of Orthopsychiatry*, 73(1), 78–90.

Warner, M. (2000). *The Trouble with Normal: Sex, Politics, and the Ethics of Queer Life*. Cambridge, MA: Harvard University Press.

Weston, K. (1991). *Families we Choose: Lesbians, Gays, Kinship*. New York: Columbia University Press.

PART V

Special Responsibilities of Parents

10

Special Responsibilities of Parents Using Technologically Assisted Reproduction

Jamie Lindemann Nelson

Parental Responsibilities

Quite apart from ARTs, a parent's standard obligations to preserve, nurture, and socialize her or his children are "special" as a matter of course, contrasted with the general run of human moral responsibilities.[1] Most obviously, perhaps, parental obligations are special in that they aren't owed to all morally considerable beings indiscriminately. A person is not generally taken to be morally remiss if she'd rather write her book than raise whatever unclaimed children there might be about. Or if she'd rather buy a book than send that potentially life-saving amount of money to Oxfam, most would hold her blameless for her purchase.[2]

Further, parental responsibilities are not merely what theorists tend to call "positive." Most positive obligations—i.e. those that direct people to act in support of others rather than just to refrain from harming or wronging them—are regarded as limited in scope and content. On many accounts, positive obligations are usually acquired by specific, uncoerced agreements made with others, or by free choices to occupy a certain role and to discharge its associated duties. It is also commonly thought that a person may find herself obliged to assist others if such help is vital to the recipient and close to costless to the provider. But such ideas don't seem to fit parental responsibilities. One doesn't, after all, sign contracts with babies or otherwise bargain with them, and a parent's duties to her children go considerably beyond what's demanded by "easy rescue."

---

[1] I'm here accepting Sara Ruddick's summary of parental duties as the preservation, nurturing, and socialization of one's children, as developed in her (1989).

[2] I acknowledge that this claim this controversial in the context of moral theory, as work by Peter Unger (1996) and Peter Singer (e.g. 2009) attest. I maintain only that it is not controversial as a matter of common practice.

It is more plausible to see parental responsibilities as based on the voluntary adoption of a role, but only marginally. If I'm a doctor or a firefighter, I have special obligations to my patients or to those threatened by fire who live in my district. But I can at least in principle resign my position if I decide that, say, I'd really rather pursue a career as a dancer, and taking on the job in the first place was presumably a matter of my own choice. The entrance and exit conditions for parents seem rather more complex.

Consider that some people are taken aback by the discovery that they are about to be fathers or mothers—indeed, they may have made some decided efforts to avoid exactly that result. Yet to think that a woman who had used multiple methods of birth control could find herself pregnant, deliver a child, and merit no moral criticism if she simply walks away, leaving the infant to its fate, strains ordinary understandings of what people are bound to do for one another. Among male progenitors, sadly, something pretty similar to "simply walking away" is not as rare as it ought to be, nor are cases of newborn abandonment by both parents unheard of. Yet those who do abandon their children are hardly admired for it. I take this common moral response to suggest that, as ordinarily understood, parental responsibilities can be incurred otherwise than by deciding to assume them.

Children are, of course, sometimes relinquished into the custody of competent willing caregivers; explicit undertakings are surely one road to becoming a parent, and they can be one way out, too. Relinquishment is, however, a far cry from abandonment. Typically, it's the children's own interests that are key to justifying transfer of responsibility. While there are many ways for parents to be involved responsibly in rearing their children, and it isn't impossible to imagine circumstances where a parent's flatly abandoning her or his children might be defensible, you need powerful reasons for doing it if you are to escape with any credit at all. Many—perhaps most—would regard an otherwise capable person who shrugged off her or his parental responsibilities to become a dancer as having done something seriously wrong. Whatever parenthood is, it isn't a contractual arrangement entered into by roughly equal parties, whose responsibilities to each other can be unilaterally dissolved.

Parenthood is not the only case in which people are generally thought to have binding responsibilities to others that do not necessarily hinge on agreements. It may be telling, though, that many of the most plausible examples of such "nonconsensual duties"—responsibilities of children to their parents, for example, or moral ties between siblings—are also from family settings. None of these relationships need be considered absolutely indissoluble; their attendant responsibilities are not indefeasible. Yet despite the fact that they do not necessarily rest on contract, nor consent, they are extensive, enduring, focused on specific individuals, and presumptively binding.

At the same time, parents have wide latitude with many aspects of their children's physical, social, cultural, and intellectual welfare. While there are general limits on defensible forms of nurturing and socializing children, and reams of solicited and unsolicited advice, there are also many accepted ways of nurturing them, and bewilderingly many ways of socializing them. If "preservation" is taken to encompass considerations of health and health risks, as well as simply striving to keep one's kids alive, parents have fair discretion even here. As seems true of the source of parental duties, the justification for this latitude is not altogether plain. As I'll develop in this chapter, the range of parental discretion is not in practice constrained by a "best interests of the child" principle. Parents routinely make decisions involving their children that not only do not advance their interests, but also put them at some risk of bad outcomes, for reasons that can include gratifying rather minor parental desires.

If parents have wide discretion over *what* they may do, *why* they may do it is a matter where less laxity is permitted. Parents are expected to love their children, and to live their lives in ways that, at least to some extent, are structured by and reflect that love. It's widely thought as well that parents ought to convey to their kids that they are cherished for the particular individuals they are, and not simply respected as morally considerable individuals among others.

Parental responsibilities, then, seem special as a matter of routine. In sum, they are owed to specific people, persist for years and make weighty demands, can be incurred without explicit agreement and against intention, are difficult to relinquish, allow wide discretion in how they are discharged, and are to be motivated by love.

Yet this is just a start at roughing out and making sense of the broad structure of parental responsibilities. If not exclusively through consent, how does a child wind up being a given adult's responsibility in the first place? Why is it biological progenitors to whom parental responsibilities are presumptively assigned, at least initially? Just how is the extent of parental responsibilities to be weighted against the range of parental discretion? How much freedom do parents have to gratify their own interests in how they raise their children? How do their responsibilities to their offspring stack up against each other? What risks, for example, may responsible parents take with their child's preservation in order to nurture or socialize them in a fashion that seems good to those parents? What risks to their children may they run to have children at all?

Matters get even more vexed when mothers and fathers try to discern the specific content of what they must do in order to be a good or "good-enough" preserver, nurturer, and socializer to the very particular children they end up having. All parents face the possibility that the challenges involved in performing their responsibilities will outrun what their own previous experience or present resources have equipped them to handle readily; they may suddenly find themselves confronting needs or interests that may be particularly difficult or even simply beyond their ability to satisfy.

ARTs and Parental Responsibilities

Here is where the issue starts to be joined between these general considerations and assisted reproductive technologies. For example, uses of ARTs that involve gametes provided by people who are not inclined or expected to take up parental responsibilities are reasonably commonplace, but—properly understood—they may leave parents with some special problems. Current practice understands selling or giving away gametes solely as a transaction between adults.[3] Yet, given the fact that there are other interests involved, it's perfectly in order to wonder whether adult interests alone ought to determine who gets to be a mom and a dad. Jack may have an agreement with Jill that she can have some of his sperm and use it as she pleases. As far as they are concerned, that agreement excuses Jack from any special responsibilities for providing any care for, or having any relationship with, children that may arise from the arrangement. But it may not let him off the hook as far as those children might be concerned. They didn't sign any agreements excusing him from anything at all. The straightforward thought would seem to be that parental responsibilities, or at least many of them, are owed to the children themselves.

Further, some ARTs increase the chances that a woman will be pregnant with more than a single foetus. This circumstance can present greater challenges to maternal health, and makes premature delivery more likely. Prematurity, in turn, ups the odds that parental responsibilities will in some important respects be even more exacting than parents typically face, threatening that the lives of ensuing children may go less well than they would have, had they been gestated to term.[4]

Creating Vulnerabilities as a Source of Special Parental Responsibilities

Deciding who bears parental responsibilities when the standard genetic grounds for identifying a person as a parent are newly configured requires getting straight about how people pick up such responsibilities in the first place. I have observed that common moral intuitions indicate that, while one can assume parental obligation to a child as a result of choosing to do so, a person can find herself encumbered by those responsibilities just in virtue of being a progenitor. Why is that? How can a mere physical relationship carry with it such a weighty set of responsibilities?

The line I want to defend rests on this basic thought: what we cause to happen can matter morally. Accordingly, biological parents have responsibilities to preserve, nurture, and socialize their children, because otherwise those children may come to serious harm—they will in fact die if no one steps in to care for them—and as the children's

---

[3] Donating or selling gestational ability has been a somewhat more complicated matter. In some celebrated cases involving parenthood abetted by "surrogacy," there have been struggles over the status of women who have borne children on the understanding that they would be relinquished to the care of commissioning couples. Some of these women have asserted their parental status despite those agreements, and at least in some cases where those women were both genetically and gestationally related to the child, the issues were adjudicated on the basis of the child's interests, rather than the status of the agreements struck between adults.

[4] For discussion and documentation, consult the statement of the American Society for Reproductive Medicine (2006) addressing multiple pregnancy's association with infertility therapy.

progenitors, they would not be able to escape at least a significant degree of responsibility for having brought it about that those children were in harm's way.

It may be true that the progenitors did not intend for anyone to come to harm as a result of their doings. Yet intention isn't everything: anyone may find herself morally responsible for putting others at serious risk of death or damage despite her best intentions, and even despite good faith efforts to avoid doing so. Even the most acutely careful driver can strike a pedestrian, and doing so makes the driver's moral responsibilities to the injured party different from yours or mine. Like such drivers, other people may find that they have presumptive responsibilities to help those that they harm or put in harm's way, or to avoid or reduce the impact of that harm if they can.

If progenitors can find willing and competent people to take on those responsibilities for their children, they may, and sometimes ought to, transfer them. But there is at least one consideration that speaks against doing so without serious reason—that while at best we can only *predict* that others, no matter how well vetted they may be, will do an acceptable job of performing parental responsibilities with the required kinds of motivation, we have a different relationship to ourselves. At least sometimes, we can bring ourselves to *perform* what we are obliged to do (Nelson and Nelson, 1995).

Further, at least on this causally oriented picture of parental obligation, it's at least unclear how decisive any particular relinquishment of parental ties ought to be. Children, even if reared in exemplary fashion by adoptive parents whom they dearly love, may later assert some interest in relationships with their progenitors, and it isn't obvious that they have no reasonable claim to do so. If some adults can find biological relationships with children so important that it makes going through ARTs with all their costs, risks, and inconveniences a rational choice for them, why should we assume that biological ties in the other direction won't also matter greatly to some children?

Understanding parental responsibilities as grounded in one's contribution to the existence of specific valuable and vulnerable beings and to the plight they face if not cared for provides a reason for treating biological progenitors as parents in the moral sense. This reason is rooted in general moral considerations about how we may incur responsibilities by what we *do*, and not merely what we *decide*. At least leaving aside (for present purposes) uses of gametes based on force or fraud, whatever their intentions or motivations, progenitors therefore have presumptive responsibilities to respond to their children's vulnerabilities (Sidgwick, 1982; Nelson, 1991).

This view may be thought to have disturbing implications for aspiring parents who employ gamete providers, and for the providers themselves. People who make their ova or sperm available for assisted reproduction have taken up a central—indeed, essential—place in the causal chain aimed at the production of those valuable, vulnerable beings, and they've hardly been invited to reflect on the moral relationships they may thereby be putting in place.

Perhaps, though, if they did reflect, they'd find some comfort. They may have had reasonably good grounds for thinking that any responsibilities they may have had were transferred to reasonably competent hands. Further reflection might seem even more comforting. Gamete providers aren't the only ones riding this causal train. It's also true that others have done things without which no child would have been born—the physician who introduces the fertilized ova into the prospective mother's body, for example. No one thinks that the doctor is one of the child's parents.

Reasons to be sceptical about transferring one's obligations so blithely have already been reviewed. The "predict/perform" distinction discussed earlier suggests there is in general some reason not to leave to others the discharge of duties for which you are initially responsible. That reason can certainly be outweighed—on my best day, I could not help the person I hit with my car as effectively as a trained and equipped EMT, and others might have similar thoughts about their children. Yet if the predictions that others will do a better job fall through, or if there are vulnerabilities that, as it happens, only the gamete provider can meet, she or he faces a moral reason to step into the breach. The responsibilities are owed, after all, to the children, not to other adults, and this may be the most fundamental reason not to be too blithe about your responsibilities if you've supplied the gametes that ended up contributing to a new person's entry into this rough world.

This leaves the second and more serious point: sorting out different kinds of causal contributions and why they matter as they do, a notoriously difficult and complex matter. Those who facilitate reproduction—not just the doctors and the technicians, but the taxi driver who transports the participants to the clinic—all are involved in the causal backdrop of the event of a child's birth. Yet, as the nervous gamete donor would be quick to remind us, no one thinks of them as parents or anything like parents.

One reason for thinking that there is a viable distinction to be made among these different kinds of causes, even if difficult to state clearly, is that the same kinds of considerations hold for reproduction not assisted by technology. There is a vast network of causal conditions, including the actions of other people, in virtue of which conceptions occur, just as there are behind every car accident. Yet it seems quite widely held that the driver, despite her painstaking care, has a special responsibility to respond to the injured pedestrian. The same is not true for the mechanic who finished working on her fuel injector in time for her to get her car out of the garage that very evening, despite the fact that, if he had been late finishing the job, no accident would have happened.

Intuitively, then, we seem to be able to distinguish relevant forms of causation from irrelevant forms. However, we're dealing here with an extension away from clear cases of parental relationships, to consider the possibility that gamete donors or vendors may incur parental responsibilities (while e.g. inseminating physicians do not), so it would be useful if more could be said. Rivka Weinberg (2008) has recently developed a version of a broadly causal understanding of parental obligations that accounts for why progenitors, and not others in the causal chain, have a special responsibility to parent their offspring. Weinberg begins with a moral intuition she takes to be clear—that a man who forswears any responsibility for a child he helped conceive due to contraceptive failure is presumptively behaving badly. She then considers its implications for a similar case—a sperm donor—whose actions are not typically thought of as generating similar responsibilities. In what may seem initially to be a somewhat bizarre conceit, she likens gametes to "hazardous materials" that are in one's possession and control. Yet the thought behind the conceit is compelling: because of the dire consequences that might ensue were the gametes to escape one's control, one has special responsibilities to keep them harmless, just as one would in analogous contexts. If gametes happen to find their way to a site where they causally contribute to the existence of a needy, helpless, valuable individual, it is, at least initially, the responsibility of the person who had control of the "hazmat" in the first place to do what is required to avoid or at least mitigate the harms that threaten. It isn't simply because that person was a cause of the child's existence. Rather, it's because he has a special responsibility for the wise disposition of the material in which many of the relevant causal powers actually reside—his gametes. If gametes get out and cause trouble, putting people at risk of harm, their owners must answer for it, as they might if they harboured an exotic and dangerous pet that broke its bonds.

As with the account I sketched, Weinberg also has to explain why progenitors can't simply transfer their obligations to willing and competent others. She employs ideas similar to those I have discussed to show how problematic such transfer can be: the distinction between predicting what others may do, as opposed to directing one's own performance (she stresses how poorly most sperm providers are positioned for making well-informed judgements about other people's parenting skills), and the idea that the responsibilities in question are owed primarily to children, not to adults. Highlighting the affective dimension of parental duties, Weinberg adds the provocative point that the obligation to love one's children is not something that even in principle is transferable to others. Others may of course love one's children, but it is at least odd to see the responsibility to do so oneself as something one could discharge simply by handing it over to another.[5]

By analogizing our alleged responsibility for the wise disposal of our gametes to clear cases of special responsibilities to be careful with dangerous items we own, Weinberg has added further intuitive support to the position that parental responsibilities can arise as a function of the impact we actually have on others. They do not depend solely on choices we might make (or not make) to assume them. This reinforces the idea that using ARTs that involve other people's gametes may change parental responsibilities by expanding the range of those who have presumptive duties with respect to children.

---

[5] These considerations may complicate our picture of the morality of relinquishment and adoption, but not in any way that precludes or lessens the reality of becoming a parent via adoption. It may suggest that the transfer of responsibilities from progenitors to adoptive parents, in addition to serving children's needs rather than those of adults, is not absolute, and that children retain at least some defeasible claims on their progenitors. If so, that would add a compelling consideration to the case for open adoption.

On this view, gamete suppliers may at least be presumptively liable to a summons on the part of children or those acting on their behalf to provide some nurturing, social- izing, and preserving, particularly if a child's needs for such goods aren't being well met. For example, they may have a responsibility, rebuttable but nonetheless real, to be available to their offspring, just to help such children if they want to understand more about their own stories—the biological and historical narratives that connect them to a certain legacy. Socializing a child presumably encompasses helping her consolidate a rich sense of who she is, and for at least some people, understanding of where one comes from can be an important part of that process. Accordingly, the parents raising the children may have responsibilities to keep alive possibilities of connection between their children and their other parents, and to be prepared to accommodate what might turn out to be a complicated family structure.

IVF and Parental Discretion

Some uses of ARTs place ensuing children at greater risk of developing physical condi- tions that, at least arguably, may constitute intrinsic harms, and that surely can chal- lenge parents' abilities to discharge their responsibilities creditably. ARTs can increase the chances that a woman may gestate more than a single foetus at a time. Being preg- nant with twins is correlated with increases in maternal and infant mortality, with pre- term delivery, and with the incidence of cerebral palsy—and the more foetuses one carries, the stronger the correlation becomes. Healthcare providers involved in treat- ing infertility have increasingly seen gestating multiple foetuses as ARTs' most serious and most frequent iatrogenic complication (Healy, 2004).

Multiple pregnancies are, not surprisingly, correlated with inserting more than one embryo at a time into a woman's womb. The aim is to make pregnancy more likely; given the burdens and costs associated with repeated IVF cycles, keeping the number of procedures to a minimum is hardly a trivial matter. Yet ART use generally, and spe- cifically the practice of multiple transfer, has contributed to a considerable upswing in the number of children born as twins or higher-order multiples in many nations, with a corresponding increase in the rates of medical and social complications affecting infants, mothers, families, and societies (Pinborg, 2005).

In 2012, the American Society for Reproductive Medicine (ASRM) and the Society for Assisted Reproductive Technology (SART) jointly issued a report calling for "elec- tive single embryo transfer" (eSET) to be recommended to all women who meet certain criteria, including age, number of IVF cycles undergone, and quality of avail- able embryos (ASRM, 2012a). The report also took note of the socio economic fac- tors bearing on its recommendation—the very substantial and, in the US and several other countries, typically privately funded expenses associated with IVF. However, the report stopped short of recommending that SET constitute the standard of care, even for women most likely to have as favourable an outcome from single as from multiple embryo transfers; a 2012 paper from ASRM continued to resist mandating SET, and recommended practice guidelines allowing MET in women 35 years of age and older (ASRM, 2012b).

In Sweden, by contrast, SET is mandatory, apart from "extraordinary circumstances"; Saldeen and Sundström (2005) reported no difference in clinical pregnancy rates after the mandatory SET policy was put in place; a 2009 study in the US corroborated this result (Kresowik *et al.*, 2011). Several other countries take up intermediate positions between the fully discretionary and the mandated. For example, when reproductive assistance is publicly funded, New Zealand allows only SET in a woman's first or second IVF cycles, unless the ovum donor is older than 35 (Maheshwari *et al.*, 2011).

In many contexts, then, women have to weigh the chances of giving birth against increased risks to their children. Yet if transferring more than a single embryo heightens the risks to the children-to-be, how can people who are serious about preserving, nurturing, socializing—and loving—children elect such procedures?

Here's what may seem a tempting reply: no such children would even exist without their parents' use of ARTs, so unless the children's existence is completely miserable, it seems that they have little ground for complaint. I think this reply should be resisted. The point here is not what is necessary to a given child's existing at all, but whether the parents are in a position to respond well, or even adequately, to their child's ongoing needs. If someone had good reason to fear, for example, that any child she was likely to have would require kinds of care that would dwarf her own resources, or those that she could muster with the cooperation of others, she wouldn't be worried about the metaphysics of coming into existence. She would be troubled about whether she would fail to provide what her child needs day by day.

No one who doesn't exist, if such an expression is permitted, is harmed or wronged if her procreation is forgone. It may well be that anyone who values his life overall is not harmed by actions and circumstances that are strictly essential to his existence. Who *can* suffer harm are children whose needs are continually scanted even if their parents are doing their best but are genuinely out of their depth.

The question, then, is what risks of this kind of outcome can people responsibly accept in pursuit of parenthood? Do those risks lie within the range of a parent's discretion?

Here's another tantalizing but dubious answer: as parents have a responsibility to promote and protect their children's best interests, they ought not to use ARTs that may put their children in a position where their needs might exceed the parents' capacities to satisfy them optimally. There are two problems with this. First, to repeat, forgoing procreation doesn't set back the interests of the unconceived; on the other hand, knowingly worsening a child's odds of flourishing, compared to the "baseline" hazards involved in "standard" pregnancies, runs the risk of doing just that. Second, the key assumption here might seem innocuous: of course, parents must act in the best interests of their kids. Yet while "best interests" is sufficiently ill-defined to allow creative interpretations, the most reasonable way to deal with this assumption is to deny it. In family life, interests of equal significance cannot always be made to harmonize, and therefore parents must be allowed some latitude in weighing and balancing them; often enough, the interests of some must give way so that the interests of others can be promoted. Yet even beyond trade-offs that cannot be avoided, there are plenty of instances of parents' slighting their children's interests for reasons that are perfectly optional, and no one thinks the worse of them for doing so. Ordinary life is full of examples of parents taking at least small risks of very negative outcomes for their children, in order to obtain reasonably high odds of comparably very small benefits for themselves. Strapping your toddler into her car seat for a ride through snowy streets to pick up the evening's wine would fit into this category nicely. Granted competency with the presenting conditions and reasonable care as a driver, the fact that your child might die or be so seriously injured as to substantially reduce her options in life and to stretch out your own caring capacity very thinly indeed—and all because you decided you wanted some wine—is not regarded as a reason to stay home and drink tea.

What if you're going out for fertility treatments? Is it defensible to take the risks involved in transferring two or more fertilized embryos at a time? The fact that parents can take risks with children's interests in other contexts, while suggestive, does not settle the question. Suppose, for example, the justification for allowing parents to put their kids' interests at risk to gratify their own is pragmatic ("how could we possibly monitor all parent–child interactions?") or instrumental ("even if monitoring were possible, it would damage familial intimacy and hamper the good work it does"). That societies must or ought to let some risks go doesn't entail that they ought to let others go. Even if the size of the risk does not differ, the costs of intervening may. What's needed is an argument showing that reasonable parental discretion can be justified by its importance to parents.

Henry Brighouse and Adam Swift have provided just such an argument. In motivating a conception of parental rights portrayed as fundamental (though at the same time limited and conditional), Brighouse and Swift (2006; and in this volume) focus on the value that having a parental relationship with children can have for adults. Intimate relationships with children afford distinctive, rich, and intense forms of significance to parents—for example, the spontaneous and unconstrained sharing of self and love that a child offers a parent, as well as the enjoyment of one's own love of that child. Getting access to that kind of value can reasonably be seen as a necessary part of a person's conception of a good life. Expanding on the distinctive character of intimacy with one's children, they write:

The love one receives from one's children, again especially in the early years, is spontaneous and unconditional, and, in particular, outside the rational control of the child. She shares herself unself-consciously with the parent, revealing her enthusiasms and aversions, fears and anxieties, in an uncontrolled manner. She trusts the parent until the parent betrays the trust, and her trust must be betrayed consistently and frequently before it will be completely undermined. Adults do not share themselves with each other in this way: intimacy requires a considerable act of will on the part of adults interacting together. (Brighouse and Swift, 2006: 93)

For my own part, I would also be inclined to include the opportunities that parenting can provide a person to exercise possibilities within her own life: to enact, and to understand herself as enacting, the roles of inheritor, preserver, challenger, and transmitter of distinctive ways of understanding and valuing experience across generations.

Brighouse and Swift are chiefly trying to defend social tolerance of parental practices that depart from a policy of aiming at an optimally autonomous future for children: to raise a child in a given set of religious or political beliefs, for example, while it does not guarantee future fidelity to that tradition, will very likely have an ongoing impact on the child's normative starting points: how she thinks about what is meaningful to her, and about her choices and her goals, what her personal moral struggles may be. Someone strongly committed to the ideal that people should all be able to choose their norms as freely as possible might look askance at raising a child as a Mormon or a Marxist; blocking this possibility, though, harms what for many will be the most important reasons for becoming a parent at all. Yet Brighouse and Swift's work has broader implications: it expresses plausibly and powerfully why parenthood is of such great importance to some people, and what that importance justifies.

There's nothing in this general picture from which one might simply deduce just where the threshold is between defensible and indefensible amounts of discretion. Brighouse and Swift (2006: 107) themselves write loosely of there being "some room for the parent to pursue her own interest even where that may not be best for the child". Still, this approach can support an analogous strategy for getting a clearer fix on risks to which parents are at liberty to expose their kids. I take their argument to support the notion that ordinary practices that make possible the distinctive forms of parental involvement with children are presumptively permissible. If what people generally do with their kids, the freedom they have to influence the shape of their lives, are so justified, then at least the risks attending to standard practices involved in raising kids, and in having kids at all, should be supported by the same considerations. Try to have children in any fashion, and you may find yourself unable to meet their needs—there's no escaping that risk. Yet if you don't have children at all, it will be much more difficult (though not impossible) to enjoy what Brighouse, Swift, and others celebrate as so especially valuable about parent–child relationships. Those common practices, then, seem at least presumptively permissible, and as such can serve as a reasonably firm basis for finding analogies that can help clarify the range of parental discretion in situations involving ARTs.

For example, women who conceive twins without assistance are not generally taken to be obligated in virtue of their parental responsibilities to undergo foetal reduction. Or, to take what is perhaps a less distracting example, consider women who know themselves to be prone to hyperovulation and thus to a greater likelihood of having twins "naturally." Like women using MET, they may find themselves facing new and quite possibly more exacting challenges in preserving, nurturing, and socializing their children. Yet they are not typically regarded as being irresponsible in nonetheless pursuing biological parenthood; Brighouse and Swift provide a way of appreciating why not.

The importance of this result goes beyond questions concerning whether SET ought to be the mandated standard of care in IVF. Other forms of ART—e.g. the use of drugs that stimulate ovulation—also lead to higher levels of multiple pregnancy than does unassisted reproduction, and it is nowhere near clear how to reduce their tendency to do so (Fauser, 2005). Further, there is some evidence that, even when ART use eventuates in a singleton pregnancy, the chances of giving birth to a low birth weight infant are elevated, roughly to the level observed in unassisted twin births (Schieve *et al.*, 2002). Weighing extra risk of harm to children against the significance of having them at all is a dimension of the special responsibilities of parents employing ARTs that won't therefore vanish with refined or mandated SET protocols.

Parental Responsibilities, Parental Rights

I've focused here on two conspicuous features of ARTs: some can complicate the causal story of a child's conception by adding people into the mix willing to provide gametes but not inclined to provide care; some can increase risks to children's health. I've argued that ARTs can generate "special" responsibilities in both cases: by increasing the number of responsible parties in ARTs requiring gamete providers, and by possibly hampering the ability of parents to discharge their responsibilities to nurture, preserve, and socialize their children in ARTs that make premature delivery likelier. How these responsibilities ought to be routinely incorporated into social or personal practices and patterns of understanding, and what moral and material resources there may be for negotiating or sharing or deflecting them, remains to be resolved. The aim here has been to put them squarely on the agenda for reflection. Teasing out the implications of new medical technologies for how we should understand our responsibilities as parents, as patients, as families, or as citizens can be subtle enough; it pales in contrast to determining how we should incorporate responsibilities newly understood into our policies, acts, habits, and feelings.

Acknowledgements

I'm grateful to Hilde Lindemann and Laura Olmstead for conversations pertaining to this paper. I am more than usually thankful to the editors of this volume for their extensive and thoughtful response to an earlier draft. This chapter is a contribution to the work of the International Consortium on the Ethics of Families in Health and Social Care.

References

American Society for Reproductive Medicine (ASRM) (2006). Multiple Pregnancy Associated with Infertility Therapy. *Fertility and Sterility*, 86(5/suppl. 1), S106–10.

American Society for Reproductive Medicine (ASRM) (2012a). Elective Single-Embryo Transfer. *Fertility and Sterility*, 97(4), 835–42.

American Society for Reproductive Medicine (ASRM) (2012b). Criteria for Number of Embryos to Transfer: A Committee Opinion. Retrieved June 2013 from http://www.asrm.org/upload-edFiles/ASRM_Content/News_and_Publications/Practice_Guidelines/Guidelines_and_Minimum_Standards/Guidelines_on_number_of_embryos(1).pdf.

Brighouse, H., and Swift, A. (2006). Parents' Rights and the Value of the Family. *Ethics*, 117(1), 80–108.

Fauser, B., Devroey, P., and Macklon, N. (2005). Multiple Birth Resulting from Ovarian Stimulation for Subfertility Treatment. *The Lancet*, 367(9473), 1807–16. Retrieved June 2013 from http://image.thelancet.com/extras/04art6002web.pdf.

Healy, D. (2004). Damaged Babies from Assisted Reproductive Technologies: Focus on the BESST (Birth Emphasizing a Successful Singleton at Term) Outcome. *Fertility and Sterility*, 81(3), 512–13.

Kresowik, J., Stegmann, B., Sparks, A., Ryan, G., and Van Voorhis, B. (2011). Five Years of a Mandatory Single-Embryo Transfer (mSET) Policy Dramatically Reduces Twinning Rate without Lowering Pregnancy Rates. *Fertility and Sterility*, 96(6), 1367–9.

Maheshwari, A., Griffiths, S., and Bhattacharya, S. (2011). Global Variation in the Uptake of Single Embryo Transfer. *Human Reproduction Update*, 17(12), 107–20.

Nelson, H. L., and Nelson, J. L. (1995). *The Patient in the Family*. New York: Routledge.

Nelson, J. L. (1991). Parental Obligations and the Ethics of Surrogacy: A Causal Perspective. *Public Affairs Quarterly*, 5(1), 49–61.

Pinborg, A. (2005). IVF/ICSI Twin Pregnancies: Risk and Prevention. *Human Reproduction Update*, 11(6), 575–93.

Ruddick, S. (1989). *Maternal Thinking*. Boston, MA: Beacon Press.

Saldenn, P., and Sundström, P. (2005). Would Legislation Imposing Single Embryo Transfer Be a Feasible Way to Reduce the Rate of Multiple Pregnancies After IVF Treatment? *Human Reproduction*, 20(1), 4–8.

Schieve, L., Meilke, S., Ferre, C., Peterson, H., Jeng, G., and Wilcox, L. (2002). Low and Very Low Birth Weight in Infants Conceived with Use of Assisted Reproductive Technology. *New England Journal of Medicine*, 346(10), 731–7.

Sidgwick, H. (1982). *The Methods of Ethics*. Chicago, IL: University of Chicago Press.

Singer, P. (2009). *The Life You Can Save*. New York: Random House.

Unger, P. (1996). *Living High and Letting Die: Our Illusion of Innocence*. Oxford and New York: Oxford University Press.

Weinberg, R. (2008). The Moral Complexity of Sperm Donation. *Bioethics*, 22(3), 166–78.

11

Adoptee Vulnerability and Post-Adoptive Parental Obligation

Mianna Lotz

Introduction

A substantial body of research has accumulated during the past four or more decades in the field of adoption studies. Debates in the psychological adoption literature have predominantly centred on delineating factors thought to predispose adoptees and their families to potential difficulties of maladjustment and poor attachment, the likely development of which has been largely assumed from the outset. Yet the weight of evidence in relation to adoptee welfare outcomes continues to support formal adoption as a morally defensible means by which to create and expand families, especially when compared with institutional or foster-care-based childrearing. Nevertheless, while Tina Rulli (in this volume) correctly points to the unique value of adoption for parents, as well as the undeniable benefit for children of having their significant need for a stable family met, it is now well understood that the actual success of any specific adoption arrangement will be determined in the interplay of a complicated range of factors, both internal and external to the family.

Indeed, recent transitions in adoption practice in most Western countries—including most significantly from "closed" to more "open" adoption arrangements—have considerably expanded the range of complexities confronting family-making in the twenty-first century. Of particular interest to me, these transitions also pose difficult questions concerning the implications of the prevailing new "ethic of openness" in adoption (Jones and Hackett, 2011: 41) for our understanding of the nature and obligations of adoptive parenting. This chapter considers the question of post-adoptive parental obligation. My principal aim is to establish the moral grounds on which specific post-adoptive parental obligations can be defended, drawing both from a philosophical consideration of adoptee vulnerabilities and interests, and from a review of key empirical findings in psychological and sociological research. I do not aim to be comprehensive in enumerating, or specifying the precise content of, all such obligations, nor to exhaustively review the psychological evidence detailing adoption outcomes and the factors contributing to them. My aim is the more modest one of putting in place a firm philosophical foundation for the examination and clarification of the scope and limits of specific obligations of this kind.

This chapter provides an analysis of three broad domains of potential adoptee vulnerability. To do so I draw upon a conception of vulnerability that accepts that it is an *ontological* and hence *universal* human condition, namely of fragility and susceptibility to harm and suffering. Importantly, our human vulnerability is often essentially social or relational in character (as elucidated e.g. by Fineman, 2008; Butler, 2004, 2009; Turner, 2006); yet it is also born of our susceptibility to naturally occurring harms including natural disasters and unpreventable and incurable disease. The conception I draw from emphasizes in particular the *contingent* susceptibility of specific persons or groups in relation to *specific types* of potential harm or suffering, or in *specific kinds of relationships* (as elucidated by Goodin, 1985). In presenting these domains of potential adoptee vulnerability, I elucidate both the nature of these vulnerabilities and the contextual conditions that make them likely to arise. That task complete, the remaining focus in the chapter will be on the normative question arising from such an analysis, which has received considerably less attention, in the scholarly literature and in adoption work itself, than it warrants: namely that of the post-adoption obligations suggested by an adequate understanding of adoption-specific vulnerability. If the grounds for such obligations exist, further questions arise concerning their scope and limits. Potential concerns and objections will need to be addressed. One such concern, for example, concerns whether the imposition of post-adoptive obligations can be defended against a possible charge of unacceptable discrimination against adoptive parents and families. While I will provide some comments on this matter, issues such as this can only begin to be explored in this chapter, given its conceptually prior task of setting out the grounds for this class of obligations.

An important point to note from the outset is that adoption research has now conclusively established that divergent welfare outcomes amongst adopted children do not reflect specific demographic, socio economic, or cultural variables and patterns. While children adopted in later childhood typically experience greater difficulties than those adopted during infancy, this chapter's survey of empirical studies highlights that the greatest determinants of successful adjustment and family attachment are *intra-familial processes,* rather than individual attributes—demographic or psychological—of either adoptees or adopters (see e.g. Brodzinsky, 1993). Emerging originally out of the highly influential work of H. David Kirk (1964; Kirk and Mass, 1959), studies have consistently demonstrated that the way in which families manage internal processes of disclosure and communication about adoption is most highly predictive of outcomes for adoptees and their families.

Nevertheless, my discussion of adoptee vulnerability and adoptive parents' related obligations proceeds with an important caveat. While I suggest that evidence and argument *do* support the existence of a class of post-adoptive obligations for adoptive parents, such a proposal is not intended to distract from a point of crucial importance to which the greatest emphasis must be allocated: namely that it will only be by the eradication of specific broader social conditions and factors—in particular what I (and also Charlotte Witt, in this volume) refer to as the "biologistic" or bionormative pre-suppositions and biases concerning the ideal family and "real" parenthood—and the removal of adoption policies and practices built upon those, that the conditions will be created in which the needs of adopted children and their families can adequately be met.[1]

Reflecting the importance of that caveat, I proceed by situating philosophical discussion of potential adoptee vulnerability within an analysis of aspects of the social context in which adoption occurs, which contribute significantly to the development of the vulnerabilities here discussed. I examine the way in which specific norms and ideals concerning family constitution and relationships give rise to a set of "imperatives of adoption," which in turn shape the views and experiences of adoption for all concerned, and play a significant role in predisposing adoptees to the specific vulnerabilities discussed here.[2]

The Social Conditions of Adoptee Vulnerability

Onora O'Neill (1979) has argued that the foundational responsibilities of *being* (as distinct from *becoming*) a parent are those involved in *raising* a child, and range from ensuring that the child's basic needs for sustenance and care are met, to emotional nurturing of the child and support of them to develop a healthy sense of self-worth. Eva Kittay's (1997) work has been helpful in elucidating the vulnerability of children, and the fact that their distinct vulnerability qua children arises from their "inevitable dependency"—for both survival and flourishing—on the actions and choices of others (especially care-givers) as well as from their temporary lack of the full complement of skills and capacities that mitigate such dependency. At the most general level, then, the obligations incurred by all who assume the social role of "parent" are essentially the same, regardless of the precise grounds of parenthood—that is, irrespective of whether biologically, genetically, or purely socially grounded. In recognition of this

---

[1] My discussion focuses on adoption generally, and on post-adoptive obligations that apply to all adoptions, irrespective of whether parents are gay, lesbian, transsexual, or heterosexual. I see no reason for thinking that the parental obligations here referred to would differ between different adoptive family types.

[2] One final qualification bears noting, given the broader focus of this volume on diverse modes of family-making. Many points I make in relation to adoption will raise questions concerning their possible extension to, and implications for, gamete donor/provider families. These questions certainly warrant investigation, and matters relevant for such an investigation are discussed elsewhere in this volume. However, space limitations preclude an exploration of these questions in this chapter. My focus here is, of necessity, on adoption alone.

Janet Farrell Smith has argued for a uniform moral foundation for being a parent (2005: 122–4).[3]

Yet while the *fundamental* needs of adopted and non-adopted children are essentially the same, in the context of adoption particular vulnerabilities that *all* children may have are cast in a different light and may be intensified. It is for these reasons that distinct adoption-related parental obligations arise. My focus is on the potential interconnected vulnerabilities that adopted children face in three specific developmental domains: identity development; development of a sense of familial belonging and security; and development of emotional independence. Importantly, construing these as *vulnerabilities* rather than deficits, disadvantages, or harms signals that our concern here is with a set of predispositions and potentialities which *may*, under certain *contingent* conditions, pose risks and threats to adoptee welfare. Clearly, the experience of adoptees varies significantly, and not all will encounter or even recognize the potential difficulties discussed here. Nevertheless, key themes and narrative tropes recur in the philosophical adoption literature and testimony. Corroborated by results from psychological research, these are suggestive of discernible patterns of adoption-specific vulnerability of the kind to be described here.

It is important to understand the conditions in which these vulnerabilities exist. They are not, I would suggest, necessary to or inherent in adoption *per se*. They arise because of specific features of the context in which adoption takes place, most significantly the prevalence of particular deeply entrenched Western cultural assumptions, norms, and ideals concerning the family. These norms in turn generate a set of "imperatives" around adoption (Leighton, 2005). They are norms that emphasize the primacy, strength, and permanence of biological connection and, by contrast, what one writer describes as the "second-class status of 'fictive' or legal kinship," which is assumed to be fragile and impermanent (Jones and Hackett, 2011: 53). They establish a context of *bionormativity* in which adoptees and their families must struggle for legitimacy.

Social science research has revealed the significant impact of this bionormativity on how adoptees and their families construe and manage adoption. Drawing on Bartholet's (1994) work describing the bias in favour of genetic and biologically related families, Brakman and Scholz (2006) refer to the ongoing salience of the "biologic paradigm," which continues to emphasize biological relatedness, judges families formed through assisted reproduction and adoption according to how closely they approximate the "natural" (i.e. biological) family, and, as Bartholet observed, "constrains" and "conditions" the options available to and choices made by women and prospective parents generally (Brakman and Scholz, 2006: 56–7). The recent trajectory of biological bias can be traced from its relative retreat during the 1920s–1960s in favour of an emphasis on social experience, environment, and nurture—and hence closed adoptions and

---

[3] Presumably Smith would not deny that differences may arise in the parenting of different children, and that those may ground *special* obligations; her point is rather about the essential moral *equality* of various possible forms of parent–child relation.

sealed birth records—to its re-emergence in the late 1960s and 1970s as part of both the feminist and civil rights movements and significant developments in IVF and biological genetics (Brakman and Scholz, 2006; see also Modell and Dambacher, 1997). This resulted in a swing towards emphasis on biological inheritance and connection, the assertion of adoptees' and birth mothers' rights, and a push towards open adoption practices. The trend continues today with recent legislative developments in many countries and states—including the UK, Sweden, Norway, the Netherlands, Switzerland, Australia, New Zealand, and British Columbia—aimed at reducing or banning donor anonymity and promoting the entitlement of donor-conceived children to knowledge of their biological parentage.

The persistence of this "biologic paradigm" is clearly discernible in recent empirical research with adoptees and their families, in spite of the fact that, due to increasing divorce rates as well as increasing access to adoption, IVF, gamete donation, and surrogacy by traditionally excluded groups such as same-sex couples, we now live in an era in which complicated family compositions are considerably more common. Marsiglio's (1998) research found a notable persistence of the view that family forms *not* based on blood or formal marriage are regarded as less meaningful and less legitimate than those that are. In more recent research Holtzman (2008) found evidence of a broadening of cultural understandings of the "family"—at least amongst her sample of young, university-level sociology students. Yet in spite of evidence of greater acceptance of alternative family structures, Holtzman found traditional conceptions of the *biologically grounded* family still figured dominantly in young people's discussions of the parent–child relationship, seeming to coexist with rather than having been supplanted by a social conception of the family (2008: 174). Of particular interest for my purposes, discussions revealed the ongoing salience of notions of genetic connection, "bloodlines," and the physical bond of pregnancy and childbirth, giving these a pivotal role in conceptions of the family and parenting. Terms like "real," "true," and "actual" were heavily used in describing biological parents, even where the overarching narrative purpose was to deny the importance of biological connection (e.g. referring to a biological child as a "real" child even while denying that genetic links are superior to social ones; and referring to their propensity to love a non-biologically related child "as if" they were one's "own" or "real" child (Holtzman, 2008: 177–80). Clearly, adoption remains a social practice with a very powerful capacity to distil and magnify cultural attitudes concerning "family." Thus Modell and Dambacher (1997: 7) comment that "as consciously constructed and scripted kinship, adoption reveals fundamental premises about birth, blood, and contact, which themselves reflect ideologies that in Western societies enshrine the dichotomies [of] nature and culture, fictive and real." As we will see, the impact of these presuppositions and ideologies in shaping the expectations, experiences, and vulnerabilities of adoption are significant. I turn now to an analysis of specific domains of adoptee vulnerability and the corresponding social conditions, values, and norms that underpin these.

The Domains and Conditions of Adoptee Vulnerability

First, a number of philosophers (e.g. Witt, 2005; Hahn, 2005; Leighton, 2005) writing on adoption have highlighted special challenges encountered by adoptees in the development of a *sense of self and identity*. Of course, where someone possesses very little or no information regarding her biological ancestry and family, her sense of her own historically based identity and generational continuity will be severely curtailed. Yet adoption theorists have also pointed to more widespread difficulties with the establishment and maintenance of a secure and stable sense of self, and with managing the multiple aspects and layers of self-knowledge and personal identity, even for adoptees who possess some knowledge of their adoption status and biological parentage. For example, March's research found evidence that amongst adoptees uncertainty about biological background did indeed appear to translate into uncertainty about self, as well as to motivate search and reunion activities as a means by which to "place self within the biosocial context valued by their community" (1995: 657–8). The significance of knowledge of biological background becomes especially clear given the fact that, in Western countries at least, adoptions occur in the context of a deeply entrenched commitment to a *family-resemblance ideal* according to which overt resemblance to other members of one's family is regarded as highly significant. Indeed, at least one theorist has explicitly linked family resemblance to self-knowledge and identity development, even positing a strong connection between *literal* family resemblance and self-constitution:

Knowing what I am like would be that much harder if I didn't know other people like me. And if people bear me a literal family resemblance, then the respects in which they are like me will be especially important to my knowledge of what I am like, since they resemble me in ways that are deeply ingrained and resistant to change. (Velleman, 2005: 365–6)

Whether or not one can agree with Velleman's strong account of self-knowledge and known family resemblance, it is clear that, within a broader social context in which significant value is accorded to biologically related family units, family resemblance will serve as an important signifier of conformity to both the "normal" and the ideal.[4]

Strong social commitment to the family resemblance ideal has meant that, in the absence of a biological basis for family resemblance, ameliorative attempts have routinely been made to enable adoptive families to approximate, by non-biological means, what cannot be achieved biologically: namely the *semblance* of biological relatedness. Such efforts express commitment to a broader imperative of "fit," positing resemblance and similarity as the basis for both identity and familial belonging. Writers such as Modell and Dambacher (1997) and more recently Hahn (2005) have highlighted the extent to which this particular conception of "fit" (which I will refer to

---

[4] I criticize Velleman's arguments in more detail in Lotz (2008).

as "resemblance-based fit") has been highly salient in shaping adoption practice and policy as well as adoptive parents' preferences. Since the late 1940s and early 1950s in most Western countries, adoption agencies, social workers, and lawyers have gone to considerable lengths to place children in adoptive families to which they have the greatest similarity (Modell and Dambacher, 1997). The practice of "matching"—not just for "race," ethnicity, or religion, but also for physical attributes such as height, hair/eye colour, socio economic class, educational background, and even psychological traits—has been professionally sanctioned by reference to a largely untested presupposition that trait similarity will be most conducive for bonding and attachment within adoptive families, and will best support adoptees in constructing stable identities. This commitment to ensuring as far as possible the outward concealment of a child's adoption status has constituted the dominant ideology informing institutional adoption practice in Western countries from the mid-nineteenth century at least (see Pertman, 2000; Shanley, 2001; Fogg Davis, 2002, 2005). Modell and Dambacher (1997) provide an insightful analysis of the extent to which biologistic and essentialist notions of identity and kinship have informed the conception of "best interests of the child" that is the cornerstone of official adoption policy in the US, for example. They point out that, culturally, the goal that an adopted child resemble someone in the adoptive family and is therefore able to appear "*as-if-begotten*," reflects "a conviction that genealogy establishes an enduring, affectionate bond between parent and child" (Modell and Dambacher, 1997: 9–10). This amounts to what they describe as "an almost mystical notion of identification" according to which attachment between those who look and therefore are alike "just happens" (Modell and Dambacher, 1997: 16). In such cases, they note, "resemblances stand (in) for the innate quality of a parent–child attachment, the continuity of 'substance' over generations, and the unconditional, unchosen connection between those who are kin" (Modell and Dambacher, 1997: 11). While explicit reference to requirements of sameness are no longer so prevalent, the notion of "matching" continues to be operative in guidelines for placement assessment, and analysis of policy and guidelines reveal the ongoing prevalence of implicit—and not-so-implicit—assumptions concerning the benefit of a child being matched with adoptive parents to whom they are culturally and ethnically similar, and preferably related as kin.[5] The so-called "double-edged sword" of matching policy is well captured by Wegar (2000: 367) who comments that, although it might be thought that the practice of matching for physical resemblance could alleviate the difficulties of adoption by outwardly making adoption status less obvious, the unintended effect is not as might be hoped, since in trying to recreate the biological family, adoption workers in fact only end up further emphasizing the importance of the genetic connection for family bonding.

---

[5] See e.g. NSW Government (2010).

A commitment to resemblance-based "fit" has of course also facilitated denial of differences and secrecy, both within the adoptive family and in their dealings with others. In their early work on this issue, Kirk and Mass (1959: 316–19) highlighted the extent to which adoptive parents are particularly susceptible to de-emphasizing intra-familial difference. Pointing to the "incongruous role obligations" faced by adoptive parents—of trying both to *integrate* and *individuate* family members—they suggested that the fact that adoption takes place within a social culture which valorizes biological-relatedness means that the pressure to present as an integrated family is likely to dominate:

In the absence of.... biological bonds and of family line continuity, the emphasis of the... adoptive couple appears to be on integration rather than on the differentiating forces of autonomy.... Adoptive parents can be expected to respond by greater-than-ordinary protectiveness of the child and by trying, with all means at their disposal, to make inviolable the integrity of their familial unit. (Kirk and Mass, 1959: 318)

An example of this is reported in more recent research by Miall (1989). Commenting on the perception that the outward disclosure of adoption would likely increase feelings of differentiation rather than aiding adjustment, one respondent stated: "We don't like to tell others because we think it sets the child apart when our desire is to integrate the child into our family, not differentiate. We love this child. It's ours no matter what others say or think" (Miall, 1989: 296–7).

Importantly, this emphasis on sameness, resemblance, and integration must be understood as a response to a perceived need to establish the legitimacy of the adoptive family within a socio cultural context in which it is still widely regarded as falling short of the "real" or "ideal" family type. Nevertheless, while almost always well intentioned, adoptive parents' efforts to outwardly perpetuate the appearance of familial biological relatedness—to present an *"as if* family," as Modell (1994) has termed it—will typically be psychologically and emotionally fraught for adoptees. In denying or downplaying real differences in favour of emphasis on, or construction of, intra-familial resemblance and similarity, an adoptee is susceptible to coming to regard her belonging within the adoptive family as *conditional* upon similarity to its members. As Hahn (2005: 214) notes, "There is something peculiarly unstable about a love made conditional on a continued perception of sameness, in the face of salient differences." Being "passed off" as biologically related can be suggestive of an implicit shame, regret, or rejection on the part of the adoptive parent(s) regarding the child's true status and origin. Against a background of pervasive family-resemblance imperatives, such outward denial or suppression of adoption status has the potential to create difficulties for healthy identity development. It may give rise to a troubling internal bifurcation in which the adoptee suppresses elements of her self and personality for the sake of maintaining the endorsed "fictive" identity as a biologically related (and therefore more "real" or "true") family member. In the process, feelings of inadequacy, shame, insecurity, and anxiety may be provoked. A child may respond to such anxieties by accentuating, or striving to develop, certain "conforming" attributes or talents at the expense of others. Insofar as the notion makes sense, the achievement of "authentic" identity is likely to be significantly constrained as a result of the perceived imperative to "fit in" to the family by resembling it.[6]

Second, aside from the way in which the imperative to *be like* and *fit in with* the adoptive family can cause an adoptee to reject and suppress diverging attributes she identifies in herself, additional aspects of adoption potentially destabilize the adoptee's *sense of security and belonging*. As we have seen, requirements of "fit" and other criteria for "suitability" or "match" frequently comprise the governing conditions for a child's entry into—and in some (tragic) cases exit from—the adoptive family. Whatever their precise nature, the existence of these governing conditions highlights a significant feature of adoption—namely that it is, in essence, both *transactional* or *negotiated* and *conditional*. It is both mediated and facilitated by the preferences of the adoptive parents, which in turn inform "placement suitability" decisions.[7] Quite unlike in the case of the conceived biological child, then, in the adoptee's history there was both a voluntary transaction and a kind of *selection process*, in which selection was conditional upon criteria that the adopted child must meet. These may be fairly minimal (e.g. being healthy) or more substantial (being fair or dark-haired; blue-eyed or brown; of medium-statured, short, or tall parentage; black, white, or "coloured"; and so on).

Yet the knowledge of having been "selected" for/by the adopting family, on these kinds of grounds, can be hazardous for an adoptee's sense of family belonging and security, potentially giving rise to a perception of having been accepted *conditionally*—on grounds of satisfying *external* criteria—as opposed to in an unqualified or unconditional way. This in turn brings with it the spectre (perceived or real) of being rejected, and even returned, post-adoption, should one somehow not fulfil expectations or continue to satisfy criteria. The dangers are not merely imagined. Adopted children sometimes are "returned."[8] In addition, relatively recent research (Hollingsworth, 2003) indicates considerable ambivalence in public attitudes regarding whether it is ever acceptable or permissible for adoptive parents to "disrupt" or

---

[6] There is of course considerable conceptual difficulty involved in providing an account of "authentic" identity development, and of its value. It will be sufficient for my purposes, however, that we can discern degrees of heteronomy in identity development, and can agree that the more an agent seeks consciously to tailor the development of attributes and character traits to meet the expectations and perceived desires of others, the less independent and hence authentic will be her identity development.

[7] Adoption may not be entirely unique in this regard. Processes of voluntary transaction and selection also play a role in ART more generally, and concerns about potential rejection (to be discussed shortly) could likewise affect these children. Nevertheless, I judge this particular vulnerability to be more acute in adoptees. The potential for their specific attributes to influence their selection seems greater given that their attributes are already realized and instantiated, and hence an aspect of who they already are, as opposed to being mere predispositions.

[8] As highlighted by recent news reports of a senior Russian official's urging of the freezing of all adoptions by US families, in light of the case of an adopted Russian 7 year old who was put on a one-way flight back to Russia, as well as three killings of Russian children by their US adoptive parents since 2006 (Associated Press, 2010).

"dissolve" an adoption and "return" a child (e.g. where the child has severe behavioural problems).[9] Hollingsworth's (2003) research—which involved telephone surveys with 749 members of the general US adult population who identified as *not* part of or directly familiar with an adoption arrangement—revealed that 23 per cent believe adoptive parents *should* be able to change their minds about keeping a child after placement; and a further 19 per cent either *didn't know/gave no answer* (6.6 per cent) or gave the qualified answer that whether or not this was permissible would *"depend on the situation"* (12 per cent). Thus, while a majority (58 per cent) of the general public believe that adoptive parents should be required to keep a child after placement, a large minority regard giving back the child as at least sometimes or potentially permissible. Findings such as these serve to highlight that an adoptee's placement and ongoing position in the adoptive family is *negotiated* and *conditional* rather than assumed, unquestioned, and enduring. This feature of adoption may predispose adoptees to particular forms of concern and anxiety in regards to their need, and capacity, to fulfil expectations. The adoptee's sense of self-worth may be excessively structured around and dependent upon external validation. Any such anxiety will be compounded where the adoptee believes that her original relinquishment was due to some flaw or inadequacy within herself.

Notably, this sense of having been the object of a conditional and negotiated transaction is not mitigated by deliberate efforts adoptive parents may make to assure their child that their adoption is evidence of their having been "really wanted." Hahn (2005: 216–17) discusses this point eloquently, detailing the poignant unintended effects of the "we *chose* you" consolation. As Hahn explains, the difficulty confronting adopted children is to achieve some understanding of how the choice that was made in selecting them was really a choice that tracked something about *them*. The notion that adoptive parents responded to some property in them that signified them as "especially desirable" actually invests what Hahn calls a "bogus value" in the child: a valuing of properties that are purely incidental and to which the child is unlikely to be able to lay a unique claim. While Hahn points out (following Frankfurt, 2004) that love is always contingent and aimed at "an accidentally determined object," nevertheless she suggests that the "accidents and contingencies of love" are likely to be more problematic for adoptive families. As she says, "[t]he need to take the sting out of the contingency of love is more pressing in adoptive families, since adoptive children are even more a product of arbitrary circumstances external to themselves" (Hahn, 2005: 216). Thus, while a fear of inadequacy or of being the cause of parental disappointment is not

---

[9] Following Hollingsworth's explanation, "disruption" refers to removal of a child from an adoptive placement prior to adoption being finalized (Festinger, 1990; Rosenthal, 1993; Schmidt *et al.*, 1988). "Dissolution" involves the permanent cancellation of an adoption after finalization (Festinger, 1990). Hollingsworth (2003: 161) notes that there is little difference in the outcome of these, since both involve a decision that ends adoption.

a childhood anxiety unique to adoptees, for the reasons outlined the potential for it is heightened and intensified in the context of adoption.

Vulnerabilities around sense of self and of belonging dovetail with a third domain of vulnerability, specifically concerning *emotional independence*. By "emotional independence" I mean, broadly speaking, the ability to express one's own emotions and needs without being unduly inhibited by sensitivity to those of others. The particular type of vulnerability I have in mind here relates to adoptees' susceptibility to assuming excessive responsibility for the (real or perceived) emotional needs of adoptive parents. In many cases it appears that adoptees feel unable to speak openly about their curiosities and desires to investigate biological family origins.[10] A significant concern may develop over the potential for compounding the hurt caused by both infertility and adoption stigma. Parental silence, and especially outward denial, about adoption may contribute to the internalization of a sense of there being something regrettable, shameful, and painful about adoption, for both adoptees and their adoptive parents.

While closely tied to the other two types of vulnerability, the existence of a propensity for adoptees to be overly concerned and protective with regard to the emotions of their adoptive parents has been demonstrated in recent research indicating that adoptees both find it more difficult to talk about adoption, and are more interested in their birth parents, than their adoptive parents realize. From a study of 162 intercountry and domestic adoptees and their adoptive parents (Hawkins *et al.*, 2008) it was found that adoptive parents generally report their children to be fairly uninterested in most aspects of their adoptions, with over 70 per cent reporting that their children asked no questions at all in regard to the aspects of adoption the study examined.[11] Yet this result is in notable tension with two other findings of the research, namely that most of the 15-year-old adoptees reported thinking about their birth parents either "occasionally" or "frequently"; and that a majority would like some form of contact with birth parents. More significantly still, the study found that approximately 20 per cent of the adoptees believed that their parents have "some" or "great" difficulty talking about their adoption generally, where the child's perception of parental difficulty was generally not shared by the adoptive parents (amongst whom 89 per cent reported no difficulty talking about the adoption itself, and 87 per cent reported no difficulty talking about the children's birth families). Also significant for my purposes is the fact that 25 per cent of the 140 adoptees who responded to this particular survey believed that thinking about their birth parents would *definitely* (10%) or *possibly* (15%) hurt their adoptive parents' feelings if they knew, a result that was again discrepant with adoptive parents' self-reports.

Thus we see that many adoptees admit to experiencing a significant sense of actual or anticipated guilt about their interest in their biological family, as if such interest

---

[10] Such inhibition may, of course, be shared by donor-conceived children, who may equally desire to find out about, or even meet, their genetic progenitors.

[11] The studies discussed here draw on the larger, longitudinal "ERA" study.

necessarily embodies a rejection of the adoptive family, and will be unavoidably a reminder and a renewed source of emotional pain and anxiety for adoptive parents. To the extent that this occurs, the adoptee's capacity for emotional independence will likely be significantly curtailed.

Before moving on to the moral obligations that flow from acknowledgement of these adoption-specific domains of vulnerability, I want to conclude my discussion of the conditions that contribute to the development of these vulnerabilities by considering sociological and psychological analyses of infertility and adoption status as socially stigmatizing. Understanding adoption status as a stigmatizing trait motivates Miall's important suggestion that "the success or failure of adoption depends more on the ability of family members to resist devaluing societal attitudes and behaviours than on psychological adjustment per se" (1987: 38).

Occurring in an estimated 10–15 per cent of the population in Western countries, we know that the experience of infertility can be extremely emotionally painful for those affected, giving rise to feelings of failure, inadequacy, and grief (e.g. McQuillan *et al.*, 2003; Kirkman, 2003; Miall, 1986; Greil, 1997). However, to properly understand the causes of such emotional pain requires that we direct our attention beyond individual psychological characteristics in order to give due acknowledgement to the fact that infertility occurs and is experienced within a wider social context dominated by an "ideology of biology"—a *bionormativity*—that valorizes biologically related families and against which childlessness, whether voluntary or involuntary, might even constitute a form of "deviance" (Veevers, 1972: 268). As Miall points out, most adoption research has focused on examining the relationship *between* adoptive parents and adopted children, typically drawing on psychological, social work, or demographic models, and presenting psychological adjustment as "the crucial variable in explaining the success or failure of adoption" (1987: 34). As a result, analyses of maladjustment in adoptive families have tended to focus on parents' potential unresolved grief and other psychological effects of infertility. Considerably less analysis has been undertaken of the relationship between the adoptive family and the broader social community; or of broader social attitudes towards adoption and adoptive families—though this deficiency of attention to the *social* construction of infertility has begun to be corrected in the past decade or so (Greil, 1997; Greil *et al.*, 2010). Of crucial importance for my discussion, therefore, is Miall's (1987: 34) claim that "[s]ocial values surrounding adoption may have as much relevance for the success or failure of an adoption as the parents' adjustment to infertility or modes of coping."

Miall's proposition, which is both based on and reaffirms Kirk's (1964; Kirk and Mass, 1959) major sociological study of adoptive parents in the US and Canada, is particularly pertinent given Kirk's conclusion that the success or failure of adoption is most significantly a function of the family's modes of coping with adoption disclosure. Analyzed within its social context, and in particular through the theoretical model of social stigma influentially developed by sociologist Erving Goffman in 1963, the pain and grief associated with infertility must be understood not as necessary, natural, or innate, but as the very real product of a social stigma that has not been widely acknowledged. As Miall (1987: 35) states, "[i]n a society that values biological kinship ties, the lack of a blood tie between a mother and her children may be an attribute which is discrediting or stigmatizing to her." A brief examination of the results of adoption stigmatization studies illustrates the significant impact of broader social attitudes on the adoptive family, including in shaping the ways in which adoption is "managed" within and by the family. It also strongly suggests the importance, noted by Wegar, of acknowledging that "the identities, attitudes, and behaviour patterns of stigmatized individuals cannot be understood apart from the social context that shapes them" (2000: 364–5).

First, both Kirk and Miall found evidence of significant social prejudice experienced by adoptive *parents*, in particular in relation to the perceived validity of their adoptive parent–child bonds. Miall (1989: 292) claims that her research shows that a majority of respondents "are aware of negative attitudes about adoption in the larger society, attitudes which focus on the lack of a biological or blood tie." Miall claims that her research indicates three broad themes, identified by adoptive parents as negative societal beliefs that create a sense of being rendered inferior and which are, therefore, stigmatizing for them: first, that the biological tie is important for bonding and love, and therefore bonding and love in adoption are "second best"; second, that adopted children are "second-rate" because of their unknown genetic past; and third, that adoptive parents are not "real" parents (1989: 281–2). These kinds of findings are further confirmed by Wegar's analysis of the first large-scale study of community attitudes towards adoption in the US, which found that "only 32% of respondents expressed unqualified support for adoption," and that many Americans retain a view of adoption as "second-best" and "a suspect family form" (2000: 363).[12]

Second, available research with *adoptees* likewise reveals a perception of adoption as socially stigmatizing, with the absence of "blood ties" regarded as the objective basis of the "social failing" for which adoptees are discredited (March, 1995: 654). Surveyed adoptees reported experiencing social discrimination from others, expressed in certain forms of questioning (e.g. of adoptee's "rightful position within the adoptive family structure") and in comments around the adoptee's need to "be grateful," their assumed lack of inheritance, and so on. Clear evidence of adoptee awareness of being perceived and characterized as "different" enabled March (1995: 658–9) to conclude that adoption can indeed be understood as socially stigmatizing, both for adoptees themselves and for adoptive parents.

As psychologists and sociologists have explained, those who experience social stigma characteristically respond by developing a range of strategies for dealing with its impact. In particular, Kirk's (1964; Kirk and Mass, 1959) research led to his observation that adoptive parents typically develop "dichotomous patterns of defining

---

[12] A separate study analyzed by Wegar (2000: 363) also revealed that 30 per cent of adolescent adoptees believed that "people expect adopted kids to have problems."

adoption in response to the community's view of them and their children as different"—adaptive patterns which he labelled "rejection of difference" and "acknowledgement of difference" tendencies. Subsequent research has elucidated the many subtle and nuanced ways in which adoptees and their families carry out so-called "information management" tasks around adoption (e.g. Pachankis, 2007; March, 1995; Miall, 1989). Such strategies vary depending on whether the stigmatizing trait is overt or concealable; where concealable, common strategies involve "defensive" practices such as secrecy, selective concealment, fabrication, use of so-called "disidentifiers" (to mislead/confuse others regarding one's status), and outright deviance disavowal (Miall, 1989: 280–4).

But while the context of social stigmatization certainly makes deployment of "defensive" strategies of the above kinds understandable, the impact on adoptees of denial or minimization of adoption status and difference forces us to consider the moral implications for adoptive parenting. It is to this question that I now turn.

Post-Adoptive Parental Obligation

What, if anything, does the preceding analysis of domains of adoption-specific vulnerability, and the conditions in which those vulnerabilities arise, suggest about the existence of special post-adoptive obligations of adoptive parents? In addressing this question I will be particularly interested in the extent to which post-adoptive parental obligations extend beyond so-called *negative* obligations—obligations of non-interference—to encompass also *positive* obligations—namely obligations to more actively provide for and further an adopted child's interests. Negative duties of non-interference—particularly in relation to a child's efforts to seek family-of-origin information and perhaps contact—are fairly straightforwardly supported by a child's interests in knowing her biological family history. However, taking the special vulnerabilities of adoptees seriously requires us to recognize that the obligations of adoptive parents are more extensive than is commonly recognized in adoption policy and practice, and in the support and resources made available to adoptive parents. To be sure, there are important limits to post-adoptive parental obligations, about which I shall say more; but for now, what might be proposed in terms of such obligations?

We have advanced a long way in both our understanding and our socio-legal safeguarding of the rights and needs of adopted children, and in our commitment to a child-centred approach based on the principle of the adopted child's interests as paramount.[13] Basic legal entitlements and rights of adoptees include: to be told of their adoption status; access to information about origins from the time of placement; and support in their search for information or contact. Minimally, then, these entitlements

---

[13] The Hague Convention on Intercountry Adoption (in force since 1995) e.g. establishes minimum standards and rules for procedural transparency and safeguards, as well as for respect for the interests of the children in addition to those of adoptive families and families-of-origin.

translate into parental obligations to ensure that a child knows of her adoption, and to provide available family-of-origin information in line with regulatory and legal requirements. As indicated, however, I think the specific vulnerabilities arising in particular in the domains of adoptee identity development and emotional independence suggest more substantial and distinctly *moral* (as opposed to legal) post-adoptive parenting obligations, beyond allowing the adopted child freedom to pursue her adoption-related interests and inquiries. One specific class of positive obligations is especially important. Fulfilment of these will be a precondition for the alleviation of other adoptee vulnerabilities examined in this chapter. Specifically, these are obligations of intra-familial "communicative openness."

The adopted child's significant vulnerabilities around emotional independence as well as identity development suggest the need for parents to go to considerable efforts to provide assurance of their support for family-of-origin curiosities and investigations. As already noted, Kirk (1964) first propounded the view that intra-familial "communicative openness," empathy, and support are of central importance for adoptee psychological adjustment. Following Kirk, Stein and Hoopes's (1985) more recent research found that the more open the adoptive family's communication style, the fewer the identity problems faced by adopted adolescents. While their study comparing adopted and non-adopted adolescent populations showed that adoptees do not *as a class* appear to manifest greater or more frequent problems of identity or maladjustment as compared with non-adoptees, other research has shown that identity problems and low self-esteem are associated with *specific forms of adoption-related communication* within the adoptive family. Brodzinky's work (2005, 2006) has been influential in demonstrating a clear association specifically between *communicative* openness and adoptee well-being. Brodzinsky (2006) was the first to examine the relationship between "communicative" and "structural" openness, the latter referring to adoptions in which arrangements are in place for continued contact with the birth parent/family (pp. 2–3). His research led him to conclude that "although structural openness and communication openness are positively correlated, only communication openness independently predicted children's adjustment"; and that "family process variables generally are more predictive of children's psychological adjustment than family structural variables" (Brodzinsky, 2006: 1). For Brodzinsky (2005: 149), the notion of "communicative openness" includes, amongst other things,

a willingness of individuals to consider the meaning of adoption in their lives, to share that meaning with others, to explore adoption related issues in the context of family life, to acknowledge and support the child's dual connection to two families, and perhaps to facilitate contact between these two family systems in one form or another.

Jones and Hackett's (2006) review of adoption studies provides further confirmation of a connection between communicative openness and the adopted child's development of a positive identity (e.g. Howe and Feast, 2003), and higher levels of adult adoptee satisfaction with their adoption (see also Raynor, 1980).[14] Conversely, secrecy and discomfort in discussing adoption has been shown to be associated with reduction in well-being, adjustment, and identity formation in adoptees (see also Rosenberg and Groze, 1997; Triseliotis, 1973).

The potential for communicative openness to significantly mitigate adoptee vulnerability requires us to be clearer regarding precisely what is involved in, and required by, communicative openness. Is communicative openness simply a matter of adoptive parents being *responsive* to adoptee-initiated adoption discussion? Or could conduct beyond mere responsive openness be required, such that adoptive parents have obligations to *initiate* and *reinvite* adoption discussion?

It is my view that the earlier-discussed reported notable discrepancies between adoptees' and adoptive parents' perceptions of difficulty in talking about adoption and interest in biological parentage provide sufficient grounds for an obligation on the part of adoptive parents to take responsibility for *initiating* adoption-related communication with their children. In making such a proposal, however, it must be acknowledged that significant challenges exist in respect of adoptive parents' capacities to fulfil such an obligation.[15] As the above-cited research has shown, parental discomfort with discussing adoption jeopardizes communicative openness within the family, and can thereby impede the adopted child's optimal psycho-emotional development. I have argued that this discomfort arises in largest measure from the impact of social conditions that significantly shape how adoption is framed and responded to within society: namely, the extant conditions of widespread social stigmatization of both infertility and adoption. There is therefore a genuine question about the extent to which adoptive parents will be capable of effectively managing and overriding their discomfort, given that such social stigma is ongoing.

What this acknowledgement underscores, however, is not the inappropriateness of proposing parental post-adoptive obligations, but rather the necessity that adoptive parents receive appropriate counselling and support in managing the post-adoption relationships. Adoption social work and support practices will, in most cases, need to be significantly reformed so as to adequately support the capacity of adoptive parents to navigate both their own emotions around adoption as well as the social stigma they encounter, and thereby to fulfil their communicative obligations. It is not my intention here to detail the nature and extent of the required reforms. One question that does warrant some preliminary comment, however, is whether an obligation of initiated communicative openness on the part of adoptive parents ought also be supplemented by obligations of *structural* openness. As noted, Brodzinsky (2005) concluded from his research that communicative openness is more important for determining adoptee outcomes than whether or not there is structural openness in the form of regular

---

[14] These results have not always been replicated in subsequent research. See e.g. Neil (2009), but note Neil's own judgement that her results may not reliably disprove Brodzinsky's corroborated findings (p. 18).

[15] Thanks to an anonymous reviewer for prompting me to more fully acknowledge this point.

contact and involvement with the birth family. Furthermore, recent studies suggest that *generalizable* claims about the impact—positive or negative—of structurally open adoption on child outcomes cannot be made. It seems clear that whether structural openness will promote adoptee welfare will be highly case-specific and reflective of a wide range of possible factors, including most notably pre-placement experiences, age, and presence of abuse or neglect (Neil, 2009). In addition, Anita Allen (2005) is surely correct in arguing that structurally open adoption—in particular of the form involving ongoing inclusion of the birth family in the lives of the adoptive family—is neither pragmatically or morally appropriate in every case.

Nevertheless, while Allen correctly points out that discharge of the primary obligation to provide for an adopted child's welfare does not entail ongoing active inclusion of birth parents in the adoptive family's life, I would argue that both communicative openness and a willingness to revisit structural arrangements throughout the course of a child's life are morally reasonable obligations to impose. Indeed I would go so far as to propose that there ought to be a defeasible presumption in favour of adoptive parents being required to actively facilitate fulfilment of a child's expressed desire for ongoing contact with her birth parent(s), absent risks of harm to the child.[16] Accordingly I would argue that only where ongoing contact would present a risk of harm to the child, either directly or by means of significant negative disruption to the adoptive family, could an adopted child's preference for contact be justifiably overridden by the adoptive parents' obligation to ensure her welfare. But such a requirement falls well short of mandating any specific form of contact or structural openness, and the required empirical evidence simply is not available to ground any such specific moral obligations.

Limits to Parental Obligations

I have suggested that the imposition of certain *positive* parental obligations of initiated communicative openness are supported by virtue of their capacity to ameliorate adoption-specific vulnerabilities, especially those affecting identity development and emotional independence. However, it might be thought that a second set of post-adoptive obligations is strongly suggested by the foregoing analysis of problems posed by the ideals and imperatives of biological relatedness and "resemblance-based fit," as well as by consideration of adoptee and adoptive parent welfare. As I have discussed, our widespread commitment to securing the closest possible "match" (of one kind or another) between adopters and adoptees has not served the adoption

---

[16] In that vein, Australia's New South Wales Adoption Regulation (2003) includes as part of its eligibility assessment for would-be adoptive parents assessment not only of their skills and experience in relation to meeting the specific needs of an adopted child, but also of "their appreciation of the importance of, and capacity to facilitate, contact with the child's birth parents and family, and exchange of information about the child with the child's birth parents and family" (clause 12).

community well. We have seen that the valorization of familial biological relatedness—which at times motivates significant effort to obscure biological *non*-relatedness where the ideal is not attainable—has its foundations in deeply flawed assumptions of a geneticist and biological essentialist kind. And we know that, in the adoption context, such assumptions and valorizations have resulted in distorted adoption practices, as well as in false attributions of necessary identity-deficits in adoptees and attachment-deficits within adoptive families.

Moreover, compelling findings from research with groups experiencing social stigma actually point to considerable potential benefit from disclosure of the stigmatizing trait, applicable here both to adoptive parents and adoptees. Miall's (1987) research indicates that while immediate revelation of adoption status to strangers may be most likely to reflect insecurity, strategies of "passing" are experienced by adoptive parents as limited and "invariably accompanied by unease" (p. 291). Furthermore, Pachankis's important recent work highlights the extent to which concealment of a stigmatizing trait poses significant cognitive hazards, including heightened anxiety, vigilance, and suspiciousness (2007: 333). Indeed, Pachankis reports that participants with a concealable stigma—such as adoption status (at least in "matched" arrangements)—experience more negative affect than those with a visible stigma; and further research has shown that those who maintain significant secrets tend to be more lonely, shy, introverted, socially avoidant, and socially anxious (2007: 334–6). While clearly aimed at avoidance of negative evaluation and rejection, and while disclosure should not be assumed to be always positive or beneficial, nevertheless there are significant psycho-emotional costs to ongoing concealment of potentially stigmatizing traits. By contrast, as Miall discusses, research has shown that people engage in selective disclosure as both "therapeutic" and "preventive" strategies: therapeutic in the sense that disclosure can enhance self-esteem, relieve anxiety, and enable renegotiations of personal perceptions of stigma as well as a sense of connection and solidarity (Miall, 1986: 275); and "preventive" (and indeed "educative") in the sense that disclosure offers an opportunity to influence others' ideas about the supposedly discrediting status, thereby potentially preventing the need for future concealment and challenging existing social attitudes (Miall, 1989: 289).

In view of the adopted child's vulnerabilities around identity-development and self-worth, as well as potential benefits for adoptive parents from refusal to conceal adoption status, might there, then, be grounds for further parental obligations of disclosure, which will preclude as unacceptable—in any case where it *could* potentially be achieved—the outward concealment of a child's adoptive status, no matter how well intentioned?[17] That is, ought there to be positive parental obligations of *extra-familial* disclosure of a child's adoptive status?

---

[17] Such an obligation would, of course, not arise in cases of transracial adoption, where "passing" is not an option due to what are typically considered to be overt markers of non-genetic relatedness.

I have space only for a brief response to this question. To begin, I would not deny that the "passing" of a child as biological progeny is morally hazardous and therefore stands in need of justification. This is the case partly in virtue of the impact of "passing" on a child's identity development vulnerabilities, but also because of the way in which such strategies effectively acquiesce to imperatives of biological relatedness, and in so doing inadvertently reinforce entrenched "biologistic" and bionormative paradigms. To be sure, adoption does present special opportunities to disrupt those paradigms and challenge the unfounded, restrictive, and distorting imperatives and norms of bio-logical relatedness. Given those opportunities, it might be thought reasonable to sug-gest that adoptive parents incur more "activist" obligations of *extra-familial* adoption disclosure.

I believe two considerations count decisively against the imposition of such obliga-tions. The first is grounded in adoptee welfare; the second points to important limita-tions to what adoptive parenting can be reasonably expected, and asked, to achieve; and to matters of political and social efficacy.

First, it is simply not obvious that an adopted child's vulnerabilities will be amelio-rated by a moral prohibition on "passing," or by the imposition of positive obligations of extra-familial disclosure. In some situations, being "passed" as biological progeny is not only a child's expressed preference but may in fact be what they most need, at that time, given those preferences and the vulnerabilities they may have around sense of belonging and security.[18] While it may indeed be less likely that a child raised with full intra-familial communicative openness will want to be "passed" as biologically related, we must be cognizant of the limitations to what attitudes *within* adoptive families can achieve in terms of attitudes *outside* of such families. The child who does not want her adoptive status known may manifest a perfectly reasonable response to awareness of broader social attitudes towards adoption as steeped in ideals of biological relatedness. Moreover, empirical research shows that a child's needs (for information, contact, disclosure, etc.) change over time, and may be intensified at certain developmental stages. Recognition of this reinforces the need for ongoing and active communicative openness within the adoptive family, in order that the child's wishes at any given time can be understood and promoted in a way that respects her needs, vulnerabilities, and entitlements. Maintaining a primary focus on the adopted child's interests thus weighs heavily against any blanket prohibitions, permissions, or prescriptions in regards to extra-familial adoption disclosure.

Second, the struggle against "biologism" must be waged where it has the greatest chance of widespread and sustainable success. Collectively, a refusal to hide or obscure

---

[18] Note that the transracially adopted child will not have such a preference, given the impossibility of being "passed" in the first place. Of course, the fact that they cannot access "passing" strategies may well have the implication that transracial adoption intensifies particular identity-related adoptee vulnerabilities. On the other hand, it is at least conceivable that the undeniability of their genetic non-relatedness, *ab initio*, could potentially aid in the consolidation of adoptee sense of self-identity. Regardless, sound evidence is required for any claims of increased vulnerability in transracially adopted children to be substantiated.

the realities of adoption in society has the power to challenge views of its abnormality and inferiority *vis-à-vis* biologically grounded family composition. Making this kind of point in relation to transracial adoption in particular, Fogg Davis (2005: 247) notes the considerable potential such adoptions have to "challenge both the normative weight of same-race family structure and the flawed assumption that strong and meaningful family bonds require a genetic tie." Nevertheless, in correctly focusing proposals for reform on social practice and policy rather than exclusively on individual behavioural change, Fogg Davis argues convincingly that considerations of cultural pluralism, fairness, and privacy cannot ultimately sustain objections to the implementation of racial randomization procedures in adoption placement, or to the removal of the capacity for race-based selection of prospective adopters and adoptees. The same applies, I think, to the other social or physiological categorizations—of sex, hair/eye colour, physical stature, culture, religion, etc.—that serve imperatives of "fit" in less obvious ways. Importantly, this entails neither denial nor trivialization of the putative value of maintaining an adoptee's cultural identities, identifications, and connections, least of all where adoptive placement occurs after the child has actually developed such identifications and connections. Nor does it entail rejection of the value of cultural pluralism. Rather, what is entailed is rejection of the notion that race- or culture-based adoptive *placement* is necessary for maintenance of "racial" or cultural identity and connection, for the kinds of reasons so forcefully provided, for example, by Haslanger's (2005) work on race and adoption.

More socially and politically transformative, then, than the refusal of individual families to conceal adoption will be the eradication of formal adoption placement practices that base selection on criteria for securing physiological, racial, or other such forms of "resemblance fit"; and their replacement by alternative policies embracing radically different priorities—including, I would suggest, full "attribute randomization" within adoption practice. Clearly, such reform objectives must be pursued through changes to adoption policies and practices themselves, facilitated by attitude change in society more broadly, including amongst prospective adoptive parents. Whatever contribution adoptive parents and their families may choose to make to such activist goals, however, they do not arise uniquely as special obligations for adoptive parents. To insist otherwise would, I suggest, amount to the imposition of unacceptably discriminatory demands and burdens upon a group of persons already carrying the burden of belonging to a group that is socially stigmatized in our bionormative culture: adoptees and their adoptive families.

Conclusion

The importance of the role played by intra-familial communicative processes in facilitating successful adoption outcomes and promoting adoptee welfare establishes grounds upon which specific communicative obligations of openness can justifiably be imposed on adoptive parents. Beyond these intra-familial post-adoptive obligations, the further imposition of *extra*-familial adoption disclosure obligations would represent one possible means by which we might begin to respond to the considerable existing need for the kind of activism and reform that would be required in order to undermine the biologistic prejudices and biases that continue to distort and stigmatize the adoption experience for all those affected.

However, this chapter has argued that a consideration of the adopted child's interests and vulnerabilities, as well as of matters of social and political efficacy, tell against the imposition of such additional obligations. As noted, an adopted child raised with full intra-familial communicative openness is perhaps unlikely to object to having her adopted status disclosed outside the family; but in the event that she does object, a consideration of the moral imperatives that properly govern all parent–child relationships requires that the child's preferences be respected, as an extension of promoting her best interests. We may thereby miss an opportunity to challenge the many problematic social norms, biases, and stereotypes that continue to complicate adoption. But we will have more appropriately responded to the adopted child's needs and vulnerabilities, as well as helped to consolidate her sense that, to a significant extent, core aspects of her identity remain a matter for her own determination and articulation. And we will have given due weight to the fact that the ongoing stigmatization of adoption is a persistent and pervasive *social* phenomenon—one that requires widespread social redress rather than a disproportionate and potentially discriminatory further burdening of adoptive families.

Acknowledgements

I would like to thank the editors, co-contributors, and reviewers of this volume for their helpful comments on earlier versions of this chapter.

References

Allen, A. L. (2005). Open Adoption is Not for Everyone. In S. Haslanger and C. Witt (eds), *Adoption Matters: Philosophical and Feminist Essays* (pp. 47–67). New York: Cornell University Press.

Associated Press (2010). Woman Returns Adopted Grandson to Homeland by Putting him on a One-Way Flight. *Sydney Morning Herald*, 10 Apr. Retrieved June 2013 from http://www.smh.com.au/world/woman-returns-adopted-grandson-to-homeland-by-putting-him-on-oneway-flight-20100410-rza5.html.

Bartholet, E. (1994). *Family Bonds: Adoption, Infertility, and the New World of Child Productions*. Boston, MA: Beacon Press.

Brakman, S. V., and Scholz, S. J. (2006). Adoption, ART and a Re-conception of the Maternal Body: Toward Embodied Maternity, *Hypatia*, 21(1), 54–73.

Brodzinsky, D. (1993). Long-Term Outcomes in Adoption. *The Future of Children*, 3(1), 153–66.

Brodzinsky, D. (2005). Reconceptualizing Openness in Adoption: Implications for Theory, Research and Practice. In D. M. Brodzinsky and J. Palacios (eds), *Psychological Issues in Adoption: Research and Practices* (pp. 145–66). Westport, CT: Praeger.

Brodzinsky, D. (2006). Family Structural Openness and Communication Openness as Predictors in the Adjustment of Adopted Children. *Adoption Quarterly*, 9(4), 1–18.

Butler, J. (2004). *Precarious Life: The Powers of Mourning and Violence*. London: Verso.

Butler, J. (2009). *Frames of War: When is Life Grievable?* London: Verso.

Festinger, T. (1990). Adoption Disruption: Rates and Correlates. In D. M. Brodzinsky and M. D. Schecter (eds), *The Psychology of Adoption* (pp. 201–18). New York: Oxford University Press.

Fineman, M. (2008). The Vulnerable Subject: Anchoring Equality in the Human Condition. *Yale Journal of Law and Feminism*, 20, 1–23.

Fogg Davis, H. (2002). *The Ethics of Transracial Adoption*. Ithaca, NY: Cornell University Press.

Fogg Davis, H. (2005). Racial Randomization: Imagining Nondiscrimination in Adoption. In S. Haslanger and C. Witt (eds), *Adoption Matters: Philosophical and Feminist Essays* (pp. 247–64). New York: Cornell University Press.

Frankfurt, H. G. (2004). *The Reasons of Love*. Princeton, NJ: Princeton University Press.

Goodin, R. E. (1985). *Protecting the Vulnerable: A Reanalysis of our Social Responsibilities*. Chicago, IL: University of Chicago Press.

Greil, A. L. (1997). Infertility and Psychological Distress: A Critical Review of the Literature. *Social Science and Medicine*, 45(11), 1679–1704.

Greil, A. L., Slauson-Blevins, K., and McQuillan, J. (2010). The Experience of Infertility: A Review of Recent Literature. *Sociology of Health and Illness*, 32(1), 140–62.

Hahn, S. (2005). Accidents and Contingencies of Love. In S. Haslanger and C. Witt (eds), *Adoption Matters: Philosophical and Feminist Essays* (pp. 195–218). New York: Cornell University Press.

Haslanger, S. (2005). You Mixed? Racial Identity without Racial Biology. In S. Haslanger and C. Witt (eds), *Adoption Matters: Philosophical and Feminist Essays* (pp. 265–89). New York: Cornell University Press.

Hawkins, A., Beckett, C., Rutter, M., Castle, J., Groothues, C., Kreppner, J., Stevens, S., and Sonuga-Barke, E. (2008). Communicative Openness about Adoption and Interest in Contact in a Sample of Domestic and Intercountry Adolescent Adoptees. *Adoption Quarterly*, 10(3–4), 131–56.

Hollingsworth, L. D. (2003). When an Adoption Disrupts: A Study of Public Attitudes. *Family Relations*, 52(2), 161–6.

Holtzman, M. (2008). Defining Family: Young Adults' Perceptions of the Parent–Child Bond. *Journal of Family Communication*, 8(3), 167–85.

Howe, D., and Feast, J. (2003). *Adoption, Search and Reunion: The Long-Term Experience of Adopted Adults*. London: BAAF.

Jones, C., and Hackett, S. (2008). Communicative Openness within Adoptive Families: Adoptive Parents' Narrative Accounts of the Challenges of Adoption Talk and the Approaches Used to Manage these Challenges. *Adoption Quarterly*, 10(3), 157–78.

Jones, C., and Hackett, S. (2011). The Role of "Family Practices" and "Displays of Family" in the Creation of Adoptive Kinship. *British Journal of Social Work*, 41, 40–56.

Kirk, H. D. (1964). *Shared Fate: A Theory and Method of Adoptive Relationships*. New York: Free Press.

Kirk, H. D., and Mass, H. S. (1959). A Dilemma of Adoptive Parenthood: Incongruous Role Obligations. *Marriage and Family Living*, 21(4), 316–28.

Kirkman, M. (2003). Infertile Women and the Narrative Work of Mourning: Barriers to the Revision of Autobiographical Narratives of Motherhood. *Narrative Inquiry*, 13(1), 243–62.

Kittay, E. (1997). *Love's Labour.* New York: Routledge.

Leighton, K. (2005). Being Adopted and Being a Philosopher: An Exploration of the "Desire to Know" Differently. In S. Haslanger and C. Witt (eds), *Adoption Matters: Philosophical and Feminist Essays* (pp. 146–70). New York: Cornell University Press.

Lotz, M. (2008). Overstating the Biological: Geneticism and Essentialism in Social Cloning and Social Sex Selection. In L. Skene and J. Thompson (eds), *The Sorting Society: Ethics of Genetic Screening and Therapy* (pp. 133–48). Cambridge: Cambridge University Press.

McQuillan, J., Greil, A., White, L., and Jacob, M. C. (2003). Frustrated Fertility: Infertility and Psychological Distress among Women. *Journal of Marriage and Family,* 65(4), 1007–18.

March, K. (1995). Perception of Adoption as Social Stigma: Motivation for Search and Reunion. *Journal of Marriage and Family,* 57(3), 653–60.

Marsiglio, W. (1998). *Procreative Man.* New York: New York University Press.

Miall, C. E. (1986). The Stigma of Involuntary Childlessness. *Social Problems,* 33(4), 268–82.

Miall, C. E. (1987). The Stigma of Adoptive Parent Status: Perceptions of Community Attitudes toward Adoption and the Experience of Informal Social Sanctioning. *Family Relations,* 36(1), 34–9.

Miall, C. E. (1989). Authenticity and the Disclosure of the Information Preserve: The Case of Adoptive Parenthood. *Qualitative Sociology,* 12(3), 279–302.

Modell, J. S. (1994). *Kinship with Strangers.* Berkeley, CA: University of California Press.

Modell, J. S., and Dambacher, N. (1997). Making a "Real" Family. *Adoption Quarterly,* 1(2), 3–33.

Neil, E. (2009). Post-Adoption Contact and Openness in Adoptive Parents' Minds: Consequences for Children's Development. *British Journal of Social Work,* 39, 5–23.

New South Wales (NSW) Government, Office of the Children's Guardian (2010). Information Sheet 11. Standard 11: Initial Assessment and Placement in Out-of-Home Care, 2 Aug. Retrieved June 2013 from http://www.kidsguardian.nsw.gov.au/example-folder-5/ Accreditation/information-sheets-for-designated-agencies.

O'Neill, O. (1979). Begetting, Bearing and Rearing. In O. O'Neill and W. Ruddick (eds), *Having Children: Philosophical and Legal Reflections on Parenthood* (pp. 25–38). Oxford: Oxford University Press.

Pachankis, J. E. (2007). The Psychological Implications of Concealing a Stigma: A Cognitive-Affective Behavioural Model. *Psychological Bulletin,* 133(2), 328–45.

Pertman, A. (2000). *Adoption Nation: How the Adoption Revolution is Transforming America.* New York: Basic Books.

Raynor, L. (1980). *The Adopted Child Comes of Age.* London: George Allen & Unwin.

Rosenberg, E. B., and Groze, V. (1997). The Impact of Secrecy and Denial in Adoption: Practice and Treatment Issues. *Families and Society,* 78, 522–30.

Rozenthal, J. A. (1993). Outcomes of the Adoption of Children with Special Needs. *The Future of Children: Adoption,* 3, 77–88.

Schmidt, D. M., Rosenthal, J. A., and Nombeck, B. (1988). Parents' Views of Adoption Disruption. *Children and Youth Services Review,* 10, 119–30.

Shanley, L. M. (2001). *Making Babies, Making Families: What Matters Most in an Age of Reproductive Technologies, Surrogacy, Adoption, and Same-Sex and Unwed Parents.* Boston, MA: Beacon Press.

Smith, J. F. (2005). A Child of One's Own: A Moral Assessment of Property Concepts in Adoption. In S. Haslanger and C. Witt (eds), *Adoption Matters: Philosophical and Feminist Essays* (pp. 112–31). New York: Cornell University Press.

Stein, L. M., and Hoopes, J. L. (1985). *Identity Formation in the Adopted Adolescent*. New York: Child Welfare League of America.

Turner, B. S. (2006). *Vulnerability and Human Rights*. University Park, PA: Penn State University Press.

Triseliotis, J. (1973). *In Search of Origins: The Experience of Adopted People*. London: Routledge & Kegan Paul.

Veevers, J. (1972). The Violation of Fertility Mores: Voluntary Childlessness as Deviant Behaviour. In C. Boydell, C. Grindstaff, and P. Whitehead (eds), *Contemporary Families and Alternative Lifestyles* (pp. 571–92). Beverly Hills, CA: Sage.

Velleman, J. D. (2005). Family History. *Philosophical Papers*, 34, 357–78.

Wegar, K. (2000). Adoption, Family Ideology, and Social Stigma: Bias in Community Attitudes, Adoption Research, and Practice. *Family Relations,* 49(4), 363–70.

Witt, C. (2005). Family Resemblances: Adoption, Personal Identity, and Genetic Essentialism. In S. Haslanger and C. Witt (eds), *Adoption Matters: Philosophical and Feminist Essays* (pp. 135–45). New York: Cornell University Press.

12

The Political Geography of Whites Adopting Black Children in the United States

Heath Fogg Davis

The adoption of black children[1] by white parents in the United States takes place in a geographical context that is marked by high levels of residential segregation between whites and blacks (Logan and Stults, 2011). The public debate over these adoptions has focused on the moment of adoptive placement—the policy question of whether whites should be permitted to adopt black children. Thus, the future-oriented question of *where* these biracial families will reside has often receded from explicit scrutiny. From a law and public policy perspective, this is justified, given that a core tenet of all constitutional democracies is that neither the government nor other individuals may dictate where someone lives. But legal questions are not the only relevant questions to consider in adoption. In this chapter, I expand the locus of the debate to consider some of the moral aspects of residential freedom as it pertains to the adoption of black children by whites in American political geography. Are white adoptive parents of black children *morally* obligated to search for and make homes for their biracial families in neighbourhoods that are not predominately white, and where they are likely to have at least some black neighbours?

I believe they do. The moral ought here is not based on the contention or speculation that a black adopted child cannot develop a healthy self-concept in a predominantly white community. Instead, the moral ought is a more magnified version of the general moral responsibility that we all have to make residential decisions that do not perpetuate long-standing patterns of racially segregated housing. Because whites by and large have much greater social and economic power than blacks do to enact their race-based

---

[1] In this chapter I will use the term "black" to include biracial individuals in adherence to the long-standing American cultural, and to some extent legal, norm of the "one drop" rule. According to this rule, anyone with one black ancestor (one drop of "black blood") is considered to be black.

housing preferences, it follows that whites have a moral duty to *lead* processes of desegregation, even and especially when such action brings about real and imagined economic and social sacrifice. White and black adults often invoke their children's welfare as a primary reason for choosing to live in predominantly white communities. "We would love to live in (gentrified parts of) the city, but we moved to the suburbs for the schools." This chapter troubles the status quo and culturally accepted assumption that parents must live in predominantly white neighbourhoods "for the sake of their children." While the prospect of depriving one's child of resources such as a high or even decent quality of education is cause for real concern, this strikes me as a very narrow way of conceiving of the short- and long-term costs and benefits of racial integration. When parents with higher levels of economic and social capital move to areas where their neighbours have lower levels of economic and social capital, they and their children will incur costs, both real and imagined. But it is also true that these parents bring their resources with them to impart to their children in these new locations. These resources can also bolster their new communities, especially when it comes to the funding of public schools, which is generated by local property taxes in most US school districts. Non-quantifiable positive outcomes flow from such integration too. For example, in living in a racially mixed neighbourhood the parents themselves are likely to be changed by their new environments, and to think of their own racial identities in different, unexpected ways that should not be thought of in strictly negative terms. How would our conception of the political geography of white–black transracial adoption change if we considered the process of white-led racial integration as yielding these benefits, instead of viewing residential racial integration as a zero-sum game in which whites and their children—adopted or not—always and only lose resources that they currently possess or may acquire in the future?

When whites parent black children, these extra-legal questions become even more pressing because the family itself, and not just a particular neighbourhood, is a place of racial integration. Blacks and whites are not equal parties to the prospect of American racial integration. And this racial power disparity is magnified in the case of whites adopting black children because children must live where their parents purchase or rent a home. By taking the time and making the effort to select a racially diverse community for the family, white adoptive parents convey an important message to their black adopted children: that the racial integration of the family is a collective endeavour led by them, and that they are willing to sacrifice the racial familiarity of living in a community that mostly reflects their white racial identity and experience. If the white parents are like most white Americans, they will have grown up and spent most of their lives in predominantly white communities. Moving out of such communities may mean physically moving outside of their racial comfort zones. It may require other compromises, too, such as having to commute greater distances for work.

Our racial comfort zones have a geographical location that we often take for granted, and thus fail to question. What feels comfortable to us can be at odds with what morality demands of us. Policy-makers and judges do not have the final word on our moral responsibilities. In my 2002 book, *The Ethics of Transracial Adoption*, I called attention to some of these extra-legal moral considerations by arguing that race-consciousness, in the form of *racial navigation*, should supplement the liberal principle of non-discrimination in adoptive placements. In retrospect, I paid insufficient attention to the actual political geography to be navigated when whites adopt black children, especially the political geography of residential location and homemaking. My intention in using the verb *navigate* was to underscore black adoptees' individual agency in constructing and revising their own racial identities, both as children and as they mature into adulthood. I did so in response to my sense that the public debate over these transracial adoptions too often conceptualized black children in need of adoption as passive receptacles of racial culture, and forever-young pawns in a more diffuse and long-standing racial debate between colourblindness and race-consciousness. In doing so, I neglected to explicitly anchor such navigation in the political geography of American race. I also paid too little attention to the parents' own racial navigation, especially the asymmetrical power that they possess in determining where the family will reside and make a home together.

I begin with a brief overview of the major philosophical differences that have shaped the public debate over the ethics of whites adopting black children in the US. I situate my race-conscious argument in support of these adoptions within this theoretical framework. Next I introduce the concept of political geography, and explain both its theoretical and empirical dimensions. I discuss some of the formal and informal factors that sustain and perpetuate high levels of residential segregation between whites and blacks. In the following section I demonstrate how the theoretical trope of racial navigation relates to political geography. I highlight the role that power and racial privilege play in normalizing the *de facto* or extra-legal racial segregation that marks out the political geography to be navigated in contemporary US society. Having discussed the macro-political geography of whites adopting black children, I conclude that section by zooming in on the micro-political geography of homemaking. I invoke Sally Haslanger's use of feminist phenomenology to describe her own experience of parenting black children as a white adoptive mother to exemplify the sort of white-led racial integration I have in mind.

The Political Debate

The political debate over white-black adoption has focused on the up or down question of whether whites should be permitted to adopt black children. To its supporters, the desire and willingness of whites to adopt black children exemplifies the principle of colourblind non-discrimination that fuelled much of the 1960s American civil rights movement (Bartholet, 1991; Kennedy, 1994). Those opposed to the practice of whites adopting black children view these adoptions in a less optimistic, even discriminatory, light. They laud the legal gains of the civil rights movement, but call attention to the structural aspects of anti-black racism that the movement's liberal legal approach left intact. The desire and willingness of some whites to adopt black children in need of homes does not address the structural racism that makes black children disproportionately in need of adoption in the first place (Roberts, 1997; Perry, 1993–41993–4). Many of these critics also question whether whites are culturally equipped to teach black children the "survival skills" necessary for coping with the everyday racism of being black in American society. More extreme critics placed these adoptions within a black nationalist framework that conceptualized black children as community resources and progenitors. On this narrative, black children belong to the black community, and their adoption out of the community by members of the white majority is tantamount to genocide (NABSW, 1994).

The preposition and definite article are instructive here because being raised *in the* black community does not require that the black adoptive parents of black children reside in a black or predominantly black neighbourhood. "The black community" refers instead to a set of cultural values and coping strategies, and the assumption is that black adoptive parents, no matter where they live, uniquely, or are at least more likely than white adoptive parents to, possess these values and strategies *and* transmit them to black children. But this assumption falters in at least two respects. First, there is no set definition of "black values" that black parents possess simply in virtue of being black (Appiah, 1992; Gooding-Williams, 1998). Second, coping strategies are manifold and situational, and not coterminous with cultural values. Conflating these two concepts and attaching a racial valence to them detracts from the more exacting question of how parents will navigate their geographical situation *with* their children in the residential location they select.

This sense of togetherness is missing in liberal defences of white–black adoption that emphasize a strong conception of individual rights. Of course, families are composed of individuals with their own set of rights, moral obligations, and life-projects. Yet families are also intimate associations that should be navigating many of life's vicissitudes together, especially the formidable challenges that race and racism pose. Feminists have pointed out that families are political institutions with internal relations of power disparity that require negotiation over time (Okin, 1989; Young, 1990). The isolation that many black adoptees experience in white adoptive families is not just about being the only black child at school or church or synagogue, but also, and perhaps more profoundly, of having to "go it alone" without the sense that the family is navigating race together as a unit or team (Smith *et al.*, 2011). My earlier work overestimated the navigational agency of children. I bypassed, or moved too quickly over, childhood, the time when we can no more racially navigate on our own than make other critical decisions tied to the acquisition of life experience, and the development of the brain's prefrontal cortex. Just as we do not expect or allow kids to traverse cities and towns on their own without parental guidance, we should not expect them to navigate the racial meaning of that geography all, or even mostly, by themselves. Nevertheless, in doing so, we should not lose sight of the adopted person's increasing autonomy in navigating racial meaning as they mature into adulthood.

The 1996 federal Multi-Ethnic Placement Act (MEPA, 1994, as amended by The Interethnic Adoption Provisions, 1996) reflects the liberal individual-rights position by prohibiting adoption coordinators from using racial consideration to deny or delay an adoption. This federal legislation governs the practice of adoption, a family law matter that falls under the jurisdiction of individual states. As in non-adoptive family law, no state legislation requires adopters to raise their adoptive children in a particular geographical location, and I am not advocating such legal restriction. The legal freedom to choose where one will make a home, however, does not obviate the moral duty to take critical stock of the racial inequity in housing patterns, and to consider how our legally permissible private choices sustain and exacerbate such inequity.

My book urged white adopters to consider the race-conscious environment in which their adoption of black children transpired, but I did not directly address the extra-legal moral question of residential choice. I was more concerned to expose the insincerity of a colourblind approach to non-discrimination in a nation rife with race-based thinking and race-based private decision-making. I presented a thought experiment of racially randomized adoptive placement as a way of debunking the myth of liberal colourblindness. I argued that if we apply the same liberal individual rights-based argument to children in need of adoption that colourblind liberals apply to prospective adopters, then it follows that children have an ethical, even constitutional, right not to be chosen based on their racial classification. Racial randomization generates two outcomes that betray the insincerity of liberal colourblindness. First, although whites want the opportunity to adopt a child of any race, most still want to select their adoptive child based on racial classification. Second, the likely outcome that some white children would be matched with black adoptive parents is an adoption scenario that many whites find disturbing because of its cultural strangeness.

The unfamiliarity, even absurdity, of that scenario communicates an underlying set of assumptions about where such a family is likely to reside, and the "goodness" and "badness" of racially coded neighbourhoods. The image of blacks adopting white children as a result of randomized placement challenges a racially coded rescue narrative prevalent in cultural depictions of adoption—of who needs rescue, of who qualifies as a rescuer, and what qualifies as rescue. This narrative takes place in a place. Why does the mere idea of blacks raising white children in black neighbourhoods become the antithesis of rescue: a story of reckless endangerment? To answer this question we must first understand how race factors into the political geography that many, if not most, Americans take for granted, and thus never question. Or when they do question this racial geography, they justify their residential decisions by referencing their own parental obligation to live in neighbourhoods with "good" schools for their children.

The Political Geography of Race

We often fail to notice the political boundaries of our location, and how our actions give meaning to the lighted lines on the GPS devices we use to guide us from place to place. Katherine McKittrick (2006) notes the seductiveness of "[g]eography's discursive attachment to stasis and physicality, the idea that space 'just is,' and that space and place are merely containers for human complexities and social relations." For, "that which 'just is' not only anchors our selfhood and feet to the ground, it seemingly calibrates and normalizes where, and therefore who, we are" (McKittrick, 2006: p. xi). Our racial comfort zones determine many of the decisions we make on a daily basis, from the mundane instinct to cross to the other side of the street upon seeing a young black male who seems "suspicious," to the major life decision to purchase or rent a home. Both decisions take place in places—the sidewalks lining a specific street, a particular residential neighbourhood with its history, reputation, price point, and rules—unspoken and codified, alike. Adoption involves two decisions that democratic liberalism deems quintessentially private: the freedom to "have" children and the freedom to choose where to make a home for oneself and one's children.

When whites adopt black children they do so within a context of persistently high levels of residential segregation between whites and blacks. Data from the 2010 US Census indicate that the rates of residential segregation between whites and blacks are higher than between any other racial groups. Moreover, when white–black racial integration occurs it is typically unidirectional, meaning that blacks move into predominately white communities, but whites do not typically move into predominately black communities. I do not argue that whites who adopt black children are ethically required to move into predominately black neighbourhoods, although this would certainly not be immoral. Instead, I argue that white adoptive parents are morally bound to search for and move into neighbourhoods that are not predominately white, and where the family will have at least some black neighbours. Again, this ethical responsibility is not based on the speculation that black children cannot navigate their racial identities in white environments. Clearly, many have done so. My focus is on parental leadership. Based on the principle of fairness, parents are morally obligated to convey the message to their black adopted children that the racial integration of the family will not be entirely unidirectional, that they are willing to physically move out of their geographical racial comfort zones. The parents show good faith in leading the family's process of racial integration by interrupting the unthinking cycle of selecting a neighbourhood that is racially familiar to them. The adoption of black children by white parents magnifies this moral obligation, but as I noted earlier, the moral imperative to break this cycle falls on the shoulders of all whites, given their disproportionate power to enact race-based housing preferences.

Some of the formal historical barriers that have kept blacks and whites from becoming neighbours include racially restrictive housing covenants that were used by whites in many locales to ban the successive sale of real estate to non-whites, the prevalent practice of banks and insurance companies refusing to sell mortgages and housing insurance to the residents of neighbourhoods racially coded as black (the practice of "red-lining"), and real estate agent practices of steering black buyers away from properties in white neighbourhoods. Whites intent on keeping blacks from moving into their neighbourhoods often resorted to extra-legal tactics as well, such as harassment and intimidation, sometimes subtle and sometimes overt, that were aimed at forcing black pioneering families out of white neighbourhoods, and discouraging other blacks from moving in. The civil rights movement of the 1960s succeeded in overturning formal racist policies and practices via court rulings and anti-discrimination legislation. In 1948 the US Supreme Court declared racially restrictive housing covenants unconstitutional (*Shelley v. Kraemer*), and in 1968 the Congress passed the federal Fair Housing Act, which charged the federal Justice Department with monitoring and enforcing fair housing practices. These and other governmental actions have reduced segregation between whites and blacks, but the levels of segregation between these two groups remain very high (Logan and Stults, 2011).

According to the 2010 US Census, the typical white person residing in a metropolitan area (this includes suburban neighbourhoods proximate to cities) lives in a neighbourhood that is 75 per cent white, 8 per cent black, 11 per cent Hispanic, and 5 per cent Asian. In stark contrast, the same data show that the typical black person living in a metropolitan area resides in a neighbourhood that is 45 per cent black, 35 per cent white, 15 per cent Hispanic, and 4 per cent Asian. Whites have lower levels of toleration for living in the same neighbourhoods as blacks than blacks do of white neighbours. After the demise of legally enforceable restrictive housing covenants, the major factor in maintaining high levels of residential segregation between whites and blacks has been the extra-legal practice of white flight. African-Americans with the economic wherewithal have been able to move into wholly or predominately white communities. But their white neighbours start to move out of such communities once the number of black neighbours reaches a "tipping point." Most whites prefer to live in neighbourhoods with no more than 20 per cent black residents. At the same time, the majority of whites personally disavow racist attitudes. Most blacks would prefer to live in racially integrated neighbourhoods that are at least 50 per cent black, but also know that such racial balancing is extremely unlikely, given whites' lower threshold for tolerating black neighbours. It will be difficult, if not impossible, for blacks to secure housing in a neighbourhood that is racially balanced 50-50 between blacks and whites, leaving them the option of either residing in an entirely black community or a white community in which they and other blacks comprise no more than 20 per cent of the population.[2]

---

[2] The cycle that maintains racial segregation in housing thus goes as follows: "Whites stay in their neighborhoods when a few Blacks move in; other Blacks follow the pioneers into a desirable neighborhood, but continue to move in past the comfort zone of some Whites. More Whites move away, and the people who buy their houses are Black. The Whites who are un comfortable with a larger number of Blacks living near them then decide to move out as more Blacks enter. This cycle continues until there are few Whites left in the neighborhood" (Spinner-Halev, 2010: 114).

These data convey important information about the political geography in which the adoption of black children by whites occurs. First, they substantiate the descriptive basis of the moral argument I am developing by showing that whites generally have much greater power than blacks to enact their race-based housing choices. The normalization of this one-way integration makes it likely that whites adopting black children will choose neighbourhoods that are predominately white. Indeed, the data show that most white adopters raise their black adoptive children in predominately white neighbourhoods (Smith *et al.*, 2011).

Navigating the Political Geography of Race

Racial navigation is the active process of interacting with one's environment and deciding for oneself how race should factor into one's self-concept. It differs from the passive notion that we inherit or get our racial identities from our parents. Philosopher Naomi Zack puts a fine point on this critical distinction by delineating between racial *identity*, a third-person imposition, and racial *identification*, which she interprets as a first-person and ongoing process of active decision-making in light of the imposition of racial identity from sources external to oneself (2007: 102). My earlier work stressed the importance of this active first-person racial navigation for adoptees, but said too little prescriptively concerning the practical steps that white adoptive parents of black children should take in leading the family's racial integration amid high levels of white–black residential segregation. When the parents choose to reside in a predominantly white neighbourhood, they enact a passive racial choice that evades moral scrutiny because it is so utterly common. By contrast, active racial navigation means being attentive to the "background rules," both formal and informal, that function in residential living patterns to accommodate and sustain racial segregation and inequality. These background rules are neither abstract nor unchangeable, but they are deeply ingrained.

Black adopted children come to their adoptive families from another place—the place(s) where the original family members resided. This geographical history, no matter how brief, matters ethically because it is part of the story of where the adopted person "came from." Upon meeting someone, we are culturally primed to ask, "Where are you from?" This can seem like a superficial question, and in many instances it is perfunctory. But its commonness belies its social importance. For adopted people, the antecedent question of the original family's geographical location complicates this social custom, even when this information is wholly or partially unavailable to her or him. What adjustments, even sacrifices in racial comfort, did the parents make in light of their black adopted child's relocation? The decision of where to live will mark out the racial terrain that must be navigated, so it should be questioned rather than assumed. The residential location of the biracial adoptive family will become an integral part of the black adopted person's origin story, and hence their identity narrative.

We grow accustomed to seeing and sensing the legal territories we live in and move within, and those we avoid, as facts of life that just are, and that we cannot alter. But in fact the rules of jurisdiction are a set of identifiable rules and practices that we collectively reinforce via our everyday individual behaviour. Legal theorist Richard Thompson Ford analogizes jurisdiction to dance notation to draw out this point concerning individual agency. Just as people learn to dance the tango by studying its diagrammed movements and observing how others perform it, we learn to "dance the jurisdiction" by mimicking peer behaviour, looking at maps, and reading descriptions of where danger lurks, what counts as "prime real estate," and what criteria to use in deciphering the goodness and badness of neighbourhoods. Charles Mills (1997) theorizes that "the racing of space" involves "the depiction of space as dominated by individuals of a certain race." In turn, individuals become "imprinted with the characteristics of a certain kind of space" (Mills, 1997: 46). The metaphor of dance is helpful because it breathes life and human agency into housing decisions that are often theorized as structural—a term that can connote ossification and permanence. The lines on maps appear fixed in time and space, a done deal, but the actions of this "deal" must be performed over and over again by individuals in order for the agreement to carry social and personal meaning. Tango dancers sustain and vivify tango notation, just as residents sustain and vivify jurisdictional mapping by heeding and performing political boundaries via their housing preferences and decisions.

When whites fail to acknowledge and make residential choices for their biracial families that are not based solely on what is racially familiar to them, they "dance the jurisdiction" too. In other words, they fail to exercise their capacity to make housing decisions that disrupt the *de facto* segregationist cycle described. The racial comfort zones of white adopters are potentially at odds with the racial comfort zones of their black adoptive children, yet they must make a way together, as a family in a racially coded neighbourhood or setting. I say "potentially" because one important consequence of being an isolated black kid in a white family and community setting is that one is likely to absorb many of the racial and racist stereotypes from that home environment. These children learn to dance the same jurisdiction as their family and friends, even as they are made acutely aware of being racially out of place where they live and attend school. In interviews with young black adults who had been raised by white adoptive parents, Darren Smith and colleagues (2011: 112) report being "stunned by the adoptees' beliefs and how they, with no apparent awareness, reinforced the stereotypically negative views of African Americans as represented in the media." One might find similar cultural absorption among blacks raised in black families and black neighbourhoods. I would thus draw a different conclusion from such interviews. The uniqueness of these families is found not in the presence of a white racial knowledge framework in a black child's life, but rather in the absence of a black racial knowledge framework in that child's home and hometown. This reality begs an explanation that goes beyond the optimistic civil rights narrative that many whites tenaciously grip.

Most Americans denounce racial segregation as immoral, but at the same time reject the proposition that whites and blacks should be forced to share residential space, or attend the same schools. As Ford puts it, most Americans claim to live by the logic of *Brown v. Board of Education*, the landmark US Supreme Court decision that declared racially segregated public schools inherently unequal and unconstitutional. Yet they live by the logic of *Plessey v. Ferguson*, the notorious 1896 Supreme Court ruling that the *Brown* decision be overturned, which asserts that racial co-mingling must not be forced. Even as Justice Harlan in *Plessey* vociferously disagreed with the idea of a legal racial caste system, he threw his full support behind the *social* concept of racial separation based on what he described as the "natural" superiority of the white race, which he predicted to last for all time. The current high levels of *de facto* segregation between blacks and whites in public schools did not "just happen," but can instead be traced to specific court rulings during the 1970s and 1980s. To name just a few critical Supreme Court decisions: *Milliken v. Bradley* (1974) exempted suburban communities from school desegregation mandates. *San Antonio Independent School District v. Rodriquez* (1973) held that property tax-based school funding was constitutional in spite of vast tax burden–expenditure ratio disparities among local school districts. And *Arlington Heights v. Metropolitan Housing Development Corp.* (1977) ruled that a facially race-neutral local zoning ordinance barring multifamily housing did not violate the equal protection clause of the fourteenth amendment to the federal constitution in spite of the racially exclusionary impact of the zoning ordinance.

These decisions might have been decided differently, which would have changed the race-class segregated living patterns that the adoption of black children by whites takes place within. Their little-known status contributes to the current sense of what Ford calls "racism without racists" (2011: 235). He traces the power cord to state legislatures because these political bodies are responsible for drawing up and approving the mapping of political localities such as municipalities, which control the relationship between taxation and the provision of local public services, lines of residency that determine who votes in which locality, and zoning rules that function as *de facto* immigration laws at the local level. "Local government boundaries are simply another set of state laws subject to the state political process. It then follows that the only relevant political process occurs at the statewide level" (Ford, 2011: 226). Municipalities are state actors subject to the federal constitution's equal protection clause. This means that the rules they enact governing taxation, residency, and land use must not use racial distinctions to deprive those residing in the *state* of the equal protection of the law.

It is safe to assume that whites making the decision to adopt black children are not overtly racist. In some sense the decision to parent a black child can be read as a strong commitment to the principle of *Brown*, and a vehement rejection of *Plessey's* principle of natural racial separation. But private non-racist decisions can, and often do, effectively reinforce racial inequality. Adoption occurs within what Clarissa Rile Hayward and Todd Swanstrom call the "thick injustice" of contemporary American city life, and the false escapism of rural life. This thick injustice consists in "the deep historical roots of unjust power relations in America's cities and suburbs, their intersection with the institutional structure of local governance in the United States, and their embeddedness in physical places" (Hayward and Swanstrom, 2011: 9). The state no longer mandates legal racial segregation, and most Americans have edited explicitly racist language out of their personal vocabularies. But American political geography still bears the organizational structure of a racial caste system once upheld by racially restrictive housing covenants, red-lining by realtors and banks, and racist terrorism. And the organization of rural life has its own segregationist history, which persists in current patterns of rural living (White, 2007).

The Micro-Political Geography of Homemaking

Families, whether adoptive or not, are political entities that are regulated by the state and have their own internal power structures and struggles. They are also zones of intimate care that develop and change as time goes by and the members age, relocate, disperse, and reunite according to certain cultural traditions and situational circumstances. In this final section, I narrow my focus from the macro-political geography of race to the micro-political geography of race as it relates to the everyday and ongoing process of homemaking. I use Iris Marion Young's feminist theory of homemaking, and Sally Haslanger's incorporation of Young's work to describe her own residential (dis)location as a white mother of black adopted children. The word "homemaking" has traditionally been used to describe and too often denigrate "women's work" of tending to the maintenance of the domestic realm, apart from the public sphere of politics and commerce. Young redefines and revalues homemaking as political work that is both misunderstood and wrongly devalued by most people, including many women. Homemaking, on her account, differs from "housework" because it involves preservation of a certain kind. In arranging the material effects that constitute the interior design of homes, women have traditionally been responsible for preserving and thus constructing meaning for families over time and across generations. The fact that women have been traditional homemakers does not mean that homemaking is something that men cannot or should not undertake. Both men and women, regardless of familial configuration, should explicitly and self-consciously incorporate homemaking into their leadership in helping the family navigate racial meaning. Young stresses that homemaking is not about fixing meaning or identity in time and place (the business of museums). Instead, homemaking is an active, fluid process of interpreting and telling stories that comprise the self and group identities of those who dwell together.

The concept of dwelling is distinct from merely sharing time and space with one's immediate family members under a shared roof. Young (2005: 139) borrows D. J. Van Lennep's definition of "dwelling": to dwell is to acknowledge and feel one's "spatial existence." Phenomenology breaks down the mind/body division, and brings awareness to subjective feeling that often cannot be adequately described in words. Sally

Haslanger draws on Young's "materialist feminism" to note the ways in which the intimate care of her black adopted children has affected her own sense of being white:

This empathetic extension of body awareness, this attentiveness to the minute signals of another's body, does not in any metaphysically real sense make the other body part of your own. But taking on the needs and desires of another's body as *if* your own, perhaps especially if the other's body is marked as different, alters your own body sense, or what some have called (following Lacan) the "imaginary body." (Haslanger, 2005: 279)

She goes on to describe how such caretaking has altered her own racial comfort zones, but is quick to acknowledge that her white social privilege remains intact no matter these alterations.

When whites adopt black children, their sense of race at the level of subjective feeling is likely to shift regardless of the family's residential location. But the political geography of particular places sets the historical and material background for racial navigation. So, it is not enough to say that one sees and feels race differently as a consequence of parenting a black adopted child; parental leadership calls for critical attention to familial emplacement. White parents should take the time and make the effort to locate their biracial adoptive families in neighbourhoods that are not predominately white, and where the family will have at least some black neighbours. Haslanger and her white husband decided to make a home for their biracial adoptive family in a predominately black neighbourhood. They also negotiated the terms of their open adoption to involve frequent visits and interaction with several of the children's black birth relatives. Both of these features produce a kind of racial dislocation that challenges assumptions about the geography of racial belonging. Of moral concern for Haslanger are the cases of transracial adoption "in which the parents' identity does not shift" (2005: 288). I would amend this slightly to add that the cases of moral concern are also those in which the parents fail to shift residential *location* in light of the shifting racial composition of their family.

Conclusion

Legal narratives are necessary, but they do not give voice or feeling to the pragmatic concerns of navigating racial meaning within the political geography of the places wherein we dwell. The active, individual, and familial project of racial navigation presented herein will in many ways be easier to initiate and sustain in a community that is not all or mostly white. It is harder to "dance the jurisdiction," to obfuscate racist decision-making, when one is no longer in the overwhelming racial majority. No matter where homemaking takes place, white–black adoptive families will be racially dislocated because these adoptions take place within the "thick injustice" of a political geography deeply scarred by "racism without racists."

Young distinguishes between *nostalgia* ("longing for elsewhere") and *remembrance* ("the affirmation of what brought us here") (2005: 143). For her, "Home as the materialization of identity does not fix identity but anchors it in physical being that makes a continuity between past and present. Without such anchoring of ourselves in things, we are, literally, lost" (2005: 140). I find Young's words helpful in rethinking racial navigation as a relational, self-conscious process that is tethered to the materiality of residential location and homemaking. With Haslanger I reject liberal models of identity formation that focus too much on individual intentionality. I find fault, too, in structural accounts of racism that ignore, or give short shrift to, our individual intentions. Missing from this binary is a somatic self-understanding and understanding of others *in place* that is also, and simultaneously, cognitive.

References

Appiah, K. A. (1992). *In my Father's House: Africa in the Philosophy of Culture*. New York: Oxford University Press.

Arlington Heights v Metropolitan Housing Development Corp., 429 US 252 (1977).

Bartholet, E. (1991). Where do Black Children Belong? The Politics of Race Matching in Adoption. *University of Pennsylvania Law Review*, 139(5), 1163–1256.

Fair Housing Act, 42 USC 3601-3619 (1968).

Fogg Davis, H. (2002). *The Ethics of Transracial Adoption*. Ithaca, NY: Cornell University Press.

Ford, R. T. (2011). The Color of Territory: How Law and Borders Keep America Segregated. In C. R. Hayward and T. Swanstrom (eds), *Justice and the American Metropolis* (pp. 223–36). Minneapolis, MN: University of Minnesota Press.

Gooding-Williams, R. (1998). Race, Multiculturalism, and Democracy. *Constellations: An International Journal of Critical and Democratic Theory*, 5(1), 18–36.

Haslanger, S. (2005). You Mixed? In S. Haslanger and C. Witt (eds), *Adoption Matters: Philosophical and Feminist Essays*. Ithaca, NY: Cornell University Press.

Hayward, C. R., and Swastrom, T. (2011). Introduction: Thick Injustice. In C. R. Hayward and T. Swanstrom (eds), *Justice and the American Metropolis* (pp. 1–29). Minneapolis, MN: University of Minnesota Press.

Kennedy, R. (1994). Orphans of Separatism: The Painful Politics of Transracial Adoption. *American Prospect*, 17(Spring), 38–45.

Logan, J., and Stults, B. (2011). *The Persistence of Segregation in the Metropolis: New Findings from the 2010 Census*. New York: Russell Sage.

McKittrick, K. (2006). *Demonic Grounds: Black Women and the Cartographies of Struggle*. Minneapolis, MN: University of Minnesota Press.

Milliken v Bradley, 418 US 717 (1974).

Mills, C. (1997). *The Racial Contract*. Ithaca, NY: Cornell University Press.

Multi-Ethnic Placement Act, MEPA, PL 103–382 (1994), as amended by the *Interethnic Provisions, IEP*, PL 104–188 (1996).

National Association of Black Social Workers (NABSW) (1994). *Preserving African-American Families (Position Statement)*. Washington, DC: NABSW.

Okin, S. M. (1989). *Justice, Gender, and the Family*. New York: Basic Books.

Perry, T. (1993-4). The Transracial Adoption Controversy: An Analysis of Discourse and Subordination. *New York University Review of Law and Social Change*, 21, 33–108.

Roberts, D. E. (1997). *Killing the Black Body: Race, Reproduction, and the Meaning of Liberty.* Boulder, CO: Westview Press.

San Antonio Independent School District v Rodriguez, 411 US 1 (1973).

Shelley v Kraemer, 334 US 1 (1948).

Smith, D. T., Jacobson, C. K., and Juarez, B. G. (2011). *White Parents, Black Children: Experiencing Transracial Adoption.* New York: Rowman & Littlefield.

Spinner-Halev, J. (2010). The Trouble with Diversity. In J. S. Davies and D. L. Imbroscio (eds), *Critical Urban Studies: New Directions* (pp. 107–20). Albany, NY: State University of New York.

White, J. A. (2007) The Hollow and the Ghetto: Space, Race, and the Politics of Poverty. *Politics and Gender,* 3(2), 271–80.

Young, I. M. (1990). *Justice and the Politics of Difference.* Princeton, NJ: Princeton University Press.

Young, I. M. (2005). House and Home: Feminist Variations on a Theme. In I. M. Young, *On Female Body Experience: Throwing Like a Girl and Other Essays* (pp. 123–54). New York: Oxford University Press.

Zack, N. (1997). Race, Life, Death, Identity, Tragedy, and Good Faith. In L. Gordon (ed.), *Existence in Black: An Anthology of Black Existentialist Philosophy* (pp. 99–110). New York: Routledge.

PART VI

Contested Practices

13

Analogies to Adoption in Arguments Against Anonymous Gamete Donation

Geneticizing the Desire to Know

Kimberly Leighton

Is it important to know to whom we are genetically related?[1] And if it is important, what makes it so? Does the value of such information come from how we might use it, for example, in diagnosing or preventing disease, or is it valuable to know to whom we are genetically related because such relatedness is, in itself, valuable? Concerns about genetic information have received extensive attention in the field of medical ethics.[2] One view about the value of genetic information that seems to be gaining significant traction beyond the clinical context is that not only is it beneficial to know the identities of our genetic relatives, but our interest in such information is so important that access to it should be protected as a *right*.

A conference on adoption I attended a few years ago provided insights into how arguments based on this purported "right to know" are specifically developing in discussions about the ethics of technologically assisted reproduction (ASAC, 2010). A key plenary focused on what adoption scholars and activists might offer to discussions about the ethics of anonymous gamete donation (AGD), the practice whereby eggs, sperm, or embryos are provided for reproductive purposes by people whose identities

[1] The argument presented here has benefited greatly from engagement with the authors in and the editors of this collection. I would like to thank especially Carolyn McLeod for her generous editorial assistance, Andrew Botterell for his comments on an early draft, as well as the anonymous reviewer whose suggestions helped improve the essay's structure.

[2] While much literature on the ethics of genetic information has focused on issues such as confidentiality and privacy, there is a growing body of work that explores ethical questions concerning the implications of genetic information for understandings of race, family, and identity. See e.g. Juengst (1998), Dorothy Roberts (1995), and Christine Hauskeller (2004).

will remain permanently undisclosed to both the recipient parent(s) and any children so conceived.[3] While the speakers on the panel presented a thorough overview of the legal, cultural, and historical contexts of adoption and donor-assisted conception, not one articulated what actually connects the topics of adoption and AGD. It became clear in the question period that each of the panellists and many in the audience made the conceptual move from adoption to AGD via the "right to know." Anonymous gamete donation, like practices of closed adoption, is fundamentally problematic, most seemed to agree, because it results in the creation of people who are permanently denied their right to know the identities of those who provided the gametes used in their conception.

The language used to make this argument, of course, was not "provider of gametes," but "genetics," "ancestry," "biological roots," "history," and "identity." When I asked the panellists how they were justifying the claim that there is a right to know that specifically protected information about gamete donors, the answer seemed to lie in the importance of knowing one's *true* origins for the sake of one's identity.[4] "You can't know who you really are if your birth certificate is a lie" was a response from an audience member that received many nods of agreement. I countered that one's origins could be understood as social, cultural, linguistic; and the truth of who one is could be cultivated through a relationship with literature, sex, work, the arts, non-human animals, and so on. How can we privilege one particular model of origin, i.e. a genetic one, given the multiplicity of models? The justification for such privileging, I was told, is the fact that a lot of people conceived through the use of anonymous donor gametes—just like adoptees—*"really want to know."*

The connection between adoption and donor-assisted conception made at this particular conference reflects a recent trend whereby organizations concerned specifically with the needs of those involved in adoption include in their missions the interests of

---

[3] In any ethical analysis, close attention must be paid to how our descriptions of the world reflect and affect our beliefs about it. Two particularly vexing language issues in discussions of AGD are what to call the act of reproduction involved and how to refer to the people conceived through that act. I have opted to use the phrase "donor-conceived" to refer to both the means by which someone was conceived using third-party gametes and the identity they have *vis-à-vis* that mode of conception. I have settled on this for three reasons: the phrase is the dominant one used in the literature; "donor-conceived" parallels "adopted" in ways that make the comparison easily intelligible; and the term reduces sentence length and awkwardness. I am not comfortable, however, with the term "donor-conceived" and see its prominent use as far from politically neutral. I resist the term because of how it emphasizes the role of the donor over all of the other agents involved, and even implies that the donor actively did the conceiving. I also consider that using the term as an adjective to describe a person or to capture something about her status as a person leads to serious question-begging. I would also like to note that gamete providers are often financially remunerated and thus are not technically "donating" anything. I use "donor" because of the term's predominance especially amongst those advocating "donor-conceived rights" (Cahn and Kramer, 2011).

[4] Scholars who have investigated the status given to genetics in discussions of adoption include Haslanger (2009), Witt (2005), Anita Allen (1993). For an investigation of the narrativity of identity in the context of adoption see Homans (2013).

people who were donor-conceived. For example, the American Adoption Congress (2007) states the following in their "Assisted Reproduction Technology Statement":

The American Adoption Congress believes that all individuals whose genetic and biological origins are different from those of their legally recognized families have the right to know those origins. This includes people created through the donation or sale of eggs or semen, the transfer of embryos, gestational surrogacies, or any other reproductive technology. Knowledge of one's origins can be vital to the psychological and physical well-being of human beings. Denying a person this information can have potentially serious consequences upon that person's family relationships, health and reproductive choices. (para. 1)

Central to this statement are two assumptions that appear throughout arguments against AGD: that the possible consequences of not knowing one's "genetic and biological origins" are comparable for all people, despite the particulars of how that not-knowing came to be; and that these consequences necessarily involve harm. But are the harms that might come to an adoptee who is unable to know the history of her adoption—including, but not limited to, the identities of her birth parents and the story of her relinquishment—actually morally equivalent to the harms that might come to someone whose conception involved anonymous gametes? How should—and *shouldn't*—we be comparing these phenomena?

As an attempt to answer this question, this chapter investigates the analogy to adoption in debates about the ethics of AGD. An analysis of the role of the analogy to adoption in this literature reveals how right to know-based arguments against AGD rest on the assumption that AGD denies donor-conceived children—and the adults they become—something critically important to their well-being: their *genetic* identity. To support this view, critics of AGD claim that we need to recognize the importance of the information lost to the donor-conceived, a recognition that can come from appreciating how adoptees have suffered by their own lack of information. "While the analogy between donor offspring and adoptees is not a perfect fit, it appears that these two groups often have similar concerns about their genetic identity" (Frith, 2007: 645). Critics such as Ruth Landau argue that harm comes to children who were donor-conceived using anonymous gametes because of what they cannot know: "As with adoption, the child's right to a genetic identity is at stake" (1998: 3268).

As I present, right to know-based arguments against AGD that specifically claim people who were donor-conceived have a right to *identifying* information about the donors hinge on the claim that not knowing the identities of those who provided the gametes used in one's conception is itself the cause of a harm that people have a right to be protected from (the Harm Claim). The analogy to adoption, I argue, functions in arguments against AGD as a means of buttressing the Harm Claim. In order for the analogy to adoption to do the work of supporting the Harm Claim, however, the experiences of adoptees regarding what they do not know (especially if theirs were closed adoptions) and the purported harm that such not knowing is thought to engender must be rendered strongly similar to those of people who were donor-conceived.[5] To accomplish this, the harm at stake must be explained in *genetic* terms, the effect of which, I argue, is to *geneticize* adoptees' relationships to what they do not know.[6] The main claim of this chapter, then, is that right to know-based arguments against AGD, in their use of the analogy to adoption, contribute to rather than ameliorate the very conditions under which people who do not know to whom they are *genetically* related experience harm.

The development of the chapter is as follows. After presenting the theoretical problems facing right to know-based arguments against AGD, I show how attempts to ground right to know arguments using human rights discourse or empirical evidence fails. I then show how the analogy to adoption in arguments against AGD has been used as evidence in support of the purported harms of AGD. I analyze the reasoning at work in uses of the analogy based on the harms of secrecy and the harm of genealogical bewilderment. For the analogy to adoption to support the claim that people who were donor-conceived have a right to know the identity of the donor(s) involved, the arguments must geneticize adoptees' experiences. I conclude by questioning the assumption that adoptees' desire to know suggests by itself the value of knowing one's genetic origins.

Challenges Facing the Right to Know-Based Arguments Against AGD

Arguments against AGD that assert that a person has a right to know and that AGD violates that right have been offered in courtrooms and in policy discussions and public debates. "Upon reaching the age of majority, someone created with transferred genetic material should have the right to learn, although not be compelled to learn, the identity of the provider, not simply medical facts or DNA profile" (Shanley, 2002: 268). Like the Statement of the American Adoption Congress, the burden on arguments in support of the right to know is to explain what makes the interest some people *might*

[5] The most obvious comparison when discussing adoption and donor-assisted conception is that between *closed* adoption and donor-assisted conception that involves *anonymously* provided gametes. Many at the conference and some adoption advocates, however, consider the harm of closed adoption as part of a broader harm that comes from not being raised by one's bio-genetic parents no matter what is known about them (Lifton, 1994). The criticisms of many against AGD use the analogy in a way that extends beyond *closed* adoption, as they too hold the view that genealogical bewilderment comes from not being raised by one's genetic parents (Velleman, 2005). I will make clear throughout the chapter when the analogy is specifically between closed adoption and AGD because I hold that the bionormative claim that families *should be* genetically related underlies the reasoning in most arguments against AGD. Hence my title refers to analogies to adoption *simpliciter*, not to closed adoption.

[6] Coining the term, Abby Lippman (1993: 40) explains "geneticization" as a growing tendency "to name things that distinguish one person from another as genetic in origin, to reduce differences between individuals to their DNA codes and to define most disorders, behaviours and physiological variations as at least partly genetic in origin."

have in knowledge of their genetic origins—such that they could feel harmed without having it—something people conceived using reproductive technologies have a "right" to.

The philosophical challenges facing right to know arguments thus include substantively defining *what* this right is a right to know, what interest it aims to protect, and how it fares with competing rights claims. Right to know arguments that endorse policies against AGD based on the alleged harms that come from *anonymity*, moreover, must explain how the right to know, if there is one, protects an interest that is harmed when someone donor-conceived specifically cannot know the identity of the gamete provider involved.

Many advocates of right to know-based views against AGD have advanced their claims by interpreting the value of knowing a "donor's identity" using the discourse of genetics. The information that the donor-conceived have a right to know, defenders of this view allege, actually provides the very reason that they have a right to know it: the identity of the donor *is* (at least half of) the genetic origins of the donor-conceived (McWhinnie, 2001).[7] Because people who were donor-conceived have a right to know their genetic origins and the donor is the source of those origins, then, the donor-conceived have a right to know the identity of the donor.

This line of argument has seemingly been effective in motivating changes in assisted reproduction policy that concerns gamete donation. Starting with Sweden in 1984, many of the countries that regulate the practice of gamete donation have reversed their policies from requiring gamete donation to be anonymous to forbidding anonymity. Though each country—or state or province—navigates differently the issues of what information should be disclosed to whom, when, and how, the current trend in gamete donation policy is to make some information—increasingly *identifying* information—about a donor accessible to a person whose conception involved gametes provided by that donor.[8]

Despite what might seem to be a sea change in attitude against anonymity, however, not one regulatory body that has ruled in favour of donor-conceived children having access to (some) information about donors has also created a mechanism by which someone who was donor-conceived will necessarily know *that* she was. Countries that have recognized a child's right to know, in other words, have not allowed that right to know to trump presumptive parents' rights to determine how, when, or even whether to disclose to their children the particulars of their conceptions. Critics of the anonymity of gamete donation such as Eric Blyth (2002), Lucy Frith (2007), and Josephine Johnston (2002) have noted that without a mechanism that would inform

---

[7] The trope of "half of" persists in the literature on the ethics of AGD. People who were donor-conceived purportedly often describe their situation as "cut off from half of their genetic identity" (Suter, 2009: 260).

[8] In the UK, parts of Australia, British Columbia, Canada, and the US state of Washington e.g. people whose conceptions involved donor gametes (and who were born after the change in policy) can currently access identifying information about the donors.

children—independently of their parents—that they were donor-conceived, ending AGD via mandatory registrations maintains anonymity *de facto*, especially as studies continue to show that a majority of prospective parents using donor gametes in their attempts to conceive indicate that they will probably never tell their future children (Johnson *et al.*, 2012). For Blyth, Johnston, and Frith such withholding of information constitutes a fundamental deception the harm of which justifies requiring that the birth certificates of people who were donor-conceived reflect this fact.

Surely letting a child believe that she is genetically related to both of her parents when she is not constitutes a deception. The question at issue here is whether the effects of this deception are harmful such that a regulatory mechanism should be put in place to prevent such harm. There are many lies parents can and do tell their children (often in the name of the child's best interest). Why should policy be created to make sure children are not lied to in relation to the sources of gametes used in their conception? Why use regulatory policy to prevent the possibility of *this* lie?

While a mandatory disclosure mechanism on birth certificates would obviate the risk of children never *not* knowing that they were donor-conceived, the argument for implementing such a practice faces at least two major challenges. First, it must be established that a child who was donor-conceived has a right to identifying information about the donor(s), and second, it must be argued that the information interests at stake for a child—as a child or as an adult—override a parent's interest in maintaining, for example, privacy regarding her mode of reproduction or parental authority regarding how her child learns such information.

A basic outline of the right to know-based argument against AGD can help us see the challenges facing this position:

(1) People who were donor-conceived have a right to know their genetic origins
(2) Gamete donation violates this right to know if it results (or is likely to result) in

    a. A child never learning that he was donor-conceived
    b. A child who was donor-conceived having no access to identifying information about the donor(s) involved.

(3) AGD can result in a child never learning that he was donor-conceived
(4) AGD results in a child having no access to identifying information about the donor(s) involved

Therefore

(5) AGD violates the right to know of a person who was donor-conceived.

This argument might seem convincing at first glance; but it is lacking in several critical ways: it assumes without argument that a right to know exists; it assumes without argument that, if there is such a right to know, it can be extended to cases involving gamete donation; and, most importantly, it assigns to something called "genetic origins" an importance such that the interest people have in knowing them should be protected as a right. Without knowing more about what it means to know one's genetic origins, it is not clear that premises (1)–(4) ground the conclusion that is (5). Furthermore, even if (5) were true, it is not obvious how the fact that AGD violates the right to know of someone who is donor-conceived necessarily grounds the moral judgement in (6) below or provides ethical justification for the policies in (7):

(6) AGD is morally unacceptable
(7) To protect a child's right to know:
    a. The birth certificates of all children who are donor-conceived will note the fact that third-party gametes were used in their conception.
    b. Policy shall provide all children who were donor-conceived, once they reach 18, access to identifying information about their donor(s).

The weak link in this argument justifying a policy of mandatory disclosure seems to be between (5) and (6): If AGD violates a child's right to know such that AGD is morally unacceptable, then the policies drafted should ensure the protection of the child's right to know given the harm of violating this right. Because such policies should reflect the nature and import of the violation of a child's right to know, the argument against AGD based on the right to know needs to establish the Harm Claim with certainty, i.e. to show conclusively *how* AGD is against a child's best interest to such a degree that it trumps possibly competing interests. This requirement expands premise (6) from "AGD is morally unacceptable" to:

(6) AGD is morally unacceptable *because protecting the right to know of the donor-conceived*
    a. Is in the best interests of children who were donor-conceived
    b. Trumps other interests that might ground competing rights claims.

Thus we are left with the challenge of substantiating what I see as three critical assumptions of the Harm Claim at work in premise (5) that could ground premise (6):

(5a)   Violation of the right to know of people who were donor-conceived constitutes a serious harm that is *against their best interests* and trumps other interests such as parents' privacy.
(5b)   Being denied access to knowing their *genetic origins* constitutes a violation of the right to know of people who were donor-conceived.
(5c)   Being denied access to *identity of donors* involved constitutes a violation of the right to know of people who were donor-conceived.

Were these assumptions behind the Harm Claim shown to be true, the argument would seem to be sound.

Defending the Right to Know: Rights and Voices

The essential difficulty in the above argument, then, is substantiating the claims that there is a right to know and that this right would require identifying information about the relevant providers of gametes to be provided to people who were donor-conceived. What needs to be shown is how the kind of harm that comes from not having this information is one that should be given moral priority. To make this argument, some critics conceptualize the harm at stake as a human rights violation while others invoke empirical data to support it. However, both of these methods of grounding the argument, as I suggest here, cannot readily justify the Harm Claim.

The most frequently cited human rights document in arguments against allowing the anonymity of gamete providers in assisted conception is the United Nations (1989) *Conventions on the Rights of the Child* (UNCRC) which asserts in Article 8 that states shall "respect the right of the child to preserve his or her identity, including nationality, name and family relations as recognized by law without lawful interference." Additionally cited in support of the right to know of people who were donor-conceived is Article 7 which includes in paragraph (1) the statement that the child shall have "the right to know and be cared for by his or her parents."[9]

Despite common references to such human rights documents, the applicability of the framework behind them to the case of AGD is unclear at best. As legal scholar Samantha Besson notes, what the UNCRC should be considered as protecting is vague. The rights detailed therein "should be interpreted broadly; the term 'parents' is said to include not only one's social or legal parents, but also one's biological or genetic parents together with one's birth parents" (Besson, 2007: 143). Also unclear, however, is how "identity" can be invoked to specifically defend the claim that a child has a right to know the identity of the person whose gametes were used in her conception. Article 8 defers not to biology as the stable ground of what is meant by "identity" but to law; it is what the law recognizes as "family relations" that is integral to a child's identity, and this recognition need not be based on genetics.

Critics of AGD such as Vardit Ravitsky and Joanna Scheib (2010) recognize the challenges of supporting a strict rights- or duty-based approach to defending a child's right to know, and emphasize instead the importance of considering the experiences of people who were donor-conceived to the argument. The authors reach their conclusions on the ethics of AGD based on studies of people who were donor-conceived regarding their experiences searching for their donors and any (so-called) siblings whose parents used the same donor.[10] People who were donor-conceived have, according to Ravitsky and Scheib (2010: para. 13), a "curiosity about the characteristics of the donor and the desire to gain a better understanding of their genetic identity." In addition to wanting medical information, many of the donor-conceived articulate "the importance of knowing their genetic or ancestral history, and the sense of frustration they felt at

---

[9] An additional international document used to support right to know-based arguments against AGD is the Council of Europe's (1950) *European Convention on Human Rights and Fundamental Freedoms*.

[10] In the US, increasing use of the term "donor-siblings" to describe people whose parents used the same donor(s) can be tied to the online group Donor Sibling Registry, an online forum founded in 2000 "with the conviction that people have the fundamental right to information about their biological origins and identities" (Beeson, 2013: para. 2).

not being able to access this information." Though the authors recognize that study-ing people who were donor-conceived is "particularly challenging" methodologically speaking, Ravitsky and Scheib nonetheless conclude that the research being done rep-resents "[t]he current U.S. reality" which is that "donor offspring indeed perceive being told the truth about their conception and having access to information about donors as important to their well-being" (2010: para. 16).[11] These perceptions are sufficient, according to the authors, to justify mandating clinics to keep ongoing records about donors and to create registries where the donor-conceived would be able to access identifying information about donors.

Ravitsky attempts to draw out the moral importance of the feelings of the donor-conceived by highlighting their desires to know and feelings of loss. "Many [people who were donor-conceived] are telling a story of psychological distress. They describe a strong need to know 'where they came from;' to know their genetic origins as an essential part of constructing their identities" (Ravitsky, 2010: 655). Ravitsky here begins to reveal the key assumption in the argument: without knowing the identity of the donor, someone who was donor-conceived cannot know her own genetic identity.

The strategy of argument offered by Ravitsky echoes the work of Alexina McWhinnie, a much-cited opponent of AGD whose argument focuses on the "voices" of the donor-conceived and how they reveal "common" feelings. "Whatever their individual experiences and the words they use to describe it, they all make clear that they consider they have been done a serious injustice and they wish the matter recti-fied" (McWhinnie, 2001: 812). To support her view, McWhinnie (2001: 812) quotes a donor-conceived adult who, she maintains, "illustrates" the injustice being done:

I long to know who my biological father is, and to meet and speak with him at least once. I search for my half-siblings in other people's faces. I want to know the missing part of my family history, but more than anything, I need to know the other half of my ethnic background. Now that some of us are adults and, in fact, older adults, it is time for our voices to matter. We have a right to know our identity and to grow up in truth (woman, aged 42).

McWhinnie recognizes the problem of using the forcefulness of the voices of the donor-conceived to ground the right to know. Rather than revealing the harm of how they were conceived, the words of the donor-conceived could be dismissed as the views of a few disgruntled people. However, we can be sure that the voices of "donor offspring" represent the true moral harm of AGD, McWhinnie asserts, because "they

---

[11] One survey cited by Ravitsky and Scheib is "My Daddy's Name is Donor," a report (Marquardt et al., 2010) released by the Commission on Parenthood's Future, a subgroup of the Institute for American Values (IAV). The IAV (2013) believes that marriage is "society's most pro-child association" in large part because "it's the only institution that brings together the three main domains of parenthood—biological, social, and legal—into one association" (Mission statement). It is important to note that "My Daddy's Name is Donor" is not published in a journal nor were its findings peer-reviewed. It can be accessed at: http://familyscholars. org/my-daddys-name-is-donor-2. Supporters of ending anonymous gamete donation Wendy Kramer and Eric Blyth (2011) have "serious reservations" about the methodology of the report. For details about the mis-sion of the IAV see http://www.americanvalues.org/intro/index.php#mission.

mirror exactly the reactions and comments of adult *adoptees* and other adults, who, for whatever reason, have been denied access to accurate information about their origins" (2001: 813; emphasis added).

The Analogy to Adoption

References to adoption—including to the institution of adoption, to the adopted family, and to the experiences of adopted people—appear throughout arguments against AGD. The force of the argument by analogy rests on this basic claim: adoption and donor-assisted conception have sufficient significant similar features such that a judgement made regarding the ethics of disclosure in adoption holds for donor-assisted conception. Many of the countries where this argument by analogy has been put forth—including the UK, Canada, Australia, and New Zealand—had, prior to recent debates about the ethics of AGD, already recognized adoptees' right to know (to varying degrees).

The analogy to adoption functions in arguments against AGD, I argue, as a justificatory supplement to right to know-based arguments to the extent that it explains how people who were donor-conceived are or will be harmed by not knowing the identities of the donors involved. After sketching the basic way the analogy is invoked in arguments against AGD, I examine how the analogy struggles to ground the Harm Claim. I present as significant and potentially undermining of the analogy existing differences between adoption and donor-assisted conception as modes of family-making, as well as differences between experiences of people who were adopted (under closed adoptions) and people who were donor-conceived (using anonymous gametes). The analogy to adoption cannot secure the claim that not knowing the identity of a donor violates the right to know of people who were donor-conceived, I show, without imposing onto adoption an assumed valuation of genetic relatedness.

According to many who advocate outlawing AGD, the harm that comes to children conceived using this practice has been foretold by the history of adoption (Elster, 2007). Critics of AGD who use the analogy invoke both adoption as a regulated mode of family-making and the experiences of people who were adopted as proof that policies about donor-assisted conception should be pro-disclosure.

The "lessons from adoption"—an almost ubiquitous phrase in such arguments by analogy—tell us that secrecy and deception are against the best interests of children. In her argument for a mandatory registry of gamete donors, Naomi Cahn, a prominent voice in the debate in the US, outlines how and why one can draw from the analogy to adoption:

[A]doption can provide instruction on best practices concerning the legal and psychological methods of disclosure and follow-up [regarding gamete donation]. In many cases, the bases for these recommendations in gamete donation *track the same reasons* for disclosure in the adoption context: allowing offspring the opportunity for knowledge, satisfying emotional and psychological needs, and providing genetic information. (2009a: 213; emphasis added)

Richard Chisholm, in his analysis of the history of adoption policy in Australia, confirms the power of the analogy: "In short, the adoption comparison provides strong support for the view that donor offspring, when adult, should have an unconditional right to information about their genetic origins" (2012: 739–40). Adoption and the experiences of adoptees, for critics like Chisholm, are able to secure the assumptions of the Harm Claim, as they show the harm that (allegedly) comes from not knowing one's genetic origins.

A key assumption in the analogy to adoption is that, just as the institution of adoption changed with the shifting attitudes of society, so too the regulation of technologically assisted reproduction can change to reflect attitudes about the importance of knowing genetic information. According to those who use the analogy, adoption policy on disclosure changed in large part in response to the desires to know of adoptees and the harms secrecy was recognized as causing. Based on adoption research that developed in the US, UK, and elsewhere beginning in the 1970s, adoption experts came to believe that keeping a child in ignorance about her adoption was harmful (Triseliotis, 1973). Developments in family therapy that foregrounded the importance of trust in family relationships were combined with research on adoption to suggest that openness—not secrecy—was the best policy for all involved.

In Canada, the UK, and to some extent the US, these changes prompted social workers to reject the once-advocated view that it was better for a child not to know she was adopted (Cahn, 2009a). Prospective adoptive parents, previously counselled to keep silent about a child's adoption, would find their applications almost certainly rejected by most present-day agencies, were there any suggestion that they might not disclose to a child the fact that she was adopted. The harms of such deception for children and for families were recognized as being clearly against the best interest of adoptees. With this focus on secrecy, the analogy to (closed) adoption suggests what harms come from the non-disclosure of genetic origins in families created through AGD and adoption.[12]

The argument in support of the right to know of people who were donor-conceived from premises (1)–(7) left us with the problem of establishing the harm of violating the right to know of people who were donor-conceived by AGD. Using the analogy, we can now develop the argument as follows:

(8) Closed adoption and AGD are significantly similar as modes of family-making.

(9) The harm of *secrecy* in adoption is significantly similar to the harm of anonymity in AGD.

(10) Adoption has moved from being closed to (more) open because of the harm that *secrecy* causes for adoptees.

(11) Given (8)–(10), policy regulating access to information in the context of gamete donation should follow that of adoption and recognize the right to know of people who were donor-conceived.

---

[12] Some authors have questioned the overall narrative of adoption's movement from closed to open offered in arguments against AGD. See Haimes (1988) and Turkmendag (2012).

If it is correct that closed adoption and AGD are sufficiently alike, the harm suffered by adoptees because of secrecy in adoption should be sufficiently similar to the harm that comes to people who were conceived through AGD, thus providing grounds for (6):

> (6) AGD is morally unacceptable because, *as secrecy in adoption has shown*, protecting the right to know of the donor-conceived
> a. Is in the best interests of children who were donor-conceived
> b. Trumps other interests that might ground competing rights claims.

Analyzing the Analogy

What is critical to the success of such arguments against AGD by analogy to adoption is whether and to what extent the harm caused by closed records in the case of adoption actually tracks the harm caused by anonymity in the case of donor-assisted conception. To judge the adequacy of the analogy, we need to consider to what extent (9) is reliably true: that the harms of secrecy and anonymity are significantly similar, and that this similarity outweighs any differences between the two cases.

The similarities between adoption and donor-assisted conception that users of the analogy point to as grounds for (9) focus on the two modes of family-making and their histories. For the parents raising the children who came to them through adoption or donor-assisted conception, the following are generally (but not always) true:

- They are not both related by genetics to their children.
- This was not their first choice in family-making methods.
- To have genetically related children was preferred.
- There was a problem with one or both partners' fertility.
- There was some discomfort or even shame about this fertility problem.
- They may have spent much money and time in trying to have a genetically related family.
- They probably spent much time, effort, and money creating their present family.

Similarities between the parents' relations to reproduction are reflected in the histories of both adoption and donor-assisted conception. Like adoption, donor-assisted conception has involved policies advocating non-disclosure of the facts to others or even to the children themselves, and these actions and policies have been justified as being in the best interests of all. Not telling how a child came into a family when the methods of their making are considered "non-traditional," "assisted," or "artificial," has been historically considered best for helping such families feel normal (Yngvesson, 1997). Are these similarities strong enough to provide the grounds for a claim such as Chisholm's that adoption supports the "unconditional right to information about their genetic origins" (2012: 739–40)?

When we consider the differences between the two cases, there are good reasons to doubt the claim that the purported harms of secrecy in closed adoption are analogous to alleged harms of anonymity in donor-assisted conception. As Haimes (1988) has noted, there are at least three important differences between adoption and assisted reproduction using third-party gametes, including the fact that, for the former, a child already exists, while for the latter, a child is created by the process; in the case of adoption, the woman who carried the child to term is not part of the family made by adoption, whereas in families created using donor-conception it is generally the case that one of the parents involved is the gestational mother; and a child who was adopted into a family via a traditional adoption is not genetically related to her adoptive parents, while a child created using donor gametes is commonly related by genetics to one parent, most often the mother (Haimes, 1988: 48).

Taken together, these differences may position families made by adoption and by donor-assisted conception quite differently. Because in the latter a woman is pregnant and gives birth, (heterosexual) couples are likely to be interpolated by others as reproducing "naturally" such that the child will be genetically related to both partners. Further encouraging a kind of "family resemblance" in families made by donor-assisted conception, details about donors provided by most reproductive centres allow people procuring third-party gametes to "match" the features of the donor to the parent(s) not contributing sperm/ova to the conception. The growth of international adoption and the outlawing of formal race or ethnicity "matching" in the adoption process in many countries means adoptive parents often have to negotiate how others perceive (real or imagined) racial and/or cultural differences between themselves and their children that (apparently) announce the fact that a family was made by adoption (Haslanger, 2005).

This last difference points to perhaps the most important difference that could threaten the view that the harms of secrecy in adoption are significantly similar to the harms of anonymity in donor-assisted conception. Policies of non-disclosure in adoption have historically aimed at protecting the adoptive family not only from the shame or embarrassment of infertility, but also from the stigma of illegitimacy (Carp, 1995). The twentieth-century history of eugenics in the US, for example, commonly portrayed adoptees as feeble-minded or degenerate. "It is scarcely necessary to point out that in none of these cases [of children available for adoption] is the ancestry likely to be up to par" (Popenoe, 1929: 243). Children available for adoption have historically been represented as inferior given their questionable heredity:

It would seem then that they must be, in many cases, the children of profligate parents, children of families who are unable to maintain their footing in the community, or even provide for the necessities of life. (Goddard, 1911: para. 3)

The differences between families made by adoption and by donor-conception suggest that their reasons for secrecy and the effects of that secrecy on the family and children are not necessarily the same. Even if the cases were analogous, the harm of secrecy suggests only that parents should tell *that* the children were conceived and/or born in the ways they were, not that anonymity in gamete donation should be ended. Without further evidence that the harm at stake comes from not knowing *particular* information, rather than from e.g. the acts of deception in general, the harm of secrecy is itself insufficient to ground a policy *requiring* disclosure or a policy that would guarantee access to the identities of donors for people who were donor-conceived.

Analogous Experiences

In order for the analogy to adoption to ground the claims that the donor-conceived have a right to know and respecting this right is in their best interests, arguments against AGD must highlight more than the harms of secrecy. What is needed is a comparison that shows how a fundamental harm comes directly from not knowing one's genetic origins, and that confirms that not knowing the identity of the donor(s) causes this harm. It is clear that secrecy in the contexts of adoption and donor-assisted conception are not identical phenomena. Critics of AGD who recognize the limits to the analogy based on the positions of the *parents* involved advocate appreciating the analogous positions of the *children* in families made by adoption and donor-assisted conception. I review here the claim that people who were adopted and people who were donor-conceived suffer an analogous harm based on what closed adoption and AGD keep them ignorant *about*. Like arguments based on the secrecy comparison, however, those based on the analogous experiences of the children struggle with securing the adequacy of the claim in premise (9) that the harm is *sufficiently similar*.

Critics of AGD have recognized the limits of the analogy to adoption, especially when the focus of the analogy is on the contexts of the families, and have determined that these differences do not undermine the strength of the analogy (McGee *et al.*, 2001). As Blyth and colleagues (2001: 296) claim, "the case for comparison lies in experiences that have, hitherto, been ignored or discounted: those of adopted people and donor offspring." Ruth Landau (1998) locates at the heart of the harm of secrecy the harm she believes AGD causes to children regarding their self-knowledge: "Donor assisted conception provides new opportunities for achieving parenthood but at the same time raises issues of secrecy, anonymity, and the management of the offspring's genetic origins. As with adoption, the child's right to a genetic identity is at stake" (p. 3268).

So, what is it that grounds the primary assumption of the analogy to adoption, i.e. that decisions about disclosure in the case of adoption can and should be extended to the case of assisted conception using third-party gametes? According to Julia Feast (2003: 42), adoptees and people who were donor-conceived "often relate the same feelings and desire about the need to complete their full family and genetic history in order to achieve a fuller sense of identity." Because of similar experiences, the argument seems to be, if adoptees have a right to "be informed of their origins"—as the law in the UK recognizes—then so too do so-called "donor offspring" have a right to know their origins (Appell, 2008: 296). Thus we can offer a revised argument against AGD that would justify mandatory disclosure as follows:

(8) Closed adoption and donor-assisted reproduction using anonymous third-party gametes are significantly similar as modes of family-making.

(9) The *experiences of not knowing* in adoption are significantly similar to the *experiences of not knowing* in the context of donor-assisted conception using anonymous gametes.

(10) Adoption moved from being closed to (more) open because of the harm that *not knowing* (vs. secrecy) causes to adoptees and that occurs because of policies of non-disclosure.

(11) Because of (8)–(10) policy regulating access to information in gamete donation should follow that of adoption and recognize the right to know of people who were donor-conceived.

Our question now is whether the experiences of people who were adopted and people who were donor-conceived—in regards to what they do not know—are, in fact, sufficiently similar to ground the argument for the right of someone who is donor-conceived to have access to identifying information about the donor(s) involved. There are surely things people who were adopted and people who were donor-conceived have in common, especially if they came into their families via policies requiring closed adoption and AGD respectively. For both groups of people the following are true:

- They are not related by genetics to at least one of their parents.
- They do not know at least one of the sources of gametes used in their conceptions.
- They may not know the details of how they were conceived.
- They may not know that they do not know any of the above.

For those who do know *that* they were adopted or donor-conceived but not *who* was involved, the following are true for some members of both groups:

- They articulate a desire to know the identities of those involved.
- They articulate feelings of suffering regarding what they do not know.
- They resent having been lied to (if they found out later in their lives).

The harmful "legacy" of secrecy in the history of adoption may give support to (5a) of the Harm Claim: the harms of secrecy could be considered a violation of the right to know of people who were adopted and people who were donor-conceived. The issue here concerns locating the source of that harm as not knowing one's *genetic origins* (5b). To support the Harm Claim, the experiences of harm of the two groups must be analogous in terms of what they do not know and how this lack of information affects them. Differences between the situations of people who were adopted and people who were donor-conceived, however, suggest that theirs might not be sufficiently analogous experiences. Unlike people whose conceptions involved AGD, for example, someone who was adopted (in a traditional closed, infant adoption):

- Is not genetically related to *either* parent.
- Was not born to her parents.
- Was conceived by people not in her family.
- Was given birth to by someone not in her family.
- Was relinquished by people not in her family.

Given these differences, it is clearly not the case that what an adoptee does not know is necessarily the same as what someone who was donor-conceived does not know.

If we examine *why* an adoptee would want to know her *genetic origins*, it is still not clear how the two cases are strongly analogous. *What* an adoptee wants to find when she searches for her birth mother, for example, could be the person *who gave birth* to her. The "knowledge" of such a discovery might then be the origin stories it makes available—stories describing what the pregnancy was like, what the delivery involved, and even how the experience of surrendering felt. Knowing one's birth mother might also entail knowing someone genetically related; but the *import* of such association can certainly be explained without recourse to a valuing of genetics. Of course, there is a possible value in learning medical information, but if this was the driving force of the desire to know, then it would not ground the right to know of the donor-conceived: medical histories and even data from genetic screenings can be rendered without divulging identifying information.

Genealogical Bewilderment: The Necessity of the Harm

The analogy to adoption based on the experiences of adoptees and the donor-conceived, I have shown, must contend with the differences between the two groups for the analogy to satisfy the Harm Claim. What would satisfy the Harm Claim despite these differences, however, is if both groups *suffered* in sufficiently similar ways because of their relative ignorance. Let me consider this possibility now.

Of special relevance in the arguments against AGD that invoke the analogy to adoption is the concept of "genealogical bewilderment." This concept is used in these arguments to accomplish two things: to secure the similarity of harm experienced by adoptees and people who were donor-conceived, and to show how the harm comes from not knowing one's genetic origins. Briefly stated, the theory of genealogical bewilderment, as first articulated by two mid-twentieth-century British psychoanalysts, Eric Wellisch (1952) and H. J. Sants (1964), holds that children who are adopted suffer from a psychological disorder due to their lack of knowledge of their genealogical origins. Wellisch and Sants believed that the sources of the symptoms of "maladjustment" adoptees allegedly showed in their clinics—symptoms such as wanting to search for their birth parents, disobedience or resentment against their adoptive parents, and the exhibition of so-called anti-social behaviours—represented the fundamental failures of adoptees to have strong identities and to feel sufficiently secure senses of belonging.

Methodological and conceptual questions about the legitimacy of the diagnosis of genealogical bewilderment have been raised by critics of the term (Humphrey and Humphrey, 1986; Leighton, 2012; Walker and Broderick, 1999). Despite this criticism, "genealogical bewilderment" continues to gain traction in courtrooms, online discussion groups, and policy debates. It is commonly referred to in contemporary discussions of both adoption and gamete donation where it is used to capture the harm of not knowing. Critics of AGD specifically attempt to bolster their arguments by analogy to adoption by transferring the concept of genealogical bewilderment from the case of adoption to the case of donor-conception:

> [S]everal studies have examined whether donor offspring experience identity problems that are similar to those of adopted children, and although the studies often conflict, they do indicate that at least some donor children experience a sense of loss for not having information about their biological pasts or being able to establish a relationship with their gamete providers. Like adoptees, children of donated gametes may feel a sense of "genealogical bewilderment," a feeling that they are confused about their identity and different from other children. (Cahn, 2009b: 126)

Like all children, adoptees and people who were donor-conceived surely might experience confusion about their identities or could feel different from other children in an alienating way.[13] While possibly unpleasant, such feelings do not establish that the harms experienced by people who were adopted and people who were donor-conceived are significantly similar. Nor does such an analogy generate the reasoning necessary to support the claim that the right to know trumps other competing interests.

Other supporters of the right to know are more bold than Cahn in their references to genealogical bewilderment as they make clear what they see as the cause and extent of the damage at stake. For Chisholm (2012: 737), adoptees and donor offspring are "in very similar positions": they lack "information about their genetic origins"—something by which they "may be disadvantaged" and that places them "equally at risk of what has been called 'genealogical bewilderment'." In their argument in support of unified laws that would protect the right to know of people who were donor-conceived, Moyal and Shelley explain that the term genealogical bewilderment "applies to adopted children" where there is "uncertain knowledge of their biological parents." Genealogical bewilderment "leads to an ensuing state of confusion and uncertainty which *fundamentally undermines* [adoptees'] sense of security and affects their mental health. This concept can be applied equally to the children of AHR [assisted human reproduction]" (Moyal and Shelley, 2010: 437; emphasis added).

Making explicit what risks of harm adoptees and people who were donor-conceived face regarding their mental health, Dr Diane Ehrensaft, an expert witness for the plaintiff in *Pratten v. British Columbia*, describes the experiences of "donor offspring" as

---

[13] Andrew Solomon's work (2012) illustrates how fundamental differences in identity within families can prompt parents to feel alienated in ways that encourage them to reconsider their expectations of what it means to be related.

being "deluded about who they are." Because of what they cannot know, offspring of anonymous donors "are left with the same sense of genealogical bewilderment that has so negatively affected adopted children's sense of self, belonging and identity and indeed led to the transformation of adoption laws" (quoted in *Pratten v. British Columbia*, 2011, para. 94).

Thus, if genealogical bewilderment is a condition that is against the best interest of adoptees, so too is it against the best interests of people who were donor-conceived. We can now strengthen the description of the harm associated with AGD needed by the right to know-based argument:

(8) Closed adoption and donor-assisted reproduction using anonymous third-party gametes are significantly similar as modes of family-making.

(9) The *experiences of genealogical bewilderment for adoptees* are significantly similar to the *experiences of genealogical bewilderment* for *people who were donor-conceived*.

(10) Adoption has moved from closed to (more) open because of the harm recognized to come to adoptees from *genealogical bewilderment*.

(11) Because of (8)–(10) policy regulating access to information in gamete donation should follow that of adoption and recognize the right to know of people who were donor-conceived.

Geneticizing Adoption and the Value of Relatedness

I have so far presented an attempt to support the argument against AGD based on the claim that AGD violates the right to know. The analogy to adoption seems to provide justification for the Harm Claim, especially when we see how both adoptees and the donor-conceived suffer genealogical bewilderment because of what they do not know. The question now is whether the argument supplies enough force to the moral judgement of AGD in (6) such that the policy recommendations on AGD recommended in (7) are justified.

The similarities listed suggest that there are some commonalities between people who were adopted and people who were donor-conceived. What a critic of AGD such as Ehrensaft contends, however, is much stronger: the psychological harm of not knowing that both groups experience because of their situations is the same because the meaning of the lack of what they do not know is the same, i.e. premise (9). But why should we assume that what someone who was adopted under a closed adoption wants to know is the same, or that what makes the information not known valuable to an adoptee mirrors what makes what is not known valuable to someone who is donor-conceived? As I show in what follows, the differences between the two cases suggest that it is far from obvious that, based on their experiences, adoptees and people who were donor-conceived necessarily suffer—*if* they suffer at all—from the same phenomena. To satisfy the Harm

Claim, the analogy *makes the cases the same* by geneticizing the experiences of adoptees. This geneticization of adoption in arguments that rely on the analogy, moreover, reveals how these arguments invest in an understanding and valuation of genetic relatedness.

The assumptions operative in the descriptions of the "genealogical bewilderment" supposedly suffered by people who were adopted and people who were donor-conceived include the following: adoptees and the donor-conceived are equally vulnerable to being genealogically bewildered; they are susceptible to the purported disorder because they do not know their genetic origins, a state of ignorance that leaves them uncertain and insecure; and without knowing for certain who their "biological parents" are, they are likely to be "deluded" about who they really are. If we recall the differences between people who were adopted and people who were donor-conceived presented earlier, however, it would seem that the strength of the analogy requires imposing a particular meaning onto the experiences of adoptees.

For the harm of not knowing actually to be analogous, the symptoms of genealogical bewilderment for adoptees must have the same aetiology as the symptoms from someone donor-conceived. It is true that people who were adopted under closed adoption and people conceived using anonymous gametes do not know certain things they may want to know. But the fundamental claim of arguments based on genealogical bewilderment is that the suffering of someone adopted who does not know provides evidence of the harms of AGD for someone who is donor-conceived. Thus what both groups do not know *must matter* to their psyches in the same way.

Someone who was adopted might surely want to know the "truth of her conception"; however, given the particulars of her adoption, the truth at stake here might focus on knowing the conditions under which she was conceived and how these circumstances connect to the decisions made about who would ultimately parent her. The value of such information might, in turn, come from how it explains to an adoptee why she was relinquished, thus relieving possible insecurities. To not know the story of one's conception and relinquishment can cause harm to an adoptee, given the imagined possibilities: a birth mother could have been raped, a relinquishment could have been forced on a birth mother, a child could have been abducted, and so on. Such cases are not unheard of in the news nor absent from representations of adoption in popular culture.

Given the circumstances of the conception of someone donor-conceived, however, there is no obvious parallel to having been relinquished for adoption.[14] Thus the desire to know of an adoptee does not obviously lend support to the right to know of someone whose parents planned and orchestrated his conception, perhaps selected the sources of gametes used, and continued the pregnancy until he was born. If there is a lesson from adoption here it would seem to be that what someone who is donor-conceived needs to know is the story of *how* his parents brought him into existence, a story that need not include the identity of the donor.

---

[14] I appreciate that some birth mothers understand their actions as "making adoption plans" rather than "giving up" or "relinquishing" children.

If the symptoms of genealogical bewilderment are an incomplete sense of identity and an uncertain feeling of belonging, it is still not necessarily the case that the harm at stake is the same. An adoptee's discomfort regarding her identity could be a reflection of the history of adoption, which, in many countries, has entailed shame about illegitimacy and even judgements of degeneracy (Goddard, 1911). In addition, an adoptee's "symptoms" of genealogical bewilderment could be a reflection of the normative requirement that only genetic relatedness can ground belonging. Her potential difficulties with her identity, then, might point to the harms that come from the view that one's identity is determined by genetics, rather than represent either the health of her psyche or the importance of knowing genetic origins. If society's expectation is that members of families necessarily physically resemble one another, an adoptee's sense of difference in terms of belonging could reflect her experience living between two families. By locating the source of genealogical bewilderment in the adoptee herself, however, possible criticisms of these social norms are stifled.

To support the claim that the genealogical bewilderment of people who were adopted justifies the right to know of someone who was donor-conceived, finally, *genetic* understandings of conception, origins, and identity are required. And once this framework that prioritizes genetics is made dominant in our interpretation of the experiences of someone who is adopted, what she wants to know and why she wants this information seem able to line up with someone who was donor-conceived. According to this framework, in wanting to know the *circumstances* of her conception, she wants to know the sperm and egg used in it. The story of her origins that she needs to learn is the sources of her genes. And the person she wants to know is not the woman who carried her to term and made the difficult decision to not remain her legal mother, but rather the woman who contributed her ova to the adoptee's conception. In other words, only by considering someone who was adopted as *not adopted* can the analogy support the right to know claim of people who were donor-conceived.

For the analogy to support the right to know argument, I offer, it must be the case that the *genetic* tie primarily drives the desire to know of people who were adopted, and not the case that the desire to know one's history and identity *as adopted* is what drives the desire to know of people who were donor-conceived. We can hear the switch in the "direction of fit" at the heart of the analogy in the conclusion reached by Blyth *et al.*: "The experiences of donor offspring therefore illustrate some parallels with the experiences of adopted people, highlighting the significance of truth and personal authenticity, the right to self-discovery, and the pervasiveness of *genetic (blood) relations*" (2001: 301; emphasis added).

Differences in experiences of not knowing had by people who were adopted and donor-conceived do not matter if the essential cause of genealogical bewilderment is necessarily the same for both groups, i.e. not knowing one's *genetic* parents. What ultimately grounds the strength of the analogy is thus what *must* cause genealogical bewilderment: not the harm of secrecy, nor even the harms that come from a weak identity, but the lack of genetic relatedness experienced in families made by adoption or by donor-assisted conception. The assumption ultimately underlying the analogy renders the differences irrelevant: people who were adopted and people who were donor-conceived are harmed by being raised by people who are not (both of) their genetic parents. According to Appell (2008: 296), genealogical bewilderment "is common among adoptees, regardless of the quality of their adoptive family relationships."

Sants (1964), in his attempt to establish the necessity of genealogical bewilderment, asserts that those who adopt—what he calls "substitute parents"—cannot provide what "natural" parents can: "'Loving care' cannot be prescribed in a set of rules and the nursing of a child not born of the nursing mother cannot be the same as the nursing of a natural child" (1964: 139). Adoptees and people who were donor-conceived are always lacking because of both what they do not know and what they cannot have. The very fact that adoptees go "in search of a biological family" is understood in this framework as proof that non-genetically related family is "less than enough" (Velleman, 2005: 360).

Arguments against AGD based on the analogy thus translate the experiences of adoptees into representations of lack, signs of disorder, evidence of the inevitable harm that comes from making families not built on genetics—i.e. not produced through so-called "natural" heterosexual sex (Somerville, 2007). Such arguments reinforce what Nordgren (2008: 265) sees as a key example of contemporary geneticization: "when an individual stresses the lack of genetic connection to family in a way that leads to psychological suffering, no matter how good life is in other respects." Once a birth parent and a donor are understood to be the same—i.e. they are the genetic parents of the children involved—then the parallel needed by the Harm Claim is satisfied.

Conclusion: My (New) Bewilderment

I have presented an analysis of the analogy to adoption that reveals how the analogy as used in arguments against AGD requires us to see the experiences of people who were adopted through the lens of genetics. The geneticization of an adoptee's relation to what she does not know, I have suggested, is motivated by a commitment to understanding identity as primarily connected to our genetic origins, and to valuing those origins because they represent genetic relatedness.

Many arguments from analogy against AGD make explicit that the harm of genealogical bewilderment is evidence of the truth that genetics are what make people related in the way that children need to know to be fully healthy and happy (Somerville, 2007). Not only does this move geneticize adoption, it naturalizes the value of genetic relatedness, converting a normative claim into an established fact. Three consequences of this geneticization, I conclude, contribute to the conditions of harm whereby *anyone* who does not know—or appears to not know—her genetic origins is marked as lacking. First, someone who does not know *cannot* flourish in the ways that someone could who has (what counts as) knowledge of his genetic heritage. Second, adults who love, nurture, and unconditionally care for a child cannot be a family that fully provides a child with what she fundamentally needs to be happy and healthy. An effect of this view is that adoptees will always be bewildered because adoptive parents can never provide their children an ingredient *necessary* to a healthy identity: a genetically related family. Third, there is an erasure from the experience of adoption of the *difference* that being adopted makes. To compare people who were adopted and people who were donor-conceived with the specific aim of proving the good of knowing *genetic origins* requires—and enacts—a foreclosure on how being adopted might engender alternative ways of understanding the meaning of relatedness.

This point brings me back to the conference I mentioned in my introduction. Once the panel on adoption and AGD was over, I decided to rewrite the talk I was giving the next day, hoping it could turn what seemed an immovable discussion into an opportunity to think through as a community how we understand and value knowing origins. I crafted a paper that asked how "relatedness" could be approached from different angles, i.e. how we might address our desires for bodily connection and our needs to recognize ourselves in and through others, without assigning genetic relatedness intrinsic value. I wanted to break through the false dilemma I see critics like David Velleman (2005) imposing onto discussions of identity, family, and the information interests of adoptees. To challenge the legitimacy of right to know arguments against AGD, I wanted to emphasize, does *not* delegitimate the moral claim adoptees have regarding access to information about their births. I offered my view as both a philosopher and an adoptee.

The session seemed to engage the participants without provoking any outrage at what could seem, to some, to represent a dismissal of adoptees' right to know. I was surprised to learn later that evening, then, about the content of some ongoing murmurings of several attendees during my talk. While I was trying to encourage visions of relatedness that resisted privileging a bionormative model of the family, the quiet discussion going on a few rows into the audience centred on my status as an adoptee. The response of these listeners to my argument turned out to be, "Ahhh, *wait 'til she finds...*" My desire for expanding our notion of origins was read by my (well-intentioned) critics as evidence of my need to "reunite with" my birth mother. No matter that I had found her years prior—even publishing an essay on my search (Leighton, 2005)—the response bothered me because it suggested that only those who haven't experienced the goodness of a genetically related family would question its value, while it excused those raised in such families from not challenging the privilege they receive.

The comment also bothered me because it seemed to reiterate the destructive claim at the heart of "genealogical bewilderment": that to be raised by people who are not one's genetic parents necessarily results in the loss of something essential to having a good life. I stand behind my call for rethinking what it means to be related, however. And, if my challenging the view that genetically related families are naturally good and my proposing new ways of valuing our connectedness evidence my suffering, then I am happy to own my bewilderment, and even to spread it around.

Acknowledgements

The argument presented here has benefited greatly from engagement with the authors in and the editors of this collection. I would like to thank especially Carolyn McLeod for her generous editorial assistance, Andrew Botterell for his comments on an early draft, as well as the anonymous reviewer whose suggestions helped improve the essay's structure.

References

Allen, Anita (1993) Does a Child Have a Right to a Certain Identity? *Rechtstheorie*, 15, 109–19.

Alliance for the Study of Adoption and Culture (ASAC) (2010). Conference: "Secret Histories, Public Policies" at MIT. Retrieved July 2013 from http://www.pitt.edu/~asac/conference2010.html.

American Adoption Congress (2007). Assisted Reproduction Technology Statement. Retrieved July 2013 from http://www.americanadoptioncongress.org/assisted_repro_statement.php.

American Society for Reproductive Medicine (1998). Ethical Considerations of Assisted Reproductive Technologies. *Fertility and Sterility*, suppl. 3(4), S29–S31.

Appell, A. (2008). The Endurance of Biological Connection: Heteronormativity, Same-Sex Parenting and the Lessons of Adoption. *BYU Journal of Public Law*, 22(2), 289–325.

Beeson, D. (2013). Donor Sibling Registry: Our History and Mission. The Donor Sibling Registry. Retrieved July 2013 from https://www.donorsiblingregistry.com/about-dsr/history-and-mission.

Besson, S. (2007). Enforcing the Child's Right to Know her Genetic Origins: Contrasting Approaches under the Convention on the Rights of the Child and the European Convention on Human Rights. *International Journal of Law, Policy and the Family*, 21(2), 137–59.

Blankenhorn, D. (2005). The Rights of Children and the Redefinition of Parenthood. Institute for American Values, June. Retrieved July 2013 from http://www.americanvalues.org/html/danish_institute.htm.

Blyth, E. (2002). Information on Genetic Origins in Donor-Assisted Conception: Is Knowing Who You are a Human Rights Issue? *Human Fertility*, 5(4), 185–92.

Blyth, E., (2007). Donor Conception: What to Do about Birth Certificates? *BioNews*, 438, 17 Dec. Retrieved July 2013 from http://www.bionews.org.uk/page_37966.asp.

Blyth, E., Crawshaw, M., Haase, J., and Speirs, J. (2001). The Implications of Adoption for Donor Offspring Following Donor-Assisted Conception. *Child and Family Social Work*, 6(4), 295–304.

Blyth, E., and Frith, L. (2008). The UK's Gamete Donor "Crisis": A Critical Analysis. *Critical Social Policy*, 28(1), 74–95.

Blyth, E. and Kramer, W. (2010). 'My Daddy's Name is Donor': Read with Caution! Retrieved Oct. 2013 from http://www.bionews.org.uk/page_65970.asp?print=1.

Cahn, N. (2009a). Necessary Subjects: The Need for a Mandatory National Donor Gamete Databank. *DePaul Journal of Health Care Law*, 12(1), 203–23.

Cahn, N. (2009b). *Test Tube Families: Why the Fertility Market Needs Legal Regulation*. New York: New York University Press.

Cahn, N., and Kramer, W. (2011). The Biological Clock: For Donor-Conceived Offspring? *The Huffington Post*. Retrieved July 2013 from http://www.huffingtonpost.com/naomi-cahn/donor-sperm-washington_b_879066.html.

Carp, E. W. (1995). Adoption and Disclosure of Family Information: A Historical Perspective. *Child Welfare*, 74(1), 217–39.

Chisholm, R. (2013). Information Rights and Donor Conception: Lessons from Adoption? *Journal of Law and Medicine*, 19(4), 722–41.

Council of Europe (1950). *European Convention for the Protection of Human Rights and Fundamental Freedoms, as Amended by Protocols Nos. 11 and 14*, 4 Nov., ETS 5. Retrieved July 2013 from http://www.unhcr.org/refworld/docid/3ae6b3b04.html.

Elster, N. R. (2007). All or Nothing? The International Debate over Disclosure to Donor Offspring. *Institute on Biotechnology and the Human Future*. Retrieved July 2013 from http://www.thehumanfuture.com.

Feast, J. (2003). Using and Not Losing the Messages from the Adoption Experience for Donor-Assisted Conception. *Human Fertility*, 6(1): 41–5.

Frith, L. (2007). Gamete Donation, Identity, and the Offspring's Right to Know. *Virtual Mentor: American Medical Association Journal of Ethics*, 9(9), 644–8.

Frith, L. (2010). Telling is More Important than Ever: Rights and Donor Conception. *BioNews*, 542, 19 Jan. Retrieved July 2013 from http://www.bionews.org.uk/page_53094.asp.

Goddard, H. H. (1911). Wanted: A Child to Adopt. *Survey*, 27, 1003–6. Also available from E. Herman (ed.), *The Adoption History Project*, 24 Feb. 2012. Retrieved July 2013 from http://pages.uoregon.edu/adoption/archive/GoddardWCA.htm.

Haimes, E. (1988). Secrecy: What can Artificial Reproduction Learn from Adoption? *International Journal of Law and the Family*, 2(1), 46–61.

Haslanger, S. (2005). You Mixed? Racial Identity without Racial Biology. In S. Haslanger and C. Witt (eds), *Adoption Matters: Essays on Adoption, Feminism and Philosophy* (pp. 265–89). Ithaca, NY: Cornell University Press.

Haslanger, S. (2009). Family, Ancestry and Self: What is the Moral Significance of Biological Ties? *Adoption and Culture*, 2, 91–122.

Homans, Margaret (2013). *The Imprint of Another Life: Adoption Narratives and Human Possibility*. Ann Arbor, MI: University of Michigan Press.

Humphrey, M., and Humphrey, H. (1986). A Fresh Look at Genealogical Bewilderment. *British Journal of Medical Psychology*, 59(2), 133–40.

Institute for American Values (IAV) (2013). Mission Statement. Retrieved July 2013 from http://www.americanvalues.org/intro/mission.php.

Johnson, L., Bourne, K., and Hammarberg, K. (2012). Donor Conception Legislation in Victoria, Australia: The "Time to Tell" Campaign, Donor-Linking and Implications for Clinical Practice. *Journal of Law and Medicine*, 19(4), 803–19.

Johnston, J. (2002). Mum's the Word: Donor Anonymity in Assisted Reproduction. *Health Law Review*, 11(1), 51–5.

Juengst, E. T. (1998). Group Identity and Human Diversity: Keeping Biology Straight from Culture. *American Journal of Human Genetics*, 63(3), 673–7.

Landau, R. (1998). The Management of Genetic Origins: Secrecy and Openness in Donor-Conception in Israel and Elsewhere. *Human Reproduction*, 13(11), 3268–73.

Leighton, K. (2005). Being Adopted and Being a Philosopher: An Exploration of the "Desire to Know" Differently. In S. Haslanger and C. Witt (eds), *Adoption Matters: Essays on Adoption, Feminism and Philosophy* (pp. 146–70). Ithaca, NY: Cornell University Press.

Leighton, K. (2010). Questioning the Right to Know One's Genetic Origins. *Bioethics Forum*, 27 Sept. Retrieved July 2013 from http://www.thehastingscenter.org/Bioethicsforum/Post.aspx?id=4891&blogid=140&terms=leighton+and+%23filename+*.html.

Leighton, K. (2012). Addressing the Harms of Not Knowing One's Heredity: Lessons from Genealogical Bewilderment. *Adoption and Culture*, 3, 63–107.

Lifton, B. J. (1994). *Journey of the Adopted Self: A Quest for Wholeness*. New York: Basic Books.

Lippman, A. (1993). Worrying—and Worrying about—the Geneticization of Reproduction and Health. In G. Basen, M. Eichler, and A. Lippman (eds), *Misconceptions: The Social Construction of Choice and the New Reproductive and Genetic Technologies* (pp. 39–65). Hull, QC: Voyageur Publishing.

McGee, G., Brakman, S.-V., and Gurmankin, A. D. (2001). Gamete Donation and Anonymity: Disclosure to Children Conceived with Donor Gametes Should Not Be Optional. *Human Reproduction*, 16(10), 2033–8.

McWhinnie, A. (2001). Gamete Donation and Anonymity: Should Offspring from Donated Gametes Continue to be Denied Knowledge of their Origins and Antecedents? *Human Reproduction*, 16(5), 807–17.

Marquardt, E., Glenn, N. D., and Clark, K. (2010). My Daddy's Name is Donor: A New Study of Young Adults Conceived through Sperm Donation. Center for Marriage and Families at the Institute for American Values. Retrieved July 2013 from http://familyscholars.org/my-daddys-name-is-donor-2.

Moyal, D., and Shelley, C. (2010). Future Child's Rights in New Reproductive Technology: Thinking Outside the Tube and Maintaining the Connections. *Family Court Review*, 48(3), 431–46.

Nordgren, A. (2008). Genetics and Identity. *Community Genetics*, 11(5), 252–66.

Popenoe, P. (1929). The Foster Child. *Scientific Monthly*, 29(3), 243–48.

Pratten v. British Columbia (Attorney General), 2011 BCSC 656 (2011). Retrieved July 2013 from http://www.courts.gov.bc.ca/jdb-txt/SC/11/06/2011BCSC0656cor1.htm.

Ravitsky, V. (2010). "Knowing Where You Come From": The Rights of Donor-Conceived Individuals and the Meaning of Genetic Relatedness. *Minnesota Journal of Law, Science and Technology*, 11(2), 655–84.

Ravitsky, V., and Scheib, J. (2010). Donor-Conceived Individuals' Right to Know. *Hastings Center Report*, 42(4). Posted 20 July 2010 to *Bioethics Forum*. Retrieved July 2013 from http://www.thehastingscenter.org/Bioethicsforum/Post.aspx?id=4811&blogid=140.

Roberts, D. (1995). The Genetic Tie. *University of Chicago Law Review*, 62(1), 209–73.

Rose and Another v. Secretary of State for Health and Human Fertilisation and Embryology Authority (2003). EWHC 2 FLR 962. Queen's Bench Division. Retrieved from LexisNexis.

Sants, H. J. (1964). Genealogical Bewilderment in Children with Substitute Parents. *British Journal of Medical Psychology*, 37, 133–41.

Shanley, M. L. (2002). Collaboration and Commodification in Assisted Procreation: Reflections on an Open Market and Anonymous Donation in Human Sperm and Eggs. *Law and Society Review*, 36(2), 257–84.

Soloman, A. (2012). *Far from the Tree: Parents, Children, and the Search for Identity.* New York: Scribner.

Somerville, M. (2007). Children's Human Rights and Unlinking Child–Parent Biological Bonds with Adoption, Same-Sex Marriage and New Reproductive Technologies. *Journal of Family Studies*, 13(2), 179–204.

Suter, S. (2009). Giving in to Baby Markets: Regulation without Prohibition. *Michigan Journal of Gender and Law*, 16, 217–98.

Sylvester, T. K. (2007). The Case Against Sperm Donor Anonymity. *Yale Law School.* Retrieved July 2013 from https://www.donorsiblingregistry.com/sites/default/files/images/docs/legal.pdf.

Triseliotis, J. P. (1973). *In Search of Origins: The Experiences of Adopted People.* London: Routledge & Kegan Paul.

Turkmendag, I. (2012). The Donor-Conceived Child's "Right to Personal Identity": The Public Debate on Donor Anonymity in the United Kingdom. *Journal of Law and Society*, 39(1), 58–75.

United Nations General Assembly (1989). *Convention on the Rights of the Child, 20 Nov.* United Nations, Treaty Series, vol. 1577, p. 3. Retrieved July 2013 from http://www.unhcr.org/refworld/docid/3ae6b38fo.html.

Velleman, J. D. (2005). Family History. *Philosophical Papers*, 34(3), 357–78.

Walker, I., and Broderick, P. (1999). The Psychology of Assisted Reproduction: Or Psychology Assisting its Reproduction? *Australian Psychologist*, 34(1), 38–44.

Walker, I., Broderick, P., and Correia, H. (2007). Conceptions and Misconceptions: Social Representations of Medically Assisted Reproduction. In G. Moloney and I. Walker (eds), *Social Representations and Identity: Content, Process, and Power* (pp. 267–300). New York: Palgrave Macmillan.

Wellisch, E. (1952). Children without Genealogy: A Problem of Adoption. *Mental Health*, 13, 41–2.

Yngvesson, B. (1997). Negotiating Motherhood: Identity and Difference in "Open" Adoptions. *Law and Society Review*, 31(1), 31–80.

14

Transnational Commercial Contract Pregnancy in India

Françoise Baylis

Introduction

In *The Mother Machine*, Gena Corea (1985) reports on conversations with entrepreneurs in the United States who pioneered the "surrogacy"[1] business at a time when artificial insemination was used to impregnate women hired as reproductive labourers. In one such conversation, John Stehura (then president of the Bionetics Foundation Inc.) speculates (as paraphrased by Corea) that "once embryo transfer technology is developed, the surrogate industry could look for breeders—not only in poverty-stricken parts of the United States, but in the Third World as well. There, perhaps one tenth the current fee could be paid women" (Corea, 1985: 215). With contract pregnancy using artificial insemination, the women labourers provide genetic material and gestational services. With contract pregnancy using embryo transfer, the women labourers only provide gestational services; they do not contribute genes to the children they will deliver and relinquish. According to Stehura, where there is no genetic contribution to the future children, the women's IQ and skin colour won't matter. The women need only be poor and (thus) seriously motivated. This speculation on the part of Stehura leads Corea to write about a possible future in which "reproductive brothels" house "non-valuable" women to work as "breeders, not mothers" for the embryos of "valuable women."[2]

Fast forward thirty years . . .

---

[1] Hereafter, the terms "surrogacy" and "surrogate" only appear when citing the work of others. Otherwise, in this chapter, I follow Susan Moller Okin (1990), Debra Satz (1992), and others in using the terms "contract pregnancy" and "pregnancy contract" interchangeably.

[2] The terms in quotation marks are introduced by Corea (1985). See ch. 14, "Breeding Brothels: A Caste of Childbearers," pp. 272–82.

Transnational commercial contract pregnancy (which uses embryo transfer technology) is now a multi-billion dollar industry[3] that, like the organs market, travels along the "modern routes of capital: from South to North, from Third to First World, from poor to rich, from black and brown to white, and from female to male" (Scheper-Hughes, 2000: 193). Wealthy and middle-class American, European, Asian, and Middle Eastern buyers travel to low-income countries such as India, Malaysia, Thailand, South Africa, Guatemala, Russia, and the Ukraine to arrange transnational commercial contract pregnancies (Bailey, 2011: 716). Among these countries, currently, India is a destination of choice (Smerdon, 2008; Centre for Social Research, 2012: 4, 15, 23): "labor is cheap, doctors are highly qualified, English is spoken, adoptions are closed, and the government has aggressively worked to establish an infrastructure for medical tourism" (Bailey, 2011: 717). This infrastructure includes an Indian Medical Travel Association to promote India as a global health destination; low interest rates on loans; eligibility for low import duties on medical equipment; tax breaks for private hospitals that treat foreign patients; and special visas to facilitate medical travel[4] (Centre for Social Research, 2012: 22). Moving from policy to practice, there are two other attractive features of transnational commercial contract pregnancy in India, at least from the perspective of contracting couples.[5] First, there is the fact that several clinics have the women labourers live in hostels during the nine months of pregnancy so that they can be monitored (Hochschild, 2009; Bailey, 2011). Second, there is the fact that the state does not confer parental rights on the Indian women who give birth. Parental rights are assigned exclusively to contracting couples (who are presumed to be the biological parents).[6]

Now, let us review current practices in India against Stehura's predictions. Originally Stehura estimated that a tenth of the usual fee for commercial contract pregnancy might be acceptable payment for poverty-stricken gestating women (whether in the

---

[3] E.g. as reported in the London *Sunday Telegraph* (26 May 2012), contract pregnancy in India is worth an estimated £1.5 billion each year (Bhatia, 2012).

[4] There is the "Medical Visa" or "M Visa" to smooth the arrival and departure of foreign patients. Foreigners who wish to visit India to purchase gestational services from an Indian woman must produce a duly notarized agreement in order to obtain the appropriate Medical Visa (Government of India, Ministry of Home Affairs, 2012). There is the "Visa on Arrival" available to tourists from select countries who come for an advance health check or short duration medical treatment. And finally, there is the "Tourist Visa" that can be used for a reconnaissance trip to check out available health facilities and services (Government of India, 2013a and b).

[5] The reference is solely to couples (and not individuals), because of the recent directive from the Ministry of Home Affairs to the Ministry External Affairs which specifies that a Medical Visa for contract pregnancy should only be granted if (among other conditions) "[t]he foreign man and woman are duly married and the marriage should have sustained at least for two years" (Government of India, Ministry of Home Affairs, 2012, pt. i).

[6] "In India, according to the National Guidelines for Accreditation, Supervision and Regulation of ART Clinics, evolved in 2005 by the Indian Council of Medical Research (ICMR) and the National Academy of Medical Sciences (NAMS), the surrogate mother is not considered to be the legal mother. The birth certificate is made in the name of the genetic parents" (Government of India, Law Commission of India, 2009, para. 1.14). As well, the 2012 directive from the Ministry of Home Affairs to the Ministry External Affairs specifies that a Medical Visa for contract pregnancy should only be granted if (among other conditions) "[t]he child/children to be born to the commissioning couple through the Indian surrogate mother will be permitted entry into their country as a biological child/children of the couple commissioning surrogacy" (Government of India, Ministry of Home Affairs, 2012, pt. ii(b)).

US or the Third World). At the present time, the usual fee for gestational commercial contract pregnancy in the United States is between US$40,000 and US$150,000, of which the gestating women receive between US$20,000 and US$30,000 (Bailey, 2011: 718). A tenth of this fee would be US$2,000 to US$3,000. In India, the usual fee for gestational commercial contract pregnancy is between US$12,000 and US$25,000, of which the gestating women receive between US$2,000 and US$10,000 (Bailey, 2011: 718).[7] Using these numbers, poverty-stricken gestating women in India might receive between a tenth and a third of what gestating women in the United States receive.

Where Stehura appears to have erred "slightly" is with his prediction that IQ and skin colour wouldn't matter because there would be no transfer of genetic material from the reproductive labourers to the offspring. Reports from India and clinic advertisments suggest otherwise. At least some contracting couples have a clear preference for lighter-skinned gestating women with some education (Schulz, 2008; Centre for Social Research, 2012: 6).

So it is that predictions about the future of commercial contract pregnancy have come to pass, with India as a prime site of reproductive outsourcing. Consistent with Corea's forecasting, we now have "reproductive brothels" that house "non-valuable" women (from the South, from the Third World, who are poor, black, and brown) to work as "breeders, not mothers" for the embryos of "valuable" women and men (from the North, from the First World, who are well off (if not rich), and often white).[8]

This chapter is in two parts. First, I argue that transnational commercial contract pregnancy involves the exploitation of impoverished and often uneducated Indian women and that this exploitation is morally objectionable. It capitalizes on existing social, material, and political inequalities and reinforces morally unacceptable structural injustices, thereby harming Indian women, as individuals and as a group. Second, I argue that transnational commercial contract pregnancy in India harms the children born of this family-making strategy—a strategy that involves both the use of assisted reproductive technologies and adoption. Of particular concern are the damaging effects on identity formation. The offspring of Indian reproductive labourers are denied the benefits of access to genetic and personal information about biological parentage and kinship relations, and they are at risk of stigma associated with not knowing biological relatives, with being birthed by a woman who was exploited, and with being commodified.

---

[7] It is difficult to know exactly what fees are charged, and what payments are made, for contract pregnancy in India. According to the Centre for Social Research in India, the usual fee for gestational contract pregnancy is between US$10,000 and US$35,000 and payments to gestating women in the range of US$2,500 and US$7,000 (Center for Social Research, 2012: 23).

[8] Note, the terms in quotation marks are those introduced by Corea (1985). See ch 14, "Breeding Brothels: A Caste of Childbearers," pp. 272–82.

Harms to Women

In an effort to present a so-called "balanced" perspective on contract pregnancy in India, media reports sometimes include statements by Indian reproductive labourers, Indian brokers, and Indian clinicians insisting that transnational commercial contract pregnancy is not exploitative and, moreover, can be mutually advantageous. Consider, for example, the often cited case of Sofia Vohra, a 35-year-old woman with "five children of her own, a husband who's a lazy drunk, and a job crushing glass that's used in making (of all things) fortified kite string, for which she earns $25 a month" (Haworth, 2007: 6). Sofia agrees to contract pregnancy to pay for her daughters' dowries. When asked about her experience she says "I'll be glad when this is over.... It's exhausting being pregnant again" (Haworth, 2007: 6) But then, lest her remarks be perceived as complaints, she adds: "This is not exploitation. Crushing glass for 15 hours a day is exploitation. The baby's parents have given me a chance to make good marriages for my daughters" (Haworth, 2007: 6). Indeed, the suggestion is that both parties (the reproductive labourer and the contracting couple) benefit from transnational commercial transactions (see e.g. Anonymous, 2007; Scott, 2007; Gentleman, 2008; Springer, 2012). On this view, the only significant ethical concern with commercial contract pregnancy in India is whether the women reproductive labourers are adequately cared for and fairly compensated for the inconvenience, discomfort, and potential physical and psychological harms.

In important respects, however, the focus on mutual advantage can be said to distort, erase, or misread the lived experiences of many Indian women involved in contract pregnancy (Bailey, 2011: 716; Gupta, 2012) for whom the crux of the matter is entrenched structural injustices. For many women reproductive labourers in India, transnational commercial contract pregnancy is a means to an end in a context where they have no (or few) other means (Centre for Social Research, 2012: 4). Some of the women are illiterate, and some only have seventh to twelfth-grade educations (Hochschild, 2009). A small study conducted in Gujarat state (in Anand, Surat, and Jamnagar) reports on the level of education of women involved in contract pregnancy. Half of the study participants only had primary school education (31.7 per cent in Anand, 54.3 per cent in Surat, and 60 per cent in Jamnagar); and in Anand 51.7 per cent were illiterate (Centre for Social Research, 2012: 31). More generally, we know that in India literacy for adults over age 15 (both sexes) is only 62.8 per cent,[9] the mean years of schooling for adults over 25 (both sexes) is 4.4 years,[10] and there is limited schooling for girls (Sen, 2011). As Amelia Gentleman (2008) reports, in at least one Mumbai

---

[9] International Human Development Indicators (2013). See http://hdrstats.undp.org/en/indicators/101406.html.
[10] International Human Development Indicators (2013). See http://hdrstats.undp.org/en/countries/profiles/IND.html.

clinic "the thumbprint of an illiterate surrogate stands out against the clients' signatures" (para. 26).

Moreover, it is important to note that most of the women reproductive labourers do not have access to "basic social goods such as housing, food, clean water, education, and medical care" (Bailey, 2011: 722). These women put their social and familial relations, their physical and mental health, and sometimes their life at risk[11] in order to provide for their family (see e.g. Bailey, 2011; Gupta, 2012; Haworth, 2007; TNN, 2012; Centre for Social Research, 2012: 4). Not only is the work often physically and psychologically demanding, it is also highly stigmatized (Centre for Social Research, 2012: 5, 18). In the words of one Indian woman:

I know I have to do this for my children's future... This is not work, this is *majboori* [a compulsion]. Where we are now, it can't possibly get any worse... in our village we don't have a hut to live in or crops in our farm. This work is not ethical—it's just something we have to do to survive. When we heard of this surrogacy business, we didn't have any clothes to wear after the rains—just one pair that used to get wet—and our house had fallen down. What were we to do? (Pande, 2009: 160)

In the words of another Indian woman:

Why do I want to do it again? Because everybody knows we are poor, we don't have money and consequently we need to do this. Those who are rich, they don't have to do it. We who don't have a house, whose husbands don't work, drink a lot, only those people have to do it. But I still have to pay debt for my house, so I will do it. It is upsetting, but one has to do it. It is our fate that we have to struggle and still do it. I feel it is good also, but sometimes you feel bad—that there is something wrong that I have done. But what can you do, it is our need.... There... there is nothing to look forward to. I have no love left for life now. How many more years will I live, I ask myself sometimes. (Springer, 2012: 18:45 min)

Contrast these words with those of the woman who paid for her gestational services: "They are doing something that is good. In their eyes they don't feel exploited" (Springer, 2012: 20:25 min). Can this statement reasonably be understood as anything other than an effort to distort, erase, or misread: (i) the lived experience of the woman she hired to bear and birth her child; as well as (ii) the lived experience of Indian reproductive labourers more generally, many of whom are pressured by their husbands, families, brokers, and personal circumstances to rent their wombs for cash?

Ethical concerns about the exploitation of reproductive labourers in India come into sharp relief when particular attention is paid to the vulnerabilities of Indian women

---

[11] The maternal mortality rate in India is exceedingly high. Indeed, nearly one quarter of the world's maternal deaths occur in India (Center for Reproductive Rights, 2008: 9). One might think that the risk of death is reduced for women reproductive labourers as they would have access to better care provided by IVF clinics. This is probably so, but it is nonetheless worth noting that recent data from wealthier nations suggest that overall maternal mortality in IVF pregnancies is higher than maternal mortality rates in the general population. So, even with very good care, there is a very real risk of death with IVF pregnancies. (See Bewley *et al.*, 2011; TNN, 2012.)

and to the unequal distribution of power and wealth between the contracting parties. One perspective on exploitation focuses on what Stephen Wilkinson (2003: 117) refers to as "wrongful exploitation" which is when one person uses another (merely or solely) as a means to his or her end. The person used (in this case the gestating woman) is treated as though she only has instrumental value, when in point of fact she has independent intrinsic value as a person. Another perspective on exploitation focuses on what Wilkinson (2003: 172) calls "unfair advantage exploitation," which is when the distribution of benefit and harm between the contracting parties (other things being equal) is unjust, and the gestating women do not provide a valid consent. In what follows, I focus on the second category of exploitation, which I consider a more promising lens through which to look at the practice of transnational commercial contract pregnancy in India.

Unfair Advantage Exploitation

According to Wilkinson "unfair advantage exploitation" is reducible to the following two questions, both of which must be answered in the affirmative for one to conclude that commercial contract pregnancy is exploitative.

(1) Is the distribution of benefit and harm between the commissioning parents and the surrogate unjust (other things being equal)?
(2) Is the surrogate's consent invalid? (Wilkinson, 2003: 179)

In response to the first of these two questions, it can be argued in very general terms that the distribution of benefit and harm between contracting couples and reproductive labourers is unjust, insofar as the wage paid to gestating women is neither commensurate with the benefit conferred on the contracting couples, nor adequately compensates for the physical and psychological harms endured by the women reproductive labourers. Further, there are those who would insist that there can be no adequate compensation (financial or other) for the "gift" of a child (leaving aside the euphemism that payment is for services rendered, not for the child that is relinquished).

Moving from the general to the particular, with specific reference to commercial contract pregnancy in India, the distribution of benefit and harm between contracting couples and reproductive labourers arguably is unjust, insofar as the contracting couples specifically aim to secure increased benefits for themselves at the expense of increased harms to the gestating women. Here, it is important to emphasize the marked difference between the working conditions of women reproductive labourers in India, and the wages paid for their gestational services, as compared with what is available to women reproductive labourers in wealthier nations. As Vida Panitch (2013: 283) notes, in comparing the situation in India with that in the United States, "[a]long with a wage, the benefits of American surrogacy seem to include the freedom to pursue other interests while under contract, health care, travel and dietary expenses, legal representation, a post-birth opt-out clause, and the potential for a rewarding relationship with the contracting parents." These potential benefits typically are not available to gestational women in India. Their primary (and sometimes sole) benefit is a wage—a wage that is considerably lower than what is offered gestational women from wealthier nations. Of note, as regards the issue at hand, is that these reduced benefits are precisely what make contract pregnancy in India attractive to contracting couples.

In principle, it should be possible to improve the working conditions of gestational women in India, but what about the discrepancy in the wages paid to Indian women as compared with women from wealthier nations—should this be improved? As noted at the outset, Indian women typically are offered a tenth to a third of what is offered to women in the United States. And yet, the purchasing power of this reduced wage significantly exceeds the typical wage paid to gestational women from wealthy nations. Moreover, this reduced wage also significantly exceeds the typical wage available to women who "are predominately uneducated, often engaged in casual work, sometimes migrants in search of better job opportunities and living in slum areas with inadequate housing facilities" (Centre for Social Research, 2012: 28). Indeed, it has been estimated that the payment for contract pregnancy is "equivalent to upwards of 10 years' salary for rural Indians" (Haworth, 2007). To increase the payment to Indian women, and further widen the gap between payment for contract pregnancy and payment for other remunerated work, could heighten concerns about the exploitation of economically disadvantaged women in India in relation to the risk of undue inducement.

To put the financial aspects of commercial contract pregnancy in India in perspective, in 2005 (the most recent data available at the time of writing) approximately 37 per cent of the Indian population lived below the national poverty line (41.8 per cent in rural areas and 25.7 per cent in urban areas) (Government of India, Planning Commission, 2011).[12] At that time, the median annual household income in rural areas was 22,400 rupees (US$370),[13] in urban areas it was 51,200 rupees (US$840), and in rural and urban areas combined it was 27,856 rupees (US$450) (Desai *et al.*, 2010: 12). These totals contrast markedly with current wages in India for commercial contract pregnancy, which range from US$2,000 to US$10,000 (Bailey, 2011: 718). The discrepancy between the payment for contract pregnancy and other wages available to poor rural women in India undoubtedly affects (skews) decision-making. Now imagine that these women were offered the same US$20,000 to US$30,000 fee offered to economically disadvantaged women in wealthier nations. How could this be anything other

---

[12] While 37 per cent of the Indian population lived below the national poverty line in 2005, 75.6 per cent of the population lived on less than US$2 a day (see World Bank, 2013). In 2010, the percentage of the population living on less than US$2 a day was reduced to 68.7 per cent (see World Bank, 2013), but in that same year the United Nations Development Programme (UNDP) reported that there were more "MPI [Multidimensional Poverty Index] poor people in eight Indian states alone (421 million in Bihar, Chhattisgarh, Jharkhand, Madhya Pradesh, Orissa, Rajasthan, Uttar Pradesh, and West Bengal) than in the 26 poorest African countries combined (410 million)" (OPHI and UNDP, 2010).

[13] An exchange rate of 1 USD = 61.1043 INR was used to obtain these figures which have been rounded up or down.

than a coercive offer? And yet, the alternative is to endorse unequal pay for equal work, which is to capitalize on (i.e. take unfair advantage of) the plight of Indian women.

In response to Wilkinson's second question about the validity of consent, there are those who insist that "[c]ommercial surrogate contracts of the sort negotiated in India are an important expression of free choice between informed adults. They fulfil a modern need in a civilized way to everyone's advantage" (Anonymous, 2007). In the words of one gestating Indian woman: "We give them a baby and they give us much-needed money. It's good for them and for us" (Prayanka Sharma, quoted in Scott, 2007). From another perspective, however, it can be argued that the gestating women's consent is invalid, insofar as there are limited alternative ways for them to meet their basic needs for food, shelter, and so on. On this view, the idea that transnational commercial contract pregnancy is a free choice in a win-win situation is seriously misguided; it ignores salient details about the decision-maker and the decision-making context. Here Panitch reminds us of the Indian woman's "fungibility, her lack of education, the extent to which her and her family members' basic needs are unmet, and the limited range of other remunerative options available to her" (Panitch, 2013: 283). On this last point, Usha Rengachary Smerdon (2008: 54) remarks:

One must question the notion of free choice and self-determination when Indian women are agreeing to surrogacy to earn money to obtain urgent medical care for loved ones, win back lost children, raise children as a single parent or as the sole breadwinner, and pay for their children's dowries, particularly when the amount of money involved is so high in relation to the woman's standard of living.

One obvious problem with the liberal argument in defence of reproductive outsourcing (which emphasizes freedom of choice and freedom to contract) is the background liberal ideal of persons as independent, rational, separate, socially unencumbered, self-aware, self-reliant, and self-interested deliberators. As both feminists and relational theorists have long argued, persons are not discrete, circumscribed, self-contained units. Persons are affective, connected, encumbered, relational beings who develop within historical, social, economic, and political contexts and whose decisions (including decisions to contract) are deeply affected by their gender, race, ethnicity, and socio economic and political status (Downie and Llewellyn, 2012). As Susan Sherwin (1998: 34, 35) reminds us,

no one is fully independent . . . the view of individuals as isolated social units is not only false but impoverished: much of who we are and what we value is rooted in our relationships and affinities with others . . . all persons are, to a significant degree, socially constructed . . . their identities, values, concepts, and perceptions are, in large measure, products of their social environment.

And Jennifer Parks (2010: 336) makes this same point in her writing on transnational commercial contract pregnancy, "Our social embeddedness means that we are not the independent, objective and impartial beings that traditional liberal accounts have posited: on the contrary, we are *inter*dependent beings, who need one another for emotional and psychological support."

Many (if not most) gestating women in India have limited education and are desperately poor, with few options for accessing the necessaries of life for themselves and their families. Given their educational and economic vulnerability, one cannot accept at face value the claim that participation in transnational commercial contract pregnancy is a free choice on the part of Indian women. Rather, it would appear that in India transnational commercial contract pregnancy qualifies as a coercive offer—an offer that cannot be refused without great cost.[14] As Jennifer Damelio and Kelly Sorensen explain in their discussion of commercial contract pregnancy, when women make "a choice they do not prefer, yet cannot refuse because the price of refusal is too high," we have good reason to think they were coerced (2008: 274). Transnational commercial contract pregnancy sometimes provides Indian women with room and board (for those required to stay in a hostel managed by the fertility clinic), medical care (at least for the nine months of pregnancy), and education (including English lessons and computer use). But most importantly, it provides them with money—for example, money to care for their family (to pay for food, clothing, and medical care), to pay down a debt, to buy a house, to purchase a child's education, or to provide for a daughter's dowry. To be sure, not all Indian women who participate in transnational commercial contract pregnancy live in grinding poverty; some are a little less poor. But, like their poorer counterparts, they find themselves with expenses they can't otherwise meet. As recounted by reporter Tanya Springer (2012), "Charha did road work in Gujarat for years. But when one of her sons fell ill with pneumonia and her husband went into the hospital with a burst appendix, she made a decision. Desperate for money, she offered her services as a surrogate. In a matter of months she was carrying a baby girl for a Canadian couple" (17:40 min).

What Indian women participating in transnational commercial contract pregnancy—the desperately poor and the not so desperately poor—have in common is the experience of significant structural injustice, which includes social stratification through India's caste system, limited access to education, marginalization in the labour market, and patriarchal structures in the social and family domains. These vulnerabilities belie the image of gestating Indian women as independent, self-interested contractors freely choosing to sell their reproductive labour. As colleagues and I have argued elsewhere, "No matter what their social position, the choices individual persons can make depend fundamentally on the options available to them. These options are often determined by policy decisions in their societies" (Baylis *et al.*, 2008: 202). What are those policy decisions in India?

Since the 1990s, the Indian government has focused on GDP growth at the expense of issues of social justice and equality. It has actively encouraged reproductive travel to India as a way to grow the economy without simultaneously (i) addressing the

---

[14] An important feature of coercive offers, as Damelio and Sorensen (2008: 274) explain, "is the vulnerability of the party being coerced." The educational and economic vulnerability of the gestating women in India is well documented.

background social, material, and political conditions against which women in grind- ing poverty make choices about employment (and in some cases survival) and without (ii) developing a robust regulatory framework for reproductive travel in order to pro- tect the health and well-being of Indian women.

As regards the first of these two points, instead of properly meeting its duty to provide its members with a baseline level of welfare, the Indian government has seen fit to provide women with the option of selling gestational services as the practicable means to improve their lot in life. Women who avail themselves of these means as part of a survival strategy can hardly be faulted; but their government surely can. The job of government is not to expand the range of exploitative work options available to its citizens, but rather to guard against exploitation and oppression. As regards the second point about the absence of a regulatory framework to promote and pro- tect the interests of Indian women, Jyotsna Gupta (2012: 43) reports that the fertility business is booming in India in no small part "due to a lack of regulatory and legisla- tive measures." At the time of writing, there is no specific legislation in India govern- ing transnational commercial contract pregnancy. There are professional guidelines for the accreditation, supervision, and regulation of fertility clinics (Indian Council of Medical Research, 2005). There is pending legislation on assisted reproductive technologies that aims to promote and normalize assisted human reproduction as a global enterprise (drafted in 2008, revised in 2010) (Indian Council of Medical Research, 2010). And, most recently, there is a directive from the Ministry of Home Affairs to the Ministry External Affairs clarifying the terms under which a medi- cal visa for contract pregnancy should be issued (Government of India, Ministry of Home Affairs, 2012).

In sum, following Wilkinson, gestating women in India are subject to unfair advan- tage exploitation. To this I would add that it is no defence to say of the status quo that transnational commercial contract pregnancy is but one of many ways in which we exploit others. No doubt some will mount a *reductio ad absurdum* response to this facet of my argument against transnational commercial contract pregnancy in India, by pointing to a range of exploitative employment practices in low-income coun- tries that the world not only tolerates, but encourages. "Your argument," they will say, requires you to reject all manner of unfair advantage exploitation that undergirds the modern world of global capitalism. "You" cannot condemn transnational commer- cial contract pregnancy in India without at the same time condemning analogously exploitative practices such as child labour in sweatshops, dangerous work decommis- sioning nuclear ships, or underpaid work crushing glass in factories.

While there are ways to draw important moral distinctions between certain exploit- ative practices, for the sake of argument I grant that transnational commercial contract pregnancy in India is not a uniquely exploitative practice. I also grant, as a matter of consistency, that if this morally objectionable practice is to be abolished, then other equally morally objectionable practices should also be abolished. There is much unfair advantage exploitation in the world and none of it should be tolerated or encouraged.

In this chapter, however, I cannot take on global capitalism. Here, I focus my efforts on the considerably more modest project of trying to build a coherent and persuasive argument against one instance of unfair advantage exploitation, namely transnational commercial contract pregnancy in India. My goal is to stop this nascent practice—which involves the exploitation of poor women in low-income countries by rich and middle-class women and men—*before* it is normalized as a legitimate capitalist activity. Other analogously exploitative practices are well entrenched and for this reason are exceedingly difficult (if not impossible) to abolish (which does not mean that they ought not to be abolished).

Harms to Children

Too often, arguments in favour of assisted reproductive technologies (including transnational commercial contract pregnancy) fail to take seriously (or consider at all) the needs and interests of the children born of these technologies. Instead, these arguments focus on the rights of adults to become parents (i.e. the right to procreate and the right to found a family), with nary a word about the children (Parks, 2010: 337). In this way, concern for the well-being of children born of assisted human reproduction is often overlooked.

As I write this, I am mindful of Debra Satz's warning that "[o]ne of the difficulties with evaluating pregnancy contracts in terms of their effects on children is that we have very little empirical evidence of these effects" (Satz, 1992: 122). As far back as the early 1970s, Robert Edwards (of the team of Steptoe and Edwards who created Louise Brown, the first human born of IVF and embryo transfer) worried that a child born of contract pregnancy "might suffer on learning the circumstances of its birth" (Edwards, 1974). Edwards recommended that careful consideration be given to the psychological demands on the child. Unfortunately, at the present time, we know very little about the psychological well-being of children born of contract pregnancy. What we do know is limited to a few qualitative reports on commercial contract pregnancy (including transnational commercial contract pregnancy), a modest number of court cases in different jurisdictions, and an increasing number of media reports. None of these sources, however, provide conclusive empirical evidence on the positive or negative effects of transnational commercial contract pregnancy on children born of such arrangements. This fact brings into sharp focus Satz's admonition that "[w]e should be wary of prematurely making abstract arguments based on the child's best interests without any empirical evidence" (1992: 122–3).

While I appreciate Satz's caution, I believe that it is both possible and proper to engage the moral imagination and to carefully reflect on the ways in which transnational commercial contract pregnancy in India might affect the identity formation of children born of this family-making strategy. To do so, we need a clear account of identity.

Relational Identity

While there are many competing understandings of identity, I advocate a relational account of identity that elucidates the ways in which persons are constituted in and through their personal relationships, public interactions, and ancestry (Baylis, 2012). This account of identity recognizes the embeddedness and interdependence of persons (their affective, connected, encumbered, and relational nature) and makes transparent the ways in which relationships (both close and distant) shape a person's identity—her traits, desires, beliefs, values, emotions, intentions, memories, actions, and experiences. It also recognizes the myriad ways in which identity is always encumbered by a past (a history both remembered and learned) that shapes and directs one's life in positive or negative ways. On this view, which builds on the work of Marya Schechtman and Hilde Nelson, identity is a dynamic thoroughly relational narrative construct fashioned in both intimate and public spaces.

For Schechtman (1996: 93), persons are projects of self-creation: "a person creates his identity by forming an autobiographical narrative—a story of his life." An important constraint on this narrative, however, is that it should cohere with what others "objectively" believe to be true about the person. For Nelson, personal identity requires social recognition: "Identities are not simply a matter of how we experience our own lives, but also of how others see us" (2001: 81–2). In this way, identities are forged in and through relationship with others (Nelson, 2002). The constraint on this narrative form of identity is not objectivity, but credibility.

For me, a person creates her narrative—the story of her life—over time "through complex social interactions involving an iterative cycle of 'self'-perception, 'self'-projection, 'other'-perception, and 'other'-reaction" (Baylis, 2012: 111). Indeed, through a series of actions, interactions, reactions, and transactions, one fashions a dynamic identity-constituting narrative in concert with others. The constraint on narrative identity is not objectivity or credibility, but equilibrium—i.e. there must be cyclical periods of stability during which there is support/endorsement of one's projected narrative from one's community of belonging. In this way, the identity-constituting narrative, though dynamic, is sustainable.

The viewing of personal identity as both narrative and relational is consonant with Alistair MacIntyre's belief that "we are never more (and sometimes less) than the co-authors of our own narratives. Only in fantasy do we live what story we please. In life…we are always under certain constraints" (1984: 213). The nature of those constraints matter morally. An identity can be empowering or damaging depending upon which stories are incorporated into one's narrative, and whether these promote or undermine self-esteem, self-respect, self-confidence, and so on.

Identity Formation: A Case Study

In 2008, Yonathan Gher and his partner Omer, a gay Israeli couple who were unable to adopt or arrange a contract pregnancy in Israel, travelled to India to "make a baby." They contracted with an Indian woman to bear and birth a child for them using the egg of a Mumbai woman and sperm from one of the partners. In discussing their family-making project, the couple reported that they planned "eventually to tell their child about being made in India, in the womb of a stranger, with the egg of a Mumbai housewife they picked from an Internet lineup.... 'The child will know early on that he or she is unique, that it came into the world in a very special way'" (Gentleman, 2008: para. 3).

The plan for full disclosure is supported by research on the role of parents' in the identity formation of children who are created using third-party gametes. Research by Maggie Kirkman (2003), for example, emphasizes the importance of accurate information about conception coupled with the opportunity to negotiate the meaning of this information with parents in constructing one's own narrative identity. This research also encourages disclosure of genetic and social histories to offspring before adolescence in order to avoid disruptions to narrative identity that can occur when children develop "a sense of not being the person they thought they were" (2239).

If we assume, for the sake of argument, that the contracting couple intend to love and nurture their child, then presumably they believe that it is in their child's interest to know that he is "unique" having been "made in India" in the womb of a "stranger." To be sure, once told his origin story (which may also include narrative details about his adoption by the parent who is not genetically related to him (see Crawford, in this volume)), the boy (Evyatar) will understand that he is "unique". But how will the details of his conception and birth (and possibly his formal adoption) shape his identity-constituting narrative? How will he answer the pivotal question "Who am I?" What will others make of his answer? And will the identity-constituting narrative be empowering or damaging?

As with children conceived using anonymous gametes, Evyatar likely will have limited information about his genetic origins. Does this matter? Available studies on the psychological development, emotional well-being, and identity formation of persons conceived using anonymous gametes are inconclusive. This is largely because most of the studies have involved young children, many of whom have not been told of their genetic origins. As Stacey Page writes, in her review of the relevant psychosocial research:

The outcome measures relating to the children's psychosocial development have been limited to proxy indicators of outcomes in the sense that the construct of identity, and acceptance of genetic history, have not been directly assessed because the children are not aware of how they came to be.... [For the time being,] secrecy regarding conception does not seem to affect negatively family relationships or the children's functioning... these conclusions may not hold [however] when the children become older, issues of identity become more salient or the children learn of their origins. (Page, 2012: 142)

Ben Gibbard (2012: 157) makes a similar point.

Almost uniformly, studies about donor conception indicate generally normal family and offspring outcomes. However, these studies are based on samples where non-disclosure is predominant. There are no true population-based longitudinal studies comparing groups of children who know their donor conception status and those who do not, in terms of psychosocial outcomes and antecedent risk and resiliency factors.

Given the inconclusive nature of the empirical studies, there is significant interest in the anecdotal narratives and personal testimonials of individuals conceived using anonymous gametes, many of whom report feelings of discontinuity, detachment, loss, and incompleteness due to missing information. (See e.g. McWhinnie, 2001; Turner and Coyle, 2000; Morrisette, 2006; *Pratten v. British Columbia*, 2011.) To be sure, experiences among individuals conceived using anonymous gametes no doubt vary, and those who report difficulties with identity formation may simply be a vocal minority. But they may also be representative of a silent majority. We can't know for sure. What we do know is that, for now, there is no equally vocal counter-story told by individuals conceived using anonymous gametes.

As reported by Vardit Ravitsky (2010: 655), the first generation of offspring of anonymous gamete providers, speaking as adults, are "telling a story of psychological distress. They describe a strong need to know... their genetic origins as an essential part of constructing their identities." As Ravitsky then carefully explains, "genetic origins" is not reducible to "genetic data." Rather, the need to know one's "genetic origin" is about the need to access medical information, personal (narrative) information, and contact information. It is also, more broadly, about the importance of truthfulness in family relationships. In the words of one woman conceived using anonymous gametes, Phoebe: "I got the impression that 'society' didn't feel I have a right to anything more than a medical history. People don't acknowledge a need/right to know traits, history, or even realize that their sense of identity might be tied up with their family history, or family stories, or remembrances about a person" (Turner and Coyle, 2000: 2047 n. 22). Clearly Phoebe is interested in more than genetics and biology. Indeed, her interest is in autobiography.

Evidence of similar interests can be found in a recent decision of the British Columbia Supreme Court on sperm donor[15] anonymity. As reported in this Canadian case, some offspring experienced problems with identity formation due to missing genetic and personal information about biological parentage and about kinship relations. For example, Olivia Pratten, a woman conceived using anonymous sperm, testified before the Court that "not knowing about her biological origins makes her feel incomplete" (*Pratten v. British Columbia*, 2011: at para. 43). Alison Davenport, for her part, testified that when she learned she had been conceived by sperm donation, "in a single moment, she felt that she had lost 50% of her understanding of herself" (*Pratten*

---

[15] I only use such terms as "donor," "donation," and "donor offspring" when gametes have been altruistically gifted. When there is a known commercial transaction, I use the term "provider" and write about "selling" or "trading." In Canada it is illegal to purchase gametes and so I use the term "donor" in this instance (though I recognize that there may have been financial transactions in cases that preceded the 2004 Assisted Human Reproduction Act).

v. British Columbia, 2011: para. 51). Yet another identity issue concerned a general malaise about one's origins: "from whence did I come, what are my origins, and how do they relate to who I am and who I will be?" (*Pratten v. British Columbia,* 2011: para. 95).

Expert reports submitted to the Court supported these testimonials. For example, according to Dr Diane Ehrensaft, a psychotherapist and a qualitative clinical researcher:

Donor offspring with anonymous donors may suffer from the psychological phenomenon referred to as genealogical bewilderment, confusion about from whence they come, along with accompanying psychological dysphoria as a result of grappling with the "missing piece" of themselves.... In Western culture, it is presumed that children will have a better sense of their identity and higher self-esteem if they know their genetic roots. Denied that information... they will have a more difficult time solidifying the foundations of their adult identity.... I have definitely observed genealogical bewilderment to be the experience in my own patients. (*Pratten v. British Columbia,* 2011: para. 95)

In light of these data about identity harms to persons born of anonymous gametes, it is reasonable to reflect critically on how Evyatar might feel about his missing genetic and contact information, and how this might affect his identity formation. For example, might Evyatar angst over whether he bears any family resemblance (physical characteristics, personality traits, talents, and so on) to his genetic mother and any possible half-siblings?[16] Might he want a face-to-face relationship with his biological kin? We can't know the answers to these questions (and the answers may change over time), but that shouldn't preclude careful reflection about possible/plausible answers, taking into consideration concerns that are being voiced by persons born of anonymous gametes. To be clear, the point here is not to contribute to the reification of the gene, or to endorse the view that biological ties are more important than other ties. *Identity is about much more than biology.* The point is simply to insist that we shouldn't ignore the experiential knowledge of persons conceived using anonymous gametes who report that contact with biological relatives is important for their identity formation (which is not to say that contact with biological relatives will contribute to a healthy identity). From another perspective, it is worth repeating a point made by Jamie Lindemann Nelson:

If some adults can find biological relationships with children so important that it makes going through ARTs, with all their costs, risks, and inconveniences, a rational choice for them, why should we assume that biological ties in the other direction won't also matter greatly to some children? (Lindemann Nelson, in this volume: 189)

---

[16] A woman who, under contract, gives birth to a child is the gestational mother of that child, and if she provided genetic material (in addition to gestational services), then she is also the genetic mother of that child. Efforts to erase these biological facts through the use of misnomers are to be resisted. For the sake of precision, on my view, a pregnant woman is not a mother to her foetus(es). Mothering relationships are between women and born children (with whom the women may or may not have a biological relationship). It follows that a pregnant woman is not a mother unless she has (or has had) children.

In writing about identity and biological ties with reference to children conceived using anonymous gametes, Sally Haslanger argues (against David Velleman) that it is not morally wrong to create a child who will not have ongoing contact with one or both biological parents. She further insists that contact with biological relatives is not a universally necessary condition for a healthy identity. Even so, Haslanger (2009: 101) allows "that children should have access to basic knowledge of origins where available." In this way, Haslanger does not discount the lived experience of persons conceived using anonymous gametes. Rather, she explicitly recognizes that because identities situate persons within social structures and cultural narratives, "lacking knowledge of one's biological family, one is often left without answers to questions that matter culturally, and this is stigmatizing" (Haslanger, 2009: 113). Haslanger's (2009: 92) response to the harm of stigmatization is to disrupt the dominant cultural schema by resisting "ideologies that entrench and naturalize the value of biological ties." She writes:

> I believe that knowing one's biological relatives can be a good thing, and that contact is valuable in the contemporary cultural context largely because this context is dominated by the natural nuclear family.... However, if we are to avoid harming our children, then rather than enshrining a schema that most families fail to exemplify and which is used to stigmatize and alienate families that are (yes!) as good as their biological counterparts, we should instead make every effort to disrupt the hegemony of the schema. (Haslanger, 2009: 114–15)

While this may or may not be an effective long-term strategy for dealing with the harms of stigmatization resulting from the absence of biological knowledge, the problem of stigmatization is exceedingly more complicated with transnational commercial contract pregnancy involving disadvantaged Indian women. Here, the stigma attaches not only to the missing biological information, but also, more generally, to the exploitation of women and the commodification of children.

With a relational account of identity, a person creates her identity-constituting narrative—her life story—in concert with others. These others contribute stories from the time of her birth (or even conception) (Ricoeur, 2001: 190). In this way, identity is not simply about how one sees oneself, but also about how one is seen by others, and how this, in turn, affects how one sees oneself. In this way, identity crucially depends on the recognition and acknowledgement of others. To quote Lawrence Thomas (1998: 365), "we are constituted through others.... the way in which we conceive of ourselves, at least in part, owes much to how others conceive [of] us, and this is necessarily so. The way in which we think of ourselves is inextricably tied to the way in which others think of us." A unique identity concern for persons born of transnational commercial contract pregnancy to economically disadvantaged Indian women is how they will be seen by others and, in turn, how these perceptions will shape their identity—their life story.

Children born of Indian reproductive labourers who know their origin story (such as Evyatar), at some time in their life, cannot help but become aware of the international public debate about the exploitation of economically disadvantaged women and the commodification of children born of transnational commercial contract pregnancy.

When Evyatar (as a young adult or maybe later) becomes aware of these debates, might he internalize the belief that he is a mere token of exploitation, and might he experience anxieties and internal conflicts that contribute to problems in identity formation? For example, might Evyatar come to believe that contract pregnancy—especially transnational commercial contract pregnancy in India where the power and wealth disparities between the contracting parties is so great—is exploitative? Might he even come to believe that his parents preyed on the vulnerabilities of his biological (genetic and gestational) mothers? Believing this, might Evyatar experience considerable existential angst? After all, "but for" the exploitation of his biological mothers by his legal parents, he would not exist.

From another perspective, might Evyatar internalize the belief that he is a mere commodity—not a unique individual desired and cherished in his own right by his birth mother, but an object of commerce? Identity, understood in dynamic, relational, and narrative terms, is an amalgam of self-ascription and ascription by others (Baylis, 2012: 118). From this perspective, it is possible that the prevailing discourse about the commodification of children might inform Evyatar's sense of himself as the "product" of a woman who sold her eggs *for a price*, and a woman who sold her labour *for a price*. As Smerdon (2008: 59) reminds us, "[p]ersonhood is harmed when it is not adequately recognized that the product of a woman's reproductive labor is some*one* not some*thing*."

If we look again to the experience of individuals conceived using anonymous gametes, we can find reports on the harm of commodification: "of being a product made to suit their parents' wishes" (*Pratten v. British Columbia*, 2011: para. 104). For example, in the Canadian court case discussed previously, Damian Adams "described feeling 'false,' like one of the experiments he conducts in his lab [he is a medical researcher], dehumanized and commodified" (*Pratten v. British Columbia*, 2011: para. 54). Similar feelings have been reported by persons born of contract pregnancy. Consider the following poignant statement by *Brian C.* (2006):

How do you think we feel about being created specifically to be given away? . . . It looks to me like I was bought and sold. . . . When you exchange something for money it is called a commodity. Babies are not commodities. Babies are human beings. How do you think this makes us feel to know that there was money exchanged for us? . . . Because somewhere between the narcissistic, selfish or desperate need for a child and the desire to make a buck, everyone else's needs and wants are put before the kids [sic] needs. We, the children of surrogacy, become lost. That is the real tragedy. (para 6 and 27)

In discussing plans for disclosure, Yonathan doesn't mention telling Evyatar what it cost his parents to travel to India, make a baby, and return to Israel as a "new" family. Will it matter to Evyatar's sense of self if the total cost is "a little" or "a lot"? Is this something he will think about in the abstract, or will he think about this in relation to other consumer goods that were forfeited in order that he might be created? Will he consider himself more or less precious than an expensive holiday, an expensive car, or an expensive home? Answers to these questions may be of pivotal importance for Evyatar's identity-constituting narrative. As markets increasingly reduce everything to the status of commodities, where "there is nothing fixed, stable, or sacrosanct about the 'commodity candidacy' of things" (Scheper-Hughes, 2000: 193), it matters what origin story Evyatar will tell himself, and others, in order to make sense of the unique circumstances of his coming into being. If a sense of self as commodity were to figure prominently in Evyatar's identity-constituting narrative, one can imagine how this might lead to a damaged identity plagued by feelings of depression and discontent (see Ehrensaft, 2007, 2008, 2012). Evyatar would be both the "means" to his parents' "end" of having a child, and the "end" of his parents' exploitative "means."

We can't yet know how children of transnational commercial contract pregnancy in India will make sense of their unusual origin story. This reproductive practice is relatively new (since the early 2000s in India) and the offspring are still young children. Flights of fancy are not required, however, to imagine the harms that might be experienced by someone who has incomplete information about his biological kin and who perceives himself, and is perceived by others, as a token of exploitation and/or a commodity. As such, notwithstanding the absence of conclusive empirical data about the harms to children who are born of transnational commercial contract pregnancy to Indian women, it is reasonable to be concerned about the problematic identities available to these children. To be clear, children born of this family-making strategy, like children born of anonymous gametes, may experience a need to know their "genetic origins," and without access to medical, personal, and contact information for their progenitors they may develop a less than healthy identity. In addition, children born of Indian reproductive labourers may come to think of themselves as "tokens of exploitation" or as "products."

Conclusion

Transnational commercial contract pregnancy preys on the vulnerabilities of economically disadvantaged and often uneducated Indian women and more particularly on the desperate nature of their poverty. It also clouds society's moral obligation to correct, not capitalize on, the myriad background social, material, and political inequalities, and it reinforces morally unacceptable structural injustices. On this point, I defer to Iris Marion Young (2007: 175), according to whom "individuals bear responsibility for structural injustice because they contribute by their actions to the processes that produce unjust outcomes.... we bear responsibility because we are part of the process." From this it follows that we all (and most especially the Indian government) have an obligation to change the background poverty conditions that generate exploitative situations. Transnational commercial contract pregnancy in India is an ethically objectionable practice that involves taking unfair advantage of existing social, material, and political inequalities, while reinforcing structural injustices.

A further ethical objection to transnational commercial contract pregnancy in India concerns the harms to children born of this family-making strategy. Of particular concern are the harmful effects of the children's unusual origin story on identity formation. Persons born of transnational commercial contract pregnancy who want access to genetic and personal information about biological parentage and about kinship relations may experience discontinuity, detachment, loss, and incompleteness due to missing medical, biographical, and contact information. As well, there is the risk of stigma associated with not knowing biological relatives, with being birthed by a woman who was exploited, and with being commodified. Those who internalize a sense of self as the product of an exploited reproductive labourer may experience depression and discontent.

Taken together, the harms to women and to children of transnational commercial contract pregnancy in India support the conclusion that this family-making strategy fails morally and should be abolished.

Acknowledgements

I am grateful to a number of contributors to this book for their helpful comments on an earlier draft. I would also like to thank my colleague Jocelyn Downie for exceedingly helpful comments on an earlier draft of this chapter, David Michels (law librarian at the Schulich School of Law, Dalhousie University) for research assistance, and Timothy Krahn for assistance with the references.

References

Anonymous (2007). Outsourcing Life Itself: What India Teaches Us. *Maclean's*, 120(25), 2 July, 4. Retrieved June 2013 from http://www.macleans.ca/canada/opinions/article.jsp?content=200 70702_107034_107034&page=2.

Assisted Human Reproduction Act, SC 2004, c. 2 (2004). Retrieved July 2013 from http://www.canlii.org/en/ca/laws/stat/sc-2004-c-2/latest/sc-2004-c-2.html.

Bailey, A. (2011). Reconceiving Surrogacy: Toward a Reproductive Justice Account of Indian Surrogacy. *Hypatia*, 26(4), 715–41.

Baylis, F. (2012). The Self *in situ*: A Relational Account of Personal Identity. In J. Downie and J. Llewellyn (eds), *Relational Theory and Health Law and Policy* (pp. 109–31). Vancouver and Toronto: UBC Press.

Baylis, F., Kenny, N. P., and Sherwin, S. (2008). A Relational Account of Public Health Ethics. *Public Health Ethics*, 1(3), 196–209.

Bewley, S., Foo, L., and Braude, P. (2011) Adverse Outcomes from IVF. *British Medical Journal*, 342, 292–3.

Bhatia, S. (2012). Revealed: How More and More Britons are Paying Indian Women to Become Surrogate Mothers. *Daily Telegraph*, 26 May. Retrieved July 2013 from http://www.telegraph.co.uk/health/healthnews/9292343/Revealed-how-more-and-more-Britons-are-paying-Indian-women-to-become-surrogate-mothers.html.

Brian, C. (2006). The Son of a Surrogate, 9 Aug. Retrieved June 2013 from http://sonofasurrogate.tripod.com.

Center for Reproductive Rights (2008). Maternal Morbidity in India: Using International and Constitutional Law to Promote Accountability and Change. Retrieved June 2013 from http://www.unfpa.org/sowmy/resources/docs/library/R414_CenterRepRights_2008_INDIA_Maternal_Mortality_in_India_Center_for_Huiman_Rights.pdf.

Centre for Social Research (2012). Surrogate Motherhood—Ethical or Commercial. Retrieved June 2013. http://www.womenleadership.in/Csr/SurrogacyReport.pdf.

Corea, G. (1985). *The Mother Machine: Reproductive Technologies from Artificial Insemination to Artificial Wombs.* New York: Harper & Row.

Damelio, J., and Sorensen, K. (2008). Enhancing Autonomy in Paid Surrogacy. *Bioethics*, 22(5), 269–77.

Desai, S. B., Dubey, A., Joshi, B. L., Sen, M., Sharif, A., and Vanneman, R. (2010). *Human Development in India: Challenges for a Society in Transition.* New Delhi: Oxford University Press. Retrieved June 2013 from http://www.ncaer.org/downloads/Reports/HumanDevelopmentinIndia.pdf.

Downie, J., and Llewellyn J. (eds) (2012). *Relational Theory and Health Law and Policy.* Vancouver and Toronto: UBC Press.

Edwards, R. (1974). Fertilization of Human Eggs In Vitro: Morals, Ethics and the Law. *Quarterly Review of Biology*, 49(1), 3–26.

Ehrensaft, D. (2007). The Stork didn't Bring Me, I Came from a Dish: Psychological Experiences of Children Conceived through Assisted Reproductive Technology. *Journal of Infant, Child, and Adolescent Psychotherapy*, 6(2), 124–40.

Ehrensaft, D. (2008). When Baby Makes Three or Four or More: Attachment, Individuation, and Identity in Assisted-Conception Families. In R. A. King, S. Abrams, A. S. Dowling, and P. M. Brinich (eds), *The Psychoanalytic Study of the Child* (vol. 63, pp. 3–23). New Haven, CT and London: Yale University Press.

Ehrensaft, D. (2012). Gametes for Sale, Wombs for Rent, Babies to Raise. In M. O'Reilly-Landry (ed.), *A Psychodynamic Understanding of Modern Medicine: Placing the Person at the Center of Care.* London and New York: Radcliffe Publishing.

Gentleman, A. (2008). India Nurtures Business of Surrogate Motherhood. *New York Times*, 10 Mar. Retrieved June 2013 from http://www.nytimes.com/2008/03/10/world/asia/10surrogate.html?pagewanted=1&_r=2.

Gibbard, W. B. (2012). The Effect of Disclosure or Non-Disclosure on the Psychosocial Development of Donor-Conceived People: A Review and Synthesis of the Literature. In J. R. Guichon, I. Mitchell, and M. Giroux (eds), *The Right to Know One's Origins: Assisted Human Reproduction and the Best Interests of Children* (pp. 151–65). Brussels: Academic & Scientific Publishers.

Government of India (2013a). *Instructions for Foreigners Possessing Medical/Medical Attendant Visa.* Bureau of Immigration, India. Ministry of Home Affairs. Retrieved July 2013 from http://immigrationindia.nic.in/medical_visa2.htm.

Government of India (2013b). Medical Visa. Chapter 3A. Indian Visa Online. Retrieve July 2013 from http://indianvisaonline.gov.in/visa.

Government of India, Law Commission of India (2009). *Need for Legislation to Regulate Assisted Reproductive Technology Clinics as Well as Rights and Obligations of Parties to a Surrogacy.*

Report No. 228, Aug. Retrieved July 2013 from http://www.scribd.com/doc/33525981/Law-Commission-of-India-on-surrogacy-2009.

Government of India, Ministry of Home Affairs (2012). *Foreign Nationals Intending to Visit India for Commissioning Surrogacy.* Memo F.No. 25022/74/2011-F.I, 9 July. Retrieved July 2013 from http://www.icmr.nic.in/icmrnews/art/MHA_circular_July%209.pdf.

Government of India, Planning Commission (2011). *Press Note on Poverty Estimates.* Retrieved July 2013 from http://planningcommission.nic.in/reports/genrep/Press_pov_27Jan11.pdf.

Gupta, J. (2012). Reproductive Biocrossings: Indian Egg Donors and Surrogates in the Globalized Fertility Market. *International Journal of Feminist Approaches to Bioethics,* 5(1), 25–51.

Haslanger, S. (2009). Family, Ancestry and Self: What is the Moral Significance of Biological Ties? *Adoption and Culture,* 2, 91–122.

Haworth, A. (2007). Surrogate Mothers: Wombs for Rent. *Marie-Claire,* 29 July. Retrieved July 2013 from http://www.marieclaire.com/world-reports/news/surrogate-mothers-india.

Hochschild, A. (2009). Childbirth at the Global Crossroads. *American Prospect,* 19 Sept. Retrieved July 2013 from http://prospect.org/article/childbirth-global-crossroads-0.

Indian Council of Medical Research (2010). *The Assisted Reproductive Technologies (Regulation) Bill 2010.* Retrieved July 2013 from http://icmr.nic.in/guide/ART%20REGULATION%20Draft%20Bill1.pdf.

Indian Council of Medical Research, National Academy of Medical Sciences (India) (2005). *National Guidelines for Accreditation, Supervision and Regulation of ART Clinics in India.* Retrieved July 2013 from http://www.scribd.com/doc/28694677/Guidelines-for-ART-Clinics-in-India.

International Human Development Indicators (2013). *UNDP [United Nations Development Programme].* Retrieved July 2013 from http://hdr.undp.org/en/statistics.

Kirkman, M. (2003). Parents' Contributions to the Narrative Identity of Offspring of Donor-Assisted Conception. *Social Science and Medicine,* 57(11), 2229–42.

MacIntyre, A. (1984). *After Virtue: A Study of Moral Theory* (2nd edn). Notre Dame, IN: University of Notre Dame Press.

McWhinnie, A. (2001). Gamete Donation and Anonymity: Should Offspring from Donated Gametes Continue to be Denied Knowledge of their Origins and Antecedents? *Human Reproduction,* 16(5), 807–17.

Morrissette, M. (ed.) (2006). *Behind Closed Doors: Moving Beyond the Secrecy and Shame* (Voices of Donor Conception, 1). New York: Be-Mondo Publishing.

Nelson, H. L. (2002). What Child is This? *Hastings Center Report,* 32(6), 29–38.

Nelson, H. L. (2001). *Damaged Identities, Narrative Repair.* Ithaca, NY: Cornell University Press.

Okin, S. M. (1990). A Critique of Pregnancy Contracts: Comments on Articles by Hill, Merrick, Shevory, and Woliver. *Politics and the Life Sciences,* 8(2), 205–10.

OPHI and UNDP [Oxford Poverty and Human Development Initiative and United Nations Development Programme] (2010). Oxford and UNDP Launch Better Way to Measure Poverty. *UNDP Human Development Reports,* 14 July. Retrieved July 2013 from http://hdr.undp.org/en/mediacentre/news/announcements/title,20523,en.html.

Page, A. A. (2012). A Review of Studies that have Considered Family Functioning and Psychosocial Outcomes for Donor-Conceived Offspring. In J. R. Guichon, I. Mitchell, and M. Giroux (eds), *The Right to Know One's Origins: Assisted Human Reproduction and the Best Interests of Children* (pp. 124–50). Brussels: Academic and Scientific Publishers.

Pande, A. (2009). Not an "Angel," Not a "Whore." *Indian Journal of Gender Studies*, 16(2), 141–73.

Panitch, V. (2013). Surrogate Tourism and Reproductive Rights. *Hypatia* 28(2), 274–89.

Parks, J. A. (2010). Care Ethics and the Global Practice of Commercial Surrogacy. *Bioethics*, 24(7), 333–40.

Pratten v. British Columbia (Attorney General), 2011 BCSC 656 (2011). Retrieved July 2013 from http://www.courts.gov.bc.ca/jdb-txt/SC/11/06/2011BCSC0656cor1.htm.

Ravitsky, V. (2010). Knowing Where you Come from: The Rights of Donor-Conceived Individuals and the Meaning of Genetic Relatedness. *Minnesota Journal of Law, Science and Technology*, 11(2), 655–84.

Ricoeur, P. (2001). *Soi-même comme un autre*. Paris: Éditions du Seuil.

Satz, D. (1992). Markets in Women's Reproductive Labour. *Philosophy and Public Affairs*, 21(2), 107–31.

Schechtman, M. (1996). *The Constitution of Selves*. Ithaca, NY: Cornell University Press.

Scheper-Hughes, N. (2000). The Global Traffic in Human Organs. *Current Anthropology*, 41(2), 191–224.

Schultz, S. (2008). The Life Factory: In India, Surrogacy has Become a Global Business. *Spiegel Online International*, 38 (part 2). Retrieved July 2013 from http://www.spiegel.de/international/world/0,1518,580209-2,00.html.

Scott, A. (2007). "Wombs for Rent" Grows in India. *Marketplace on NPR*, 27 Dec. Retrieved July 2013 from http://www.marketplace.org/topics/life/wombs-rent-grows-india.

Sen, A. (2011). Growth and Other Concerns. *The Hindu*, 14 Feb. Retrieved July 2013 from http://www.thehindu.com/opinion/op-ed/article1451973.ece?homepage=true.

Shenfield, F., and Steele, S. J. (1997). What are the Effects of Anonymity and Secrecy on the Welfare of the Child in Gamete Donation? *Human Reproduction*, 12(2), 392–5.

Sherwin, S. (1998). A Relational Approach to Autonomy in Health Care. In S. Sherwin and Feminist Health Care Ethics Research Network (eds), *The Politics of Women's Health: Exploring Agency and Autonomy* (pp. 19–47). Philadelphia, PA: Temple University Press.

Smerdon, U. R. (2008). Crossing Bodies, Crossing Borders: International Surrogacy between the United States and India. *Cumberland Law Review*, 39(1), 15–85.

Springer, T. (2012). Of Mothers and Merchants: Commercial Surrogacy. *CBC: The Current*, 28 Mar. Retrieved July 2013 from http://www.cbc.ca/thecurrent/episode/2012/03/28/of-mothers-merchants-commercial-surrogacy.

Thomas, L. M. (1998). Moral Deference. In C. Willet (ed.), *Theorizing Multiculturalism: A Guide to the Current Debate* (pp. 359–81). Oxford: Blackwell Publishers.

TNN (2012). Surrogate Mother Dies of Complications. *The Times of India*, 17 May. Retrieved July 2013 from http://articles.timesofindia.indiatimes.com/2012-05-17/ahmedabad/31748277_1_surrogate-mother-surrogacy-couples.

Turner, A. J., and Coyle, A. (2000). What does it Mean to be a Donor Offspring? The Identity Experiences of Adults Conceived by Donor Insemination and the Implications for Counselling and Therapy. *Human Reproduction*, 15(9), 2041–51.

Wilkinson, S. (2003). The Exploitation Argument Against Commercial Surrogacy. *Bioethics*, 17(2), 169–87.

World Bank (2013). Open Data. http://data.worldbank.org/?display=default [Search "By country," select India; search "By topic," select poverty].

Young, I. M. (2007). *Global Challenges: War, Self-Determination and Responsibility*. Cambridge: Polity Press.

15

Aged Parenting through ART and Other Means

Jennifer A. Parks

This chapter will focus on one aspect of contemporary family-making—advanced age parenting—with the aim of arguing that, within the realm of assisted reproductive technologies, new techniques in fertility preservation (FP) represent an overall advancement in women's reproductive options, and are therefore morally acceptable. FP technologies allow women to preserve their own ovarian tissue and/or eggs so that they may be used at a later date to achieve pregnancies; this lowers the demand for egg donors, thus reducing the number of younger women who are put at risk in supplying their eggs for use by women of advanced maternal age. Given this and other benefits I will highlight in this chapter, FP represents an improvement in the circumstances under which women achieve pregnancy at an advanced maternal age, and it is therefore morally permissible. I will take a feminist approach to aged parenting, considering this issue with particular concern for women's best interests. While critics of in vitro fertilization (IVF) and FP use by older women object to the practice because of potential harms to women and children, I will address other considerations that suggest there is no moral justification to ban older women from accessing these services. While feminists in particular should continue to discuss the morality and advisability of women reproducing at an advanced maternal age, I will offer some preliminary reasons for supporting women's reproductive use of cryopreserved eggs beyond the age of menopause.[1]

In what follows, I will first consider postmenopausal IVF as a form of aged parenting, exploring the health-based, social, and ethical concerns that are raised against it. I will then consider the new technique of fertility preservation and whether, from a

[1] FP is designated as serving medical purposes when it is applied to women and girls who are undergoing cancer treatment or other medical interventions that result in the loss of their reproductive capacity; it is designated as serving social purposes when it is applied to women for non-medical purposes, such as the case of women who are beyond menopause and want to use the technology for purposes of family formation.

feminist viewpoint, it represents a feminist technology that frees women from biological and cultural limits on motherhood, or whether it is rather the imposition of patriarchal values and beliefs about the central value of reproduction to women's lives. I will consider in turn the practices of custodial grandparenting and adoption of children by older persons, and other forms of advanced age parenting, to consider whether these may be more virtuous forms of parenting than the purposeful creation of offspring by older parents. From these considerations, I will conclude by arguing that older women should be morally permitted to use their own cryopreserved eggs in order to have children, but that there should also be concerted efforts to make aged parenting of existing children more accessible so that it may be a viable option for older persons.[2]

The Rise of Aged Parenting through Assisted Reproductive Technologies

Following the development and proliferation of assisted reproductive technologies (ARTs), the number of advanced age persons seeking reproductive assistance has been on the rise, rendering cultural understandings of "family" ever more complex. While still not a mainstream activity, aged parenting has garnered media attention because of sensationalized cases such as Harriet Stole (who in 1999 had a son at the age of 66); Adriana Iliescu (who in 2005, and at the age of 66, gave birth to twin daughters, one of which survived past birth); Omkari Panwar (who in 2008, and at the estimated age of 70, gave birth to twins); and Rajo Devi (who also gave birth to a daughter in 2008 at the age of 70). Ethicists have addressed this issue of aged parenting (but especially postmenopausal mothering) from a variety of perspectives, in some cases supporting older women's rights to form families by various means (see Smajdor, 2008) and in other cases citing concerns for the health and welfare of both older women and their offspring in order to justify preventing these women from accessing ART services (see Caplan and Patrizio, 2010). Objections to aged parenting through ART concern the likelihood of orphaning children, aged parents' inability to vigorously engage with children, and the physical and biological problems associated with age—concerns that extend to custodial grandparents and advanced age adopters as well.

Though ARTs were initially targeted to reproductive-aged women, a growing number of women of advanced maternal age began to take advantage of the technology during the mid-1980s. Initially, the demand by older women arose because ARTs were not available during their reproductive years when women discovered their (or their male partners') infertility (Paulson and Sauer, 1994); but as time went by, the reasons

[2] I will focus on women of advanced maternal age in this chapter because it is reproduction by older women, not older men, that is receiving the most criticism and debate. Since, biologically speaking, men are able to produce viable sperm throughout their lifespan, they are not reliant on technology to extend their reproductive years; women beyond the age of menopause, by contrast, must use technological means (and rely on interventions by a third party) in order to reproduce postmenopausally.

for women's aged parenting practices became more diverse—from career-based concerns (not wanting to interrupt their career momentum), to relationship changes (cases of remarriage or marrying a younger male partner who still wanted to start a family), to psychological issues, such as grief over the death of a beloved adult child (as in the case of a 62-year-old Italian women who used IVF services to have a child after her adult son was killed in a car accident) (Parks, 1999). Now, 35 years since the first IVF baby Louise Brown was born, there has been not only a surge in the development of and demand for ARTs in general, but a very notable increase in the number of women of advanced maternal age who have used it in order to achieve pregnancy. Over the past ten years, the number of women over 45 giving birth has more than doubled in the US (Martin, 2012). Furthermore, as revealed in a study by the Centers for Disease Control, there has been a 375 per cent increase in the same period for births among US women over 50 (Abraham, 2012).

The Physical and Moral Risks of IVF at Advanced Maternal Age

Recent studies on the physical risks and effects of postmenopausal IVF suggest that women over 50 who use donor eggs do not appear to face significantly higher risks of complications than their counterparts in their forties.[3] While using IVF at an advanced maternal age is certainly not without its risks, recent studies suggest that the outcomes for women between the ages of 50 and 59 versus women 42 years of age and younger may be comparable (Kort et al., 2012).[4] Of course, significantly older women like Omkari Panwar face additional health risks associated with having a much older body; but studies are increasingly suggesting that concerns regarding additional health risks in cases of donor egg IVF are not sufficient grounds for preventing women from conceiving and carrying a pregnancy into their mid-to-late fifties or even early sixties. A recent Columbia University study included 101 postmenopausal women between the ages of 50 and 59; participants were screened for cardiovascular issues, diabetes, and other complicating factors to ensure a low-risk population (Kort et al., 2012).

---

[3] While the risks are higher, the claim is that they are not statistically higher enough to reject women in their mid-to-late fifties as reasonable candidates for IVF.

[4] But about this new statistical evidence, Barbara Katz Rothman (2012) queries "What if the IVF rates, with purchased eggs... are just as good as the pregnancy rates for young women? What if we just stop arguing the data, and say 'so what?' Is it a good thing that young and healthy women who want education and good careers cannot in any way, not in time nor in money nor in energy, afford children? Is it a good thing for children to become a mid-life project? Is it a good thing to conquer the biological clock for reproduction if the rest of the biological clock—the one for diabetes, stroke, dementia—keeps ticking? We've had older men fathering children, often second-sets of them, as lovely late-life projects. But those men usually had young wives to mother the kids, care for them through the aging and death of the father. These delayed-childbearing women are less likely to have young partners to pick up the reins. What are we wishing on our children? And what are the costs for all of the women involved, the ones who delay, the ones who sell eggs, the ones who succeed in late-life baby-making and the ones who don't?" (para. 3).

Mark Sauer, a contributor to the study, notes that "None of these are accidental pregnancies. None of these 50-year-olds just woke up pregnant and (said), 'now what?' These are women of means who have good medical resources available to them" (Alberta Children's Hospital Research Institute, n.d.: para. 21). The study sample is representative of the older female demographic that seeks reproductive services for infertility—i.e. they are relatively wealthy, healthy individuals with the financial means to seek excellent healthcare.[5]

Indeed, such changes in the risk evaluation of postmenopausal IVF led the Ethics Committee of the American Society for Reproductive Medicine (ASRM) to address the question of setting age limits on the use of IVF. As the Committee notes, "[we believe] that achieving pregnancy through egg donation after age 50 is not such a significant departure from other currently accepted fertility treatments as to be considered ethically inappropriate in postmenopausal women" (Ethics Committee, 2013: 3). Their current position acknowledges the use of IVF by women of advanced maternal age, and recommends an upper age limit of 55 years.[6]

Physical risks aside, however, concerns about the *moral* risks associated with advanced age parenting persist, and this is where the greatest controversy arises. Critics of IVF at advanced maternal age are troubled by the use of ARTs among older women, since these technologies can allow for parenting, in some cases at a *very* advanced age. For example, commenting on the case of Omkari Panwar, Arthur Caplan (2008) claims the following:

surely it is of ethical concern when people in their 70s use IVF to have babies. Despite all those ads with great-looking 70-somethings playing tennis and dancing aboard cruise ships, a 70-year-old woman giving birth to twins is a super high-risk pregnancy for both mom and the babies, and also a likely route to making orphans. (para. 11)

Caplan later adds that,

Even if the babies are born healthy when one or both parents is over 70, are they really in a position to parent kids through their teen years? Shouldn't there be some limit on age that recognizes that kids should have a chance of having at least one parent who is physically capable of raising them? (2008: para. 12)[7]

---

[5] This indicates the degree of privilege enjoyed by the type of women who tend to pursue IVF at an advanced maternal age. Such women tend to have benefited from lifelong access to good healthcare, from living a healthy lifestyle, and from advanced education and the attending economic pay-off that comes with it.

[6] Note that the ASRM guidelines are just that—guidelines or recommendations—and are not legally enforceable. Critics argue that, without the "teeth" of serving a regulatory purpose, such guidelines do little to govern the actual practice of ART in the United States (see Maguire, 2010; Riggan, 2011; Asch and Marmor, 2008).

[7] Although Caplan (2008) has taken a stance against aged parenting, he has written in favour of anti-ageing medicine. It is worth nothing that there may be inconsistencies in Caplan's writings on these topics given his failure to construct an approach to technology that connects ART and anti-ageing technology. If technologies to extend a woman's lifespan and years of health are morally worthy goals according to Caplan, then it is unclear why technologies to likewise extend a woman's reproductive years are not similarly worthy goals.

Some commentators agree with Caplan (Landau, 2004; Kluge, 1994; Royal Commission on New Reproductive Technologies, 1993), while others are less willing to forbid the practice (Parks, 1999; Cutas, 2007; Smajdor, 2008).

Critics of aged parenting rightly focus on concerns for the welfare of the women who reproduce at a much older age and the children that result. There may be psychological health risks imposed on children with significantly older parents; and concern has been expressed over the prospect of children with old parents being orphaned at an early age or being saddled early on with the responsibility of intensive care-giving for an elderly and debilitated parent. Given our current socio-cultural state of affairs, these are justified concerns, and anyone who is considering parenting at an advanced age should be giving serious thought to them as part of their parenthood planning. Yet I argue that none of these arguments is decisive in undercutting the morality of advanced age parenting, especially if older parents are making appropriate arrangements for their own chronic care needs and for the future care and guardianship of their children.

Consider the argument that older parents present the risk of orphaning their children, a harm that should be avoided since their children could potentially be quite young when this happens. As I have argued in previous work (Parks, 1999), concerns about orphaning children are not consistently applied, and should not be raised solely against postmenopausal IVF. For example, fertility preservation is a means to allow younger women with cancer and other serious health issues to preserve their reproductive capacity so that, should their illness render them infertile after treatment, they will still be able to have children. But this promotes reproduction in younger women who, if they survive their illnesses, may be at risk of premature death, which will also result in partially orphaning their children.[8] In some cases, pregnant women who were deemed clinically dead due to brain death have been kept alive to gestate their pregnancies until the foetuses were able to survive independently and be delivered through caesarean section, most certainly resulting in motherless children. Single women are not prevented from reproducing, though should anything happen to them, their children will be fully orphaned. If we will countenance younger women's reproduction under such circumstances, there seems to be no ground for rejecting older women's reproduction out of hand based on the concern for creating orphans.[9]

---

[8] I say "partially" orphaning because in cases where these women are married, their partners will still be present to parent their children. But in many cases where women are reproducing at an advanced maternal age, they also have younger marital partners who will survive them, so the cases are similar in this regard.

[9] This is not to minimize the importance of addressing this concern about orphaned children. On the contrary, in my other work on ART I argue that we have not been sufficiently addressing concerns about children's best interests (see Parks, 2010) and that we need to do so in order to highlight the implications of parenting choices for the offspring that are created. Women and men who are reproducing at an advanced age have a special moral obligation to address this concern, since they are more likely to partially or fully orphan their children than persons who reproduce at a younger age.

The rise of the nuclear family has had a significant impact on the risk of orphaning children. Thus, as I argue, it is not that parenting at an advanced age *necessarily* leads to orphaned children, but rather that the constitution of families in the Western world threatens this outcome. Daniela Cutas, for example, argues that the problem isn't the age of the parents, but the family support and love that is available to any children that are born, and the quality of their family relationships. She queries: "If this factor is so important, and carries so much moral weight, then it is curious that we don't seem to encourage the formation of families with three or more parents.... Obviously children would be more protected from the loss of all parents to death if they had three or more of them" (Cutas, 2007: 461). If women who elect postmenopausal IVF take seriously these concerns about their future children's best interests, then it may be far less likely that they will orphan them, as they will be ensuring the good care of their children in the event of their deaths.

Concerns about the imposition of intensive care-giving upon very young adults who have aged parents should not be taken lightly. Even middle-aged children of elderly parents may find the strains associated with care-giving for elderly parents to be over-whelming (Holstein, 1999; Parks, 2003); the implications for very young adults who are responsible for their parents' intensive care-giving needs could, indeed, be grave. Thus, I argue that along with the need to ensure continuity of care for their children in the case of their death, older parents contemplating the creation of offspring are morally obligated to seriously consider how their own profound care needs may affect their children. In the absence of older children who may be in a position to care for their aged parents (e.g. Omkari and Charan Singh Panwar have two adult daughters, presumably both of whom will be responsible for the care of their parents and for the care of their young siblings when their parents eventually die), persons of advanced age ought not to purposefully bring children into the world that will carry this heavy physical and psychological care burden.[10]

Absent any personal benefit or good that is associated with parenting at an advanced age, postmenopausal IVF and other ART use by older women lack a clear rationale. What, then, might be the goods associated with advanced age parenting? The act of mothering another human being, as many feminist ethicists have argued, allows an individual the opportunity to richly develop and express his or her relational and affective capacities (Held, 1987; Ruddick, 1989, Noddings, 1984). For example, as Sara Ruddick describes it, maternal practice involves the responsibility of preserving a child's life; supporting his or her growth; and educating and nurturing a child for

[10] While I do not want to generally characterize care-giving for elderly family members as a *mere* burden—most adult children do so out of a genuine sense of love for their elderly parents—in the case of advanced age reproduction these care-giving demands take on a different meaning. Children in their late teens and early twenties should not be left with the sole responsibility for the care of their elderly parents, despite their love for their parents or their willingness to do so. Furthermore, the gendered burden of care implicit in the Panwar situation is a concern, though it is not one that I can address in this chapter.

acceptance within his or her community.[11] These are activities in which many older persons engage as the custodial grandparents to their grandchildren or as older adoptive parents, and responsibilities under which many such persons flourish. Older persons may seek ARTs in order to have these experiences.

Furthermore, adults who parent at an advanced age may arguably find deep meaning in their relationships to their children. This "meaning making" may have further implications for their sense of purpose in life, something that persons may lose as they age and their associations with others erode over time.[12] As existential psychotherapist Viktor Frankl (1967) observed in his writing on the human search for meaning in life, "A man [sic] who becomes conscious of the responsibility he bears toward a human being who affectionately waits for him, or to an unfinished work, will never be able to throw away his life. He knows the 'why' for his existence, and will be able to bear almost any 'how'" (p. 127). When one has a child who "affectionately waits" and especially a younger child who is, metaphorically speaking, the "unfinished work" to which Frankl refers, one may find special meaning and purpose in life, and a reason to invest passionately and purposefully in life. One might argue that persons who would find such purpose in advanced age parenting should not be prevented from it, especially given the moral significance of mothering as Ruddick describes it.

As critics like Patricia Smith point out, however, it may be possible for individuals to engage in Ruddick's version of maternal practice by developing relationships with children without being their parents. Smith (1993: 96–7) claims that:

You do not need to be a parent to share your life with a child. Aunts, uncles, grandparents, and friends can have an enormous effect on a child's life, and in today's busy world can be a real boon both to the child and the parents. Watching the blossoming of a child in whom you have invested time and love is always uniquely rewarding, whether it is your own child or someone else's. If none of these options appeal to you—if it is not enough to spend your time and energy on an adopted child or on someone else's child, perhaps you should examine your reasons for wanting a child. Perhaps you don't really want to spend time with a child. You just want to own one...

Smith raises an important point regarding the intentions and purposes of older individuals who desire offspring. If the desire is to connect with, love, and nurture a child, then it is not clear why adoption would not be an obvious choice.[13] Alternatively, as she

---

[11] Note that Ruddick's maternal thinking applies to men and women equally, and that the subject of the maternal thinking isn't necessarily a child, but can be other adults with whom we are in relation. Ruddick's account, however, is rooted in the experiences and virtues that come with rearing children, and it is children that I specifically have in mind here.

[12] I am not suggesting that old persons *necessarily* lack relationships and meaning in their lives, but studies have repeatedly shown and theorists have argued that, as individuals age, their opportunities to connect with others, to touch and be touched, and to find meaning in relationships is reduced just because of the way our social arrangements have been set up. (See Bartky, 1999; Parks, 2003.)

[13] While I do not have the space to pursue this question in this chapter, I refer readers to the previous section of this volume, where the authors address the moral issues associated with adoption versus biological parenting. Also, note the problematic use of Smith's language of raising "your own child or someone else's." This notion of having "one's own" child assumes a certain degree of biologism and does not take into account that adopted children are their parents' "own" children. For more on biologism, see Petropaganos (2013).

suggests, one might seek these experiences as an aunt, uncle, grandparent, or friend. Regarding this alternative, however: in cultures like the US that emphasize the privacy rights of families, the value of the nuclear family, and parental rights over children, those who are not legally and socially recognized as parents to particular children may find it extremely difficult to find children with whom they are allowed to develop that kind of close, loving relationship. Mothering—and parenting in general—constitutes a very special relationship with particular children that arguably may not be had by "spending time" with others' children (see Brighouse and Swift, this volume).

In the section that follows I will consider the advent of fertility preservation (FP) technologies and their implications for women of advanced maternal age. These technologies, while intended to serve the needs of young women who are rendered infertile due to medical conditions such as cancer, can also possibly be used by women to preserve their fertility so that they may biologically reproduce at a much later date. Thus, there may be both medical and social reasons for the use of FP; and while there has generally been moral approval for the use of FP in reproductive-aged women who are undergoing cancer treatment, its use by women for the social purpose of extending their ability to reproduce has received a degree of criticism (Harwood, 2008; Jones, 2009; Schermers, 2009). Following my discussion of FP technology, I will pursue the question of whether it constitutes a "feminist" technology. This is significant to my feminist evaluation of aged parenting, since technological advancements like FP cannot simply be assumed to always be of benefit to women; and even if they are of some benefit, one must consider how those benefits weigh against their harms. As I will argue, there are considerable arguments on both sides regarding the feminist implications of FP technologies.

The Use of Fertility Preservation Technologies by Women of Advanced Maternal Age

Until recently, when applied to postmenopausal women, IVF treatment has required the additional step of securing donor eggs[14] (i.e. "healthy" eggs donated by much younger women) to facilitate their successful implantation and gestation. Although with the use of fertility-enhancing drugs, postmenopausal women are capable of gestating and carrying a foetus to term, the quality of their eggs has declined significantly due to age. However, recent developments in ART have resulted in the possibility of fertility preservation (Loren *et al.*, 2013), where women seek assistance to preserve their own gametes and tissue for later use.

---

[14] Note that in cases of egg donation, the young women who "donate" the eggs in the US are paid, and sometimes paid very well, for the eggs they supply. The issue of whether to pay for eggs at all, and how much constitutes fair remuneration, is still being debated in bioethics (Steinbock, 2004; Baylis and McLeod, 2007).

Fertility preservation has been researched with the goal of helping reproductive-aged cancer patients to retain their procreativity (Woodruff *et al.*, 2010); but as with most technologies with multiple applications, it is being extended beyond cancer care to serve social purposes (Goold and Savulescu, 2009; Harwood, 2008). This means, for example, that women who seek IVF treatment following FP will be able to use their own ovarian tissue or eggs, cryopreserved during their peak reproductive years, to reproduce at the time of their choosing. This application of FP means that in future more women are likely to use it for the purpose of extending their reproductive years.

The form of FP technology that concerns us in the case of advanced age mothering involves "the removal, cryopreservation and subsequent storage of [eggs] for future use" (Petropaganos, 2013: 1).[15] While sperm cryopreservation has been successfully practised since the 1950s, and embryo cryopreservation has been available for several decades, these have been the only FP options available until very recently (Petropanagos, 2013). The most recent forms of FP, however, have targeted women and children with cancer diagnoses in an attempt to avoid the infertility that comes with cancer treatment.

Fertility preservation technologies involve different stages. "Stage one FP" for women involves the removal and freezing of eggs or reproductive tissues (Petropanagos, 2013). Once frozen, these materials are stored for an indefinite amount of time, until (in the case of cancer treatment) a woman's cancer care is complete and she has been determined to be in remission. This stage of FP must take place before a woman undergoes radiation or chemotherapy, both of which can render her infertile. Notice, however, that the option to cryopreserve one's eggs or ovaries is not limited to women undergoing cancer treatment: stage one FP may be practised on women for any number of reasons, especially (in the case at hand) on women who may use FP technology for the social purpose of extending their timeframe for biological reproduction.[16]

Women who use stage one FP may elect to use their cryopreserved eggs or other reproductive materials in stage two FP, which encompasses thawing these materials for use in various ART procedures. The goal in doing so is to produce a genetically related child. For a variety of reasons, not all women who undergo stage one FP continue with stage two FP: they may decide not to reproduce at all, or may conceive unassisted, or may decide to adopt rather than seeking genetic reproduction. Women who undergo FP in their early twenties may wait decades before using their cryopreserved materials, and may even wait until well beyond the age of menopause before they decide to use

---

[15] Other FP techniques include ultrasound-guided transvaginal eggs aspiration, and the surgical removal of ovarian tissue that is either cryopreserved or matured using in vitro maturation (IVM). With the last option (IVM), any resulting eggs are cryopreserved (Georgescu *et al.*, 2008; Uzelac *et al.*, 2012; Petropanagos, 2013). I will not consider these other types of FP technology, since they relate more to the cancer treatment context, and are not relevant to FP use by women of advanced maternal age.

[16] As Angel Petropaganos (2013: 2) notes, "these FP technologies could be used by individuals who wish to guard themselves against infertility that is caused by other diseases, infertility due to age (what I call 'age-related infertility'), employment or environmental hazards, or gender reassignment surgeries." The term "age-related infertility" as used in this chapter comes from Petropanagos.

them. Given that stage one FP for women involves the storage of their reproductive materials for an indefinite period of time, as the technology is perfected and risks are even further reduced, it is more likely that women will elect for stage one FP as a type of "insurance policy" against concerns such as partner selection, educational or career plans, or menopause.

Yet using FP for this social purpose raises a number of criticisms and concerns. To begin, stage one FP normally takes place during a woman's reproductive years;[17] yet as noted she may cryopreserve her reproductive materials for decades before she decides to move to stage two FP, using ART to create offspring. It is unclear how long these cryopreserved materials remain viable, and some organizations (like the Human Fertilization and Embryology Authority, 2013) recommend storing them for no more than ten years in order to prevent potential harms to women or their offspring. Second, as the argument goes for postmenopausal IVF, it might be unfair for women of advanced maternal age to use their cryopreserved materials in order to procreate when there is a good chance they will not see their children into adulthood (Parks, 1999; Petropaganos, 2013). Third, one might criticize this use of FP technology by pointing out that, despite recent studies finding no significant harms associated with IVF use by women of advanced maternal age, ageing affects women's "energy levels, mobility, health, and life expectancy" (Petropaganos, 2013) such that reproduction in the later stages of life poses too many inherent risks. Finally, feminist critics like Barbara Katz Rothman (1986, 2012) may worry about the degree of autonomy with which women can choose to reproduce given the presence of social expectations surrounding women's reproduction. The ability to reproduce serves as a signal in youthist culture that a woman is still young and fertile enough to serve this function; some older women may view reproducing as a way of denying old age and remaining relevant in an ageist society. While FP and other forms of ART are presented as increasing options for women, feminist critics might raise worries about the way they close off the option of accepting ageing gracefully and moving on to other (non-reproductive) stages of life.

Despite these concerns, FP technologies also present clear benefits to the women who use them. For example, from an ethical perspective, the technologies potentially eliminate concerns associated with the practice of egg donation, which carries risks for the young women who agree to sell their eggs.[18] The technology extends a woman's reproductive timeframe and allows the individual to use her *own* eggs, thus making egg donors unnecessary. I will develop this point later in my discussion of

---

[17] FP can be applied to young children, using techniques to collect eggs/sperm even before the age of puberty (Uzelac *et al.*, 2012; Georgescu *et al.*, 2008). See n. 15. I will not address this particular use of FP in this chapter.

[18] In some cases, the young women who supply the eggs may suffer from ovarian hyperstimulation syndrome (OHSS) which can result in morbidity or, in some extreme cases, mortality. On these grounds alone, some ethicists have rejected the practice of egg donation, since the young women are taking serious health and reproductive risks that they may be unfairly induced to accept based on the prospects of making money (Cohen, 1996).

FP as a feminist technology, but it is worth noting here that, by significantly reducing the demand for egg donors, FP technologies present an improvement over postmenopausal IVF, which is premised on access to younger women's eggs. In addition to the reduction of risks to others, FP also avoids concerns about scarcity of resources, since women of advanced maternal age who practised stage one FP would be using their own gametes to reproduce at a later date, thus obviating any concerns about the scarcity of reproductive materials.

Furthermore, from a feminist point of view, technologies such as FP that challenge our traditional notions of motherhood and family-making by extending parenthood to persons who have until recently been barred from parenting (such as lesbians, gay men, and single, disabled, or postmenopausal women) may be morally acceptable or even desirable (Parks, 2009). Attempts to limit or control parenthood to the two-parent, heterosexual norm have been undercut by various forms of ART, which have opened up definitions of what it means to be a family and who may parent. Gay or lesbian couples who use surrogates or sperm donors to meet their reproductive ends may include those third parties as part of their family, thus extending the notion of "family" beyond the two-parent norm. Single women who parent on their own may call on their brothers, fathers, uncles, or other male family members to serve as father figures to their children. And women or couples of advanced maternal age may also alter conceptions of family by extending them beyond the standard nuclear family unit. While courts have generally tried to resist these applications of ART, struggling to maintain laws that *only* recognize biological connections and that uphold the heterosexual, two-parent, nuclear family unit, the advent of ARTs like fertility preservation and IVF have required courts to acknowledge these families as legitimate (Parks, 2009). Insofar as FP extends reproduction beyond traditional conceptions of "normal" or "appropriate" motherhood, it potentially loosens the patriarchal chokehold over the ways that families are formed. This leads to a consideration of whether ARTs (and specifically, FP as a form of ART) can or should be viewed as "feminist" technologies that serve women's interests and ends; it is to this question that I will now turn.

ARTs as Feminist Technologies?

The question of whether ARTs are "feminist" technologies has not received much attention to date. Yet in their recent anthology, *Feminist Technology*, Linda Layne, Sharra Vostral, and Kate Boyer consider what makes a technology "feminist" by addressing a variety of recent technologies that have been applied to women, such as menstruation-suppressing birth control pills, home pregnancy tests, breast pumps, and tampons (Layne *et al.*, 2010). The working definition of feminist technology offered by Layne *et al.* (2010: 3) is "tools plus knowledge that enhance women's ability to develop, expand, and express their capacities." According to Layne *et al.* (2010), what makes a technology "feminist" is not that it is an artefact designed by women, for women (p. 6), since the gender of the designer and the intended users are not definitive of whether a technology develops, expands, and expresses women's capacities. Indeed, as is often the case with any technology, the way in which it is taken up may be vastly different from the intentions of the designer, meaning that even some technologies that have sexist origins may ultimately benefit women (Layne *et al.*, 2010: 10).[19] Although I agree that Layne's definition is a good starting point for an account of feminist technology, her consequentialist approach does not address ways in which a technology by its very creation may violate women's dignity. Layne's emphasis could be criticized for focusing too much on how technologies are taken up and used by women, with not enough attention to how in their very creation they represent or characterize women. However, in previous work (Parks, 2009) I have suggested that technology is not easily reducible to a single representation of women or "motherhood," so to reject a technology purely on such symbolic grounds is problematic if in practice it expands or extends women's capacities in ways that are welcomed by them.

There is a persistent problem, then, in coming to agreement about whether a technology should or can be considered feminist. This may be in part due to the nature of technology, since it often has complex and, indeed, contradictory effects (Parks, 2009). But the problem is also difficult because feminists adopt very different social and political worldviews, with liberal feminists often taking vastly different approaches from their radical or socialist feminist peers. As Layne *et al.* (2010: 18) note, "feminism is 'a many splendor'd thing.' The feminist approaches by which we evaluate technologies are diverse. A technology may appear feminist in light of one type of feminism and antifeminist through a different feminist lens." Despite its inherent difficulties, it is valuable to consider whether the practice of ART use by women of advanced maternal age will enhance women's capacities, empowering them and serving their well-being, or whether such practice is a setback to these feminist aims. By considering this issue, one can gather a fuller picture of the harms and benefits to women of FP technology, thus making it possible to better determine whether it is a technology that feminists can and should support.

As previously noted, Layne *et al.*'s criteria for determining whether a technology is feminist are consequentialist: the concern is not their intended purpose or who created them, but how they affect and are taken up by women (2010: 11). As one might observe over the past few decades in which assisted reproduction has developed, ARTs arguably enhance women's freedom from reliance on men. Consider the contemporary use of IVF or donor insemination (DI) by lesbians, disabled, single, and minority women, and women of advanced maternal age. The use of reproductive technologies by these groups has extended reproduction to women who have historically

---

[19] Layne *et al.* (2010: 11) mention the example of the dial pack for birth control pills, invented in 1964 by David Wagner to "help his wife remember to take her pills and to overcome the marital rows that emerged around her taking the pill... So even though the ring pack allowed the male inventor (and potentially other men) to engage in surveillance to see for himself whether his sexual partner was being 'medically compliant,' the product has enhanced women's ability to control their fertility."

been denied access because of their "difference" or because, as Valerie Hartouni (1997) has noted, they represent "monstrous" mothers. Similarly, in its social use to artificially extend a woman's reproductive years, FP technology may be used by women in ways that expand their options beyond reliance on men to achieve their reproductive ends. For example, an older woman who never married for career or personal reasons might use FP plus sperm donation in order to have a family; here, FP appears as a feminist technology that has been used to support feminist agendas, challenging and changing our traditional conceptions of motherhood and family.

Let us consider the arguments on both sides concerning whether ARTs that assist mothering at an advanced maternal age are feminist technologies. As Layne notes, feminists do not agree on this question. According to some feminist accounts, the application of technology to assist significantly older women in reaching their reproductive goals helps to develop, enhance, and extend their capacities. Feminist critics, on the other hand, are wary of the extension of reproduction to these older women.

Consider, for example, Patricia Smith's position on reproduction by women of advanced maternal age. Smith addresses many of the assumptions built into the growing practice of postmenopausal reproduction: that it fulfils a basic "need" for women to experience pregnancy and childbirth[20]; that it is the fulfilment of womanhood; and that it addresses the desire to have genetically related children of one's own. Smith (1993: 93) opens her essay by asking the following questions:

Is twenty-first century medicine perpetuating nineteenth-century roles for women? It certainly wouldn't be the first time. If feminist historical research has demonstrated anything, it has shown clearly that the medical establishment, especially in the past two hundred years, has been extremely conservative about what it considers good for women, healthy for women, normal for women. Always it has assumed that the central (or only) function, fulfillment, and happiness for normal women is pregnancy, childbirth, and motherhood.

Smith does well to point out the way in which IVF technology for postmenopausal women may not enhance women's well-being, but may rather serve to reinforce long-standing views held by the medical establishment concerning what is "good" for women, and what is their central function. Rather than creating new options and choices for women, the practice may actually limit their options and choices, and at a time in their lives when women are supposed to be free of the pressures to reproduce. Indeed, one of the clear motivating factors for women to reproduce at an advanced maternal age is the desire to provide their husbands with genetically related children. As Smith (1993: 97) claims,

By definition of the circumstances, genetic offspring are not a possibility for the postmenopausal woman. Presumably, she participates vicariously in the biological reproduction of her husband and another woman through the experience of pregnancy. It is a simulation of genetic

---

[20] On Smith's account, the technology more specifically fulfils a woman's role of providing biological offspring for her husband.

reproduction for her.... the question is, Why is this important? From the point of view of biological reproduction, a postmenopausal woman is not reproducing herself or her family line,
nor is she joining her family line with her mate's. So all that can be important from a biological
point of view is the reproduction of the father's genetic line, his family, or perhaps by extension,
his name.[21]

Let us return to the case of Omkari Panwar, mentioned earlier in this chapter. She had
only ever given birth to daughters, so had never delivered to her husband the son that
would carry on the family name and the family farming business. Her husband's comments to the media are telling, and echo Smith's concerns about women reproducing
to provide a biological heir for their husbands: "At last we have a son and heir. We
prayed to God, went to saints and visited religious places to pray for an heir. The treatment cost me a fortune but the birth of a son makes it all worthwhile. I can die a happy
man and a proud father" (*Mail Online*, 2008). Under these sorts of conditions, it is
difficult to see how a practice like postmenopausal IVF could be viewed as a feminist
technology, when the point is to support a system of patrilineage. While not all cases of
postmenopausal IVF resemble Panwar's situation—Adriana Iliescu, for example, was
a single Romanian woman in her late sixties when she underwent IVF and eventually
gave birth to her daughter, and she claimed she did it for herself[22]—one must consider
the practice within the broader landscape of ARTs, where "female fertility has been
industrialized, legalized, and 'medicalized,' keeping it in the hands of the conservative male establishment" (Smith, 1993: 99). According to critics like Smith, postmenopausal IVF does not present itself as a feminist technology, one that ultimately serves
women's well-being.

Supporters of the technology, by contrast, consider the ways in which it may directly
benefit individual women, and the kinds of social and familial changes the technology
may lead to that may ensure *all* children are protected from parental loss. As previously
noted, Daniela Cutas (2007) argues that the problem isn't the age of the parents, but the
family support and love that is available to any children that are born, and the quality
of their family relationships. And Anna Smajdor (2008) and Jennifer Parks (1999) have
argued that postmenopausal women have not necessarily "had their chance" to be
mothers at a younger age since, as Smajdor (2008: 174) observes, "*not deciding to have a
child* is not logically, morally, or experientially the same as *deciding not to have a child.*"

---

[21] While Smith's feminist point here is well taken, she builds into her argument an assumption regarding
the "real" purpose of having children: to carry on one's (patriarchal) family line through biological offspring.
From the viewpoint of many contributors to this volume, Smith's assumption is objectionable, and should
be questioned. At the same time, the general spirit of her argument suggests that other means of being in a
relationship with children, perhaps including adoption, but also other non-biological forms of relationship,
would be morally good.

[22] Daniela Cutas considers Iliescu's particular case by raising the question of what matters most when
bringing children into the world—the age, social environment, level of education of the parents, or economic status of the family? As she questions, "Why should we think that age is the most relevant among the
above criteria, especially in the case here under scrutiny, where the mother is a professor of literature and
author of literature for children?" (Cutas, 2007: 460).

Fertility Preservation as a Feminist Technology?

Beyond general considerations of ARTs as feminist technologies, one may question whether FP constitutes a feminist technology. As a branch of assisted reproduction, FP faces the same feminist criticisms and concerns that have been levelled against other ART practices, such as IVF. This practice may arguably re-establish the centrality and even the *necessity* of reproduction to women's lives, since it extends reproduction to women in cases where it was not possible (i.e. women undergoing radiation for cancer treatment and women of advanced maternal age). FP closes off to these women the option of simply accepting one's childless (or child-free) future, instead encouraging women to pursue what may be seen as their biological (and social) end of reproducing and parenting biological offspring.

This leads to another concern regarding FP technology: that it also overemphasizes the value and significance of having not just any children, but biologically related children. If genetic relatedness to one's children was not considered valuable, a technology such as FP would not be marketable in the first place. Indeed, since there are other less invasive means of family formation (by, for example, adopting children), the risks that women undertake by using a technology like FP suggest that they are "worth it" for the privilege of having genetically related offspring. Some oncofertility researchers note "a deep desire to propagate our own germ line" as "part of who we, as people, are" (Gardino and Emanuel, 2010: 447). This bias is evident in much of the medical literature on FP, which justifies the technology on the ground that it serves a strong and natural desire for biological reproduction (McLeod, 2010). Yet for decades, feminists have been highly critical of this notion that there is a natural urge for biological reproduction and that various forms of ART are morally good in that they open up reproductive options for women (see Corea, 1986, 1987; Raymond, 1998; Overall, 1987).

However, the use of FP for medical and social purposes, though not without the feminist concerns just mentioned, may answer other feminist worries about the way in which women's reproductive lives are determined and controlled by factors such as male-patterned career trajectories, the pressure to reproduce that comes with women's "biological clock," and the need to rely on other people's gametes for reproduction. I will address these considerations and suggest that FP can be considered a feminist technology because it liberates women in important ways from these controlling factors, and because it can be used to undercut some of these patriarchal expectations surrounding women's reproduction.

Historically, gendered role expectations long kept women out of the public sphere of work, and so a male career trajectory developed that emphasizes upward mobility and uninterrupted work life (largely because men had wives whose jobs were to raise their children and keep their homes) (Gallos, 1989; Evetts, 2000). Though contemporary women are now a major part of the workforce, they have inherited the career trajectory and expectations that are associated specifically with men—a trajectory that does not leave room for maternity leaves, nursing, or child care, especially during the career-building stage. Even once a woman's career is established, she risks losing status if she takes time off to raise children (a task, it is worth noting, that still falls almost exclusively on women). Unsurprisingly, more women are suffering the effects of this male career trajectory by putting off child-bearing until their later lives, precisely when their reproductive capacity has declined or expired. The advent of technologies like FP arguably offers women a means of dealing with these unyielding career expectations by removing the biological limitation placed on their ability to reproduce, thus increasing the window of time that women have to make their reproductive choices, especially in a masculinist culture.[23]

This leads to feminist concerns about reproductive pressures that are associated with what is referred to as women's "biological clock." As Angel Petropaganos notes,

Making egg freezing options available... can help alleviate some of the financial pressure of trying to have a family at a young age, the emotional stress of finding "Mr. Right" or the guilt and anxiety experienced when having to choose between a higher education and a career versus starting a family... egg freezing can help lessen anxiety about reproducing by offering her some security (or increased hope) for the future and for giving her some level of reproductive control. (2010: 225)

While in an ideal world women would not face reproductive pressures of any sort, in our actual world these pressures do exist and derive from both the societal expectations that women will reproduce and from the limited timeframe in which they may do so. Fertility preservation serves to reduce such reproductive pressures by extending the period during which women may choose to reproduce, thus alleviating concerns about melding education and career goals with reproductive goals or finding "Mr Right" in time to start a family. Indeed, given that FP can be practised in stages, completing stage one (which involves freezing one's eggs or preserving one's ovaries) may give women the time needed to decide that they ultimately do not want to become mothers. FP extends the option of reproducing beyond a woman's biological limits, but it is an option that some women may decide against pursuing.[24]

Finally, FP represents an important advancement over the use of IVF at an advanced maternal age because, as previously mentioned, it allows women to preserve their own eggs, thus making it unnecessary to rely on the eggs of younger women who serve as egg vendors. Some critics have pointed to the risks associated with egg "donation," arguing that women should not be encouraged to take risks with their own bodies for other people's reproductive ends (Lahl, 2010). Furthermore, given that women in

---

[23] Of course, changing this career trajectory and the expectations associated with it is another way of dealing with this problem, but in the time it takes to make these changes women will remain limited by male-based career expectations.

[24] If a woman freezes her eggs or utilizes another FP technology in her twenties, she will have to pay for the storage and maintenance of her reproductive materials for an extended period, which raises financial issues. But unlike frozen embryos, which some women will see as more than just "reproductive materials," women may find it easier to decide not to pursue FP beyond stage one, or to even destroy their frozen eggs/ovaries if they decide they do not want to have a family.

the US are paid for their eggs, FP may reduce concerns about the coercion or undue inducement that accompanies payment. While some women may still choose to use donor egg IVF at an advanced maternal age, FP represents a likely significant reduction in the number of women who will do so, which means fewer young women will be taking unnecessary risks as egg vendors.

Recall Layne's definition of a feminist technology as "tools plus knowledge that enhance women's ability to develop, expand, and express their capacities" (Layne *et al.*, 2010: 3). If we take this to be a reasonable definition of feminist technology, then given the reduction of risks to other women and the reduction of pressure on women to reproduce within a certain timeframe, FP can be seen as a feminist technology. I argue that, as such, it should be deemed acceptable for use by women of advanced maternal age. While there may be other, preferred, family-building options for older persons to pursue, the considerations listed suggest that the use of FP by women to extend their reproductive years is at least morally acceptable. Indeed, given the prevalence of patriarchal norms and values that still take reproducing as an important (if not *the* most important) part of being a woman, FP for age-related infertility may have an added degree of moral acceptability. I now turn briefly to other forms of advanced age parenting—custodial grandparenting and adoption—as a basis for comparison to the use of FP and other forms of ART by women of advanced maternal age.

Unlike the use of ARTs to support parenting at an advanced age, practices of custodial grandparenting and adoption do not involve the *creation* of offspring to be raised by significantly older parents. Rather, custodial grandparents and older adoptive parents take responsibility for the full-time care and rearing of children who are already in existence and who need a home. This marks an important difference between these types of aged parenting and the use of ARTs by older persons, and may explain why IVF and FP by women of advanced maternal age is so highly scrutinized and criticized while custodial grandparenting and older adoptive parenting are viewed as more acceptable forms of parenting. Purposefully creating children to be raised by significantly older parents may be seen as morally bad because the need for aged parenting in such cases could be avoided by simply not creating those children. By contrast, custodial grandparents and older adoptive parents are responding to a need that already exists, and though the children they raise are also subjected to the risks cited by those who criticize postmenopausal motherhood (i.e. the risk of orphaning children, or of saddling them with the care of an older parent[25]), the potential harm to children raised by grandparents or older adoptive parents is arguably far less than the harms of being parentless, especially for older children who are harder to place through adoption, or are being placed in the foster care system. Furthermore, in the case of

---

[25] Since custodial grandparents and older-age adopters are more likely to be (though not necessarily) parenting children who are older, the concern about orphaning children at a young age may also not be as pressing a concern as it is where persons in their sixties or seventies are raising children from birth.

custodial grandparenting children are not only placed in a stable home, but they are placed with their grandparent(s), who presumably already have a special bond with their grandchildren.

Yet the practices of custodial grandparenting and advanced age adopting are different, for obvious reasons. Custodial grandparents do not necessarily seek the opportunity to parent and may not even relish or welcome it; certainly, they do not have the same opportunity to plan ahead for their parenting responsibilities or to emotionally or economically prepare to care for their (grand)children. Older adoptive parents, by contrast, seek out parenthood willingly and with advance consideration of their ability to be good parents (Korkki, 2013; Brenoff, 2012). The level and degree of screening for adoptive parents versus custodial grandparents[26] also differs, presumably because the familial connections between grandparents and their grandchildren have already been established and they are biologically related to those children, while adoptive parents are being freshly introduced into a child's life. But these two forms of advanced age parenting do not involve the purposeful creation of infants who will be raised by older persons, which is exactly why IVF and FP use by women of advanced maternal age is harshly criticized. I will first consider the practice of custodial grandparenting, how and why it arises, and then move on to consider older adopters as aged parents.

The Phenomenon of Custodial Grandparenting

Custodial grandparenting is a complex issue because of the circumstances under which it takes place (Strutton and Leddick, 2005). First, in most cases the grandparents do not choose to take on full responsibility of raising their grandchildren: it usually comes about because of what have been referred to as the "four Ds"—divorce, drugs, desertion, and death (Yorkey, 1993). Second, the parenting relationship is often taken up without any accompanying financial assistance, so that many grandparents find themselves with only their pensions as income to support themselves and their grandchildren. Third, in the US context custodial grandparenting has significant racial implications, since a disproportionate number of African-American and Hispanic grandparents are taking on this responsibility (Kelch-Oliver, 2008). Although custodial grandparents come from all racial groups and social classes, single African American grandparents from urban, low-income households are represented at a higher rate than any other racial or ethnic group. Fourth, many custodial grandparents are single women who have no one else on whom they can rely for assistance or relief from their care-giving duties. Thus, although the care offered by custodial grandparents may be exemplary, the conditions under which the need for that care arises are

---

[26] In some cases, custodial grandparents may adopt their grandchildren, in which case they are both a custodial grandparent and an older-age adopter. Here, they still do not undergo the screening that people do when they embark on non-family member adoptions. Adoptions by grandparents count as family member adoptions, which generally do not involve screening. See McLeod and Botterell in this volume.

clearly not ideal. Yet while most custodial grandparents may not choose these arrangements, the value of the relationship must still be acknowledged.

Older Adopters as Parents: A Better Option?

Adoption of children by older parents is a growing phenomenon, though it may not be considered by some state adoption agencies to be an ideal situation in which to place younger children (James, 2009). Some persons of advanced age may face significant hurdles in finding agencies that will approve them as adoptive parents because of age-based concerns. As one commentator notes, "Older adopters may have difficulty in finding an agency in their area that will accept them for a home study (pre-adoption preparation) from a licensed agency, which is mandated by federal law" (Langing, 2006). Furthermore, older prospective parents also face more questions regarding their health and expected longevity. These additional challenges are put in place because agencies want to avoid adopted children facing yet another loss should an aged parent become seriously ill or die (Langing, 2006). At the same time, it is increasingly common to see persons in their fifties or sixties being considered as legitimate adopters of children, since the placement needs for children in foster care is so high (Brenoff, 2012; Bamberger, 2013). As a recent *New York Times* article notes,

Some of these older parents are empty-nesters who apparently didn't have their fill of child-rearing the first time around. Others are grandparents or older blood relatives of parents unable to care for their own children. Still others never had children, and finally have the time, desire and means to give it a go. In most cases, the children are older and have special needs; it is rare for an older adult to adopt an infant, according to national adoption groups. (Korkki, 2013: 2)

Adopting already-existing children, especially older children, allows older persons to have the experience of parenting, but not the experience of gestating, birthing, and raising a child from birth. In such cases (except when grandparents legally adopt their grandchildren) there is also not a biological connection between adoptive parents and their adopted children. One question that the option of adopting a child raises, then, is what it is an older woman is really seeking in her family-making choice: a parental relationship with a child, or the experience of gestating, birthing, (possibly) nursing, and raising an infant from birth? If the latter is the most important issue, then adoption will seem like a less appealing option because it does not offer such opportunities. But if the main goal is to love, be in a relationship with, and engage with a child, then adoption is arguably a better option for older–age women, since it avoids *any* potential risks associated with pregnancy and birth at an advanced age.

As noted above, the choice by older couples to purposefully create children for rearing is often judged, on social and moral grounds, to be a selfish and potentially harmful one, mostly for reasons relating to the best interests of the children involved. One might question whether one's reproductive desire as a person of advanced age is a sufficient ground for bringing a child into the world under relatively riskier circumstances, since there is a good chance that at 60-plus years a mother may end up orphaning (or partially orphaning) the offspring she creates. Yet those who defend FP and other forms of ART for older women (Goold and Savulescu, 2009) argue that the potential harms to offspring are not great enough to justify regulations to prevent older women from accessing the technologies. Furthermore, custodial grandparenting is not a viable route for most older persons who wish to parent, since it only arises as an option in cases where the biological parents are not able to parent their children; and short of going to court to label biological parents as being unfit to raise their children, those interested in custodial grandparenting will not be able to access this as an option.[27] Additionally, adoption may be more difficult to set up than advanced age parenting through FP, IVF, and other forms of ART, and so may be a less certain way of building a family at an older age. It is therefore unsurprising that some older persons who wish to parent go to the lengths of technological interventions like FP and IVF in order to achieve their ends. On the accessibility of these technologies to older persons, consider that in the US, despite recommendations by the ASRM to limit IVF to persons under the age of 55, clinics set the scope for their own practice, and ability to pay is often sufficient grounds to receive services. Furthermore, in nations where ARTs are more closely regulated, individuals may (and do) access services abroad. Given the realities of the new global market in reproduction,[28] ARTs may present an easier option for prospective older parents than the adoption route. Ironically, then, the parenting options that are the least risky and arguably the most virtuous (i.e. caring for already-existing, older children who need parents to step in) are perhaps the least available to older persons.

Conclusion

So how *should* feminists receive the use of FP and other ARTs by women of advanced maternal age, given the different feminist frameworks for addressing it and their different foci? To return to concerns raised in this chapter, I argue that one must adopt a feminist approach that questions the value of technology for women: in this case, asking whether FP and other ARTs constitute feminist technologies. While some feminists are concerned that the use of FP and other forms of ART to extend reproduction

---

[27] Perhaps if, for economic and care-giving reasons, more intergenerational families are formed where grandparents live with and help care for their grandchildren, the option to serve as a custodial grandparent will be more available to older persons. But as the law currently stands, a grandparent's access to her/his children can be limited by the parents, since the courts have established that grandparents do not have a legal right to see or spend time with their grandchildren. While in some US states grandparents may petition courts for visitation rights, the standard for awarding visitation may require grandparents to show it will "harm" the child if there is no relationship with their grandchildren (Goyer, 2009).

[28] Reproductive travel and tourism is a growing phenomenon, with countries like India offering significantly cheaper prices and liberal guidelines surrounding IVF and surrogacy (see Baylis, in this volume). For articles that detail the new global market in assisted reproduction, see Parks (2010), Ramskold and Posner (2013), and Donchin (2010).

to women of advanced maternal age may uphold suffocating, traditional, sexist norms for women, I suggest that feminists must not lose an important opportunity to address the way in which these technologies open up new understandings and practices of family-making. Furthermore, since as noted above, other aged parenting options are difficult for these potential parents to access, ARTs present themselves currently as the best way (and, in some cases, the only way) for older women to form families. Furthermore, since FP technology entails fewer risks to women as egg donors, extends women's timeframe for making reproductive decisions, and challenges strict notions of "legitimate" family formation, it represents a feminist technology that may be justified in its use by older women.

Beyond the feminist defences already cited in this chapter, I suggest that legislation denying FP or the use of other ART forms by women of advanced maternal age raises serious concerns about patriarchal interventions into women's reproductive lives. As I have suggested, concerns cited against older-age parenting also apply to some reproductive-aged persons. For example, stage two FP technology for women who have received cancer care places offspring at risk of being orphaned or partially orphaned, or subject to an extreme care burden should a child's mother develop cancer again, in much the same way as children born of stage two FP technology to women of advanced maternal age. Yet few feminists would, I think, support legislation that would reject cancer survivors as legitimate candidates for FP because of their risk of cancer; few feminists would likewise support legislation that would prevent single women from accessing sperm donation in order to have children, since the inability to find, or lack of interest in finding, "Mr Right" hardly seems like grounds for intervening to protect those women and their offspring. Regulations to reject outright the use of FP and other forms of ART by women of advanced maternal age, or setting an absolute age limit, face similar concerns about the role of the state in regulating women's reproductive lives. It is unclear how FP would be successfully regulated in any case, since at stage one FP a woman may not know how many years she will be cryopreserving her reproductive materials, so it is difficult to monitor the practice from the outset. Furthermore, though a government could regulate stage two FP by rejecting older women as candidates, this will not prevent aged parenting through FP, since women can take their frozen eggs to other countries to either find a surrogate or to have the eggs fertilized and transferred for implantation so they can still achieve pregnancies. Strict regulation that prohibits stage one or two FP and other ART use by women of advanced maternal age is not the answer to this complicated issue.

Rather, I argue, we need more serious conversation surrounding the purposes and obligations of parents to their offspring or potential offspring, to gain some clarity surrounding: who can/should parent; when parenting choices are unfair or risky to the offspring that will be created; the value (and what, specifically, it might be) of genetic connections and why we value them so intensely, in so many cultures; the conditions under which parenting ought not to be pursued; the broadening of family formation from the nuclear, heterosexual, two-parent model; and other options that could be made available (but are not currently easily accessible) for persons to parent in the ways mentioned by Patricia Smith and others who recommend adoption or other parenting possibilities.[29] The goal should be to encourage virtuous parenting choices for *all* persons who are contemplating parenting, whether as biological and social or merely social parents to children.

A robust ethical examination of FP and other ARTs for the social purpose of aged parenting addresses the visions or ideals for later life that these technologies create. To some extent, what we take to be moral virtues and what we perceive as ethical obligations derive from our lived environment. The virtues we might have admired in the past will (and should) give way to ones that reflect a new moral system that is based on new ways of living. With regard to my argument, these new ways of living include changes in family formation such that what it means to be a "good" mother, father, or grandparent, or what constitutes a "family," will be redefined in significant ways in the years to come (Overall, 2004).

References

Abraham, T. (2012). Pregnancy in IVF Patients Over the Age of 50 "Carries No More Risk than in Women Under 42." *Mail Online.*, 4 Feb. Retrieved June 2013 from http://www.dailymail.co.uk/femail/article-2096285/Pregnancy-IVF-patients-age-50-carries-risk-women-42.html.

Alberta Children's Hospital Research Institute (n.d.). Women Over 50—Pregnancies with Donor Eggs. Retrieved June 2013 from http://research4kids.ucalgary.ca/women_over_50-donor_egg_pregnancies.

Asch, A., and Marmor, B. (2008). Assisted Reproduction. In M. Crowley (ed.), *From Birth to Death and Bench to Clinic: The Hastings Center Bioethics Briefing Book for Journalists, Policymakers, and Campaigns* (pp. 5–10). Garrison, NY: Hastings Center.

Bamberger, J. C. (2013). Older, Wiser, and Warming Bottles. *Adoptive Families.* Retrieved June 2013 from http://www.adoptivefamilies.com/articles.php?aid=1004.

Bartky, S. (1999). Unplanned Obsolescence: Some Reflections on Aging. In M. U. Walker (ed.), *Mother Time: Women, Aging, and Ethics* (pp. 61–74). New York: Rowman & Littlefield.

Baylis, F., and McLeod, C. (2007). The Stem Cell Debate Continues: The Buying and Selling of Eggs for Research. *Journal of Medical Ethics,* 33(12), 726–31.

Brenoff, A. (2012, July 3). "Too Old to Adopt?" Not the Case for These Parents. *The Huffington Post,* 3 July. Retrieved June 2013 from http://www.huffingtonpost.com/2012/07/03/too-old-to-adopt-child-adoption_n_1606942.html.

[29] It would have been preferable had Rajo Devi and Omkari Panwar considered all of the above in making their reproductive decisions. It is worth considering how such a deeply patriarchal culture might have propelled these women (along with their husbands) to make the decision to biologically reproduce at the age of 70. We need broader (less accusatory and ageist) discussions about the conditions under which such parenting options are chosen in order to consider what limits (if any) should be placed on women's reproductive choices. The kinds of questions raised in this chapter deserve more serious and sustained public (and policy) debate.

Caplan, A. (2008). New IVF Dilemmas Make Old Fears Seem Quaint. *NBC News*, 24 July. Retrieved June 2013 from http://www.nbcnews.com/id/25837220/ns/health-health_care/t/new-ivf-dilemmas-make-old-fears-seem-quaint/#.UdBNTjugXDO.

Caplan, A. L., and Patrizio, P. (2010). Are You Ever Too Old to Have a Baby? The Ethical Challenges of Older Women Using Infertility Services. *Seminars in Reproductive Medicine*, 28(4), 281–6.

Cohen, C. B. (ed.) (1996). *New Ways of Making Babies: The Case of Egg Donation*. Bloomington, IN: Indiana University Press.

Corea, G. (1986). *The Mother Machine: Reproductive Technologies from Artificial Insemination to Artificial Wombs*. New York: HarperTrade.

Corea, G. (1987). *Man-Made Women: How Reproductive Technologies Affect Women*. Bloomington, IN: Indiana University Press.

Cutas, D. (2007). Postmenopausal Motherhood: Immoral, Illegal? A Case Study. *Bioethics*, 21(8), 458–63.

Donchin, A. (2010). Reproductive Tourism and the Quest for Global Gender Justice. *Bioethics*, 24(7), 323–32.

Ethics Committee of the American Society for Reproductive Medicine (2013). Oocyte or Embryo Donation to Women of Advanced Age: A Committee Opinion. *Fertility and Sterility*. Retrieved June 2013 from http://www.socrei.org/uploadedFiles/ASRM_Content/News_and_Publications/Ethics_Committee_Reports_and_Statements/postmeno.pdf.

Evetts, J. (2000). Analysing Change in Women's Careers: Culture, Structure and Action Dimensions. *Gender, Work and Organization*, 7(1), 57–67.

Frankl, V. E. (1967). *Man's Search for Meaning*. New York: Simon & Schuster.

Gallos, J. (1989). Exploring Women's Development: Implications for Career Theory, Practice and Research. In M.B. Arthur, D. T. Hall, and B. S. Lawrence (eds), *Handbook of Career Theory* (pp. 110–32). New York: Cambridge University Press.

Gardino, S., and Emanuel, L. (2010). Choosing Life When Facing Death: Understanding Fertility Preservation Decision-Making for Cancer Patients. In T. K. Woodruff, L. Zoloth, L. Campo-Engelstein, and S. Rodriguez (eds), *Oncofertility: Ethical, Legal, Social, and Medical Perspectives* (pp. 447–58). New York: Springer. Available from *Cancer Treatment Research*, 156, i-519. Retrieved June 2013 from http://link.springer.com/book/10.1007/978-1-4419-6518-9/page/1#page-1.

Georgescu, E. S., Goldberg, J. M., du Plessis, S. S., and Agarwal, A. (2008). Present and Future Fertility Preservation Strategies for Female Cancer Patients. *Obstetrical and Gynecological Survey*, 63(11), 725–32.

Goold, I., and Savulescu, J. (2009). In Favour of Freezing Eggs for Non-Medical Reasons. *Bioethics*, 23(1), 47–58.

Goyer, A. (2009). Grandparents' Visitation Rights. American Association of Retired Persons (AARP), 28 May. Retrieved June 2013 from http://www.aarp.org/relationships/grandparenting/info-05-2009/goyer_grandparent_visitation.2.html.

Harwood, K. (2008). Egg Freezing: A Breakthrough for Reproductive Autonomy? *Bioethics*, 23(1), 39–46.

Hartouni, V. (1997). *Cultural Conceptions: On Reproductive Technologies and the Remaking of Life*. Minneapolis, MN: University of Minnesota Press.

Held, V. (1987). Non-Contractual Society: A Feminist View. In M. Hanen and K. Nielsen (eds), *Science, Morality and Feminist Theory* (pp. 111–37). Calgary, AB: University of Calgary Press.

Holstein, M. (1999). Home Care, Women, and Aging: A Case Study of Injustice. In M. U. Walker (ed.), *Mother Time: Women, Aging, and Ethics* (pp. 227–44). New York: Rowman & Littlefield.

Human Fertilisation and Embryology Authority (2013). Freezing and Storing Eggs. Retrieved June 2013 from http://www.hfea.gov.uk/46.html.

James, S. D. (2009). Old Mom: Dies at 69, Leaves Orphan Twins. *ABC News*, 16 July. Retrieved June 2013 from http://abcnews.go.com/Health/ReproductiveHealth/story?id=8098755&page=1#. UcN8Y_mkqPx.

Jones, B. (2009). Lord Winston Labels Egg Freezing an "Expensive Confidence Trick." *BioNews*, 6 July. Retrieved June 2013 from http://www.bionews.org.uk/page_46074.asp.

Katz Rothman, B. (1986). *The Tentative Pregnancy: Prenatal Diagnosis and the Future of Motherhood*. New York: Viking Press.

Katz Rothman, B. (2012). Yet another technological fix to a social problem. 22 July. Retrieved June 2013 from http://www.barbarakatzrothman.com/2012_07_01_archive.html.

Kelch-Oliver, K. (2008). African American Grandparent Caregivers: Stresses And Implications for Counselors. *Family Journal*, 16(1), 43–50.

Kluge, E. H. (1994). Reproductive Technology and Postmenopausal Motherhood. *Canadian Medical Association Journal*, 151(3), 353–5.

Korkki, P. (2013). Filling up an Empty Nest. *New York Times*, 14 May. Retrieved June 2013 from http://www.nytimes.com/2013/05/15/business/retirementspecial/some-older-adults-are-adopting-children.html?pagewanted=all&_r=0.

Kort, D. H., Gosselin, J., Choi, J. M., Thornton, M. H., Cleary-Goldman, J., and Sauer, M. V. (2012). Pregnancy After Age 50: Defining Risks for Mother and Child. *American Journal of Perinatology*, 29(4), 245–50.

Lahl, J. (2010). *Eggsploitation*. A Documentary Film by the Center for Bioethics and Culture. Available from http://www.eggsploitation.com.

Landau, R. (2004). The Promise of Postmenopausal Pregnancy. *Social Work in Health Care*, 40(1), 53–69.

Langing, B. (2006). When are You "Too Old" to Adopt? *Rainbowkids.com—the Voice of Adoption*, 1 June. Retrieved June 2013 from http://www.rainbowkids.com/ArticleDetails.aspx?id=66.

Lawrence, N. M. (2004). Grand Rounds: What's the Best Approach to Spontaneous Premature Ovarian Failure? *Contemporary OB/GYN*, 1 Nov., 46–55. Available from http://poi.nichd.nih.gov/pdf/Contemp_ObGyn_2004_POF.pdf.

Layne, L. L., Vostral, S. L., and Boyer, K. (2010). *Feminist Technology*. Chicago, IL: University of Illinois Press.

Loren, A. W., Mangu, P. B., Beck, L. N., Brennan, L., Magdalinski, A. J., Patridge, A. H., Quinn, G., Wallace, W. H., and Oktay, K. (2013). Fertility Preservation in Patients with Cancer: American Society of Clinical Oncology Guideline Update. *Journal of Clinical Oncology*, 31(19), 1–12.

Maguire, M. (2010). ASRM Report Denies Regulatory Reality. *Biopolitical Times*, 14 July. Retrieved June 2013 from http://www.biopoliticaltimes.org/article.php?id=5296.

Mail Online (2008). World's Oldest Mother Gives Birth to Twins, 5 July. Retrieved June 2013 from http://www.dailymail.co.uk/news/article-1031722/Worlds-oldest-mother-gives-birth-twins-70.html.

Martin, D. (2012). Number of Babies Born to Women of 45 and Over Triples in Just Ten Years. *Mail Online*, 27 Jan. Retrieved June 2013 from http://www.dailymail.co.uk/news/ article-2092971/Number-babies-born-women-45-trebles-just-years.html.

McLeod, C. (2010). Morally Justifying Oncofertility Research. In T. K. Woodruff, L. Zoloth, L. Campo-Engelstein, and S. Rodriguez (eds), *Oncofertility: Ethical, Legal, Social, and Medical Perspectives* (pp. 187–94). New York: Springer. Available from *Cancer Treatment Research*, 156, i-519. Retrieved June 2013 from http://link.springer.com/book/10.1007/978-1-4419-6518-9/ page/1#page-1.

Noddings, N. (1984). *Caring: A Feminine Approach to Ethics and Moral Education*. Berkeley, CA: University of California Press.

Overall, C. (1987). *Ethics and Human Reproduction: A Feminist Analysis*. New York: Routledge.

Overall, C. (2004). Longevity, Identity, and Moral Character: A Feminist Approach. In S. G. Post and R. H. Binstock (eds), *The Fountain of Youth: Cultural, Scientific, and Ethical Perspectives on a Biomedical Goal* (pp. 286–303). New York: Oxford University Press.

Parks, J. A. (1999). On the Use of IVF by Postmenopausal Women. *Hypatia*, 14(1), 77–96.

Parks, J. A. (2003). *No Place Like Home? Feminist Ethics and Home Health Care*. Indianapolis, IN: Indiana University Press.

Parks, J. A. (2009). Rethinking Radical Politics in the Context of Assisted Reproductive Technology. *Bioethics*, 23(1), 20–7.

Parks, J. A. (2010). Care Ethics and the Global Practice of Commercial Surrogacy. *Bioethics*, 24(7), 333–40.

Paulson, R. J., and Sauer, M. V. (1994). Pregnancies in Post-Menopausal Women: Oocyte Donation to Women of Advanced Reproductive Age: How Old is Too Old? *Human Reproduction*, 9(4), 571–2.

Petropaganos, A. (2010). Reproductive "Choice" and Egg Freezing. In T. K. Woodruff, L. Zoloth, L. Campo-Engelstein, and S. Rodriguez (eds), *Oncofertility: Ethical, Legal, Social, and Medical Perspectives* (pp. 223–35). New York: Springer. Available from *Cancer Treatment Research*, 156, i-519. Retrieved June 2013 from http://link.springer.com/book/10.1007/978-1-4419-6518-9/ page/1#page-1.

Petropanagos, A. (2013). *Fertility Preservation Technologies for Women: A Feminist Ethical Analysis*. University of Western Ontario, Electronic Thesis and Dissertation Repository. Paper 1299. Retrieved June 2013 from http://ir.lib.uwo.ca/etd/1299.

Raymond, J. (1998). *Women as Wombs: Reproductive Technology and the Battle over Women's Freedom*. Melbourne: Spinifex Press.

Ramskold, L. A., and Posner, M. P. (2013). Commercial Surrogacy: How Provisions of Monetary Remuneration and Powers of International Law Can Prevent Exploitation of Gestational Surrogates. *Journal of Medical Ethics*, 39(6), 397–402.

Riggan, K. (2011). Regulation (or Lack Thereof) of Assisted Reproductive Technologies in the U.S. and Abroad. Center for Bioethics and Human Dignity, 3 May. Available online at http://cbhd.org/content/regulation-or-lack-thereof-assisted-reproductive-technologies- us-and-abroad.

Royal Commission on New Reproductive Technologies (1993). *Proceed with Care: Final Report of the Royal Commission on New Reproductive Technologies*. Toronto: Federal Publications, Inc.

Ruddick, S. (1989). *Maternal Thinking: Toward a Politics of Peace*. Boston, MA: Beacon Press.

Schermers, J. (2009). No "Wishful Medicine" Please. *NRC Handlesblad*, 16 July. Retrieved June 2013 from http://vorige.nrc.nl/article2302404.ece.

Smajdor, A. (2008). The Ethics of Egg Donation in the Over Fifties. *Menopause International*, 14(4), 173–7.

Smith, P. (1993). Selfish Genes and Maternal Myths: A Look at Postmenopausal Pregnancy. In J. Callahan (ed.), *Menopause: A Midlife Passage* (pp. 92–119). Indianopolis, IN: Indiana University Press.

Steinbock, B. (2004). Payment for Egg Donation and Surrogacy. *Mount Sinai Journal of Medicine*, 71(4), 255–65.

Strutton, J. N., and Leddick, G. R. (2005). Grandparents as Parents: A Growing Phenomenon. In G. R. Walz and R. K. Yep (eds), *VISTAS: Compelling Perspectives on Counseling* (pp. 111–13). Alexandria, VA: American Counseling Association.

Uzelac, P., Christensen, G., and Nakajima, S. (2012). In T. K. Woodruff, L. Zoloth, L. Campo-Engelstein, and S. Rodriguez (eds), *Oncofertility: Ethical, Legal, Social, and Medical Perspectives* (pp. 77–89). New York: Springer. Available from *Cancer Treatment Research*, 156, i-519. Retrieved June 2013 from http://link.springer.com/book/10.1007/978-1-4419-6518-9/page/1#page-1.

Woodruff, T., Zoloth, L., Campo-Englestein, L., and Rodriguez, S. (eds) (2010). *Oncofertility: Ethical, Legal, Social, and Medical Perspectives*. New York: Springer. Available from *Cancer Treatment Research*, 156, i-519. Retrieved June 2013 from http://link.springer.com/book/10.1007/978-1-4419-6518-9/page/1#page-1.

Yorkey, M. (1993). Picking up the Pieces. *Focus on the Family*, 17, 13. Available from http://www.safamily.org.za/?showcontent&global%5B_id%5D=594.

Index

Printed and bound by CPI Group (UK) Ltd, Croydon, CR0 4YY